Extensive Air Showers

Extensive Air Showers

M. V. S. Rao

Tata Institute of Fundamental Research
India

B. V. Sreekantan

National Institute for Advanced Studies
India

World Scientific
Singapore • New Jersey • London • Hong Kong

Published by

World Scientific Publishing Co. Pte. Ltd.

P O Box 128, Farrer Road, Singapore 912805

USA office: Suite 1B, 1060 Main Street, River Edge, NJ 07661

UK office: 57 Shelton Street, Covent Garden, London WC2H 9HE

British Library Cataloguing-in-Publication Data
A catalogue record for this book is available from the British Library.

EXTENSIVE AIR SHOWERS

ISBN 981-02-2888-0

This book is printed on acid-free paper.

Printed in Singapore by Uto-Print

We are thankful to Prof. V. Singh, Director, Tata Institute of Fundamental Research, Mumbai, Prof. R. Ramanna, Director, National Institute of Advanced Studies, Bangalore, and the Homi Bhabha Fellowship Council for their support and for allowing us to use the facilities of the respective institutes. Our thanks are also due to Dr. B. S. Acharya, who has painstakingly gone through the manuscript and made several suggestions.

To
Vijaya
&
Ratna

Preface

The Extensive Air Shower is a grandiose phenomenon that takes place in the atmosphere whenever a very high energy cosmic ray primary—proton or heavy nucleus—arriving from depths of space is incident on it. The primary particle, being a hadron, collides with air nuclei several times as it travels down and in each collision through strong interaction produces a large number of charged and neutral pions, kaons, nucleon-antinucleon pairs etc., which being all hadrons interact further down and give rise to a nuclear cascade. Some of the unstable hadrons decay instead of interacting and give rise to the very penetrating particles—μ mesons and neutrinos—in addition to hadrons, electrons, positrons as the case may be. The π^0 mesons produced in the various interactions decay almost instantaneously into high energy gamma rays which initiate huge electromagnetic cascades. Thus by the time the shower, enriched by a variety of particles, reaches mountain altitude or sea level the single primary would have multiplied into thousands to billions of particles depending upon its energy, the dominant component being electrons and positrons. In this development all the three forces of nature would have played a significant role. The other force namely gravitation too has an important role to play in the development of the shower and in making it a very valuable tool for the study of very high energy cosmic rays. Because of the nature of the gases of the atmosphere and the gravitational force of the Earth, the atmospheric density decreases with height in an exponential manner with a scale height of about 7 km. This exponential character has two important effects on the development of the shower. Since the distance between successive collisions becomes large at the higher heights in the atmosphere, the unstable particles have a chance to decay and secondly this same nature of the atmosphere allows the secondaries to spread laterally. The lower energy particles with larger opening angles will spread more compared to the higher energy particles. This gives a unique character to the air shower. There will be a dense core of high energy particles and a characteristic decrease of density with distance from the core. The core direction defines the angle of incidence of the primary at the top of the atmosphere. The lateral distribution of the density of particles, which is different for the different species becomes an important parameter for measurement. The extended nature of the atmosphere also introduces a characteristic time structure in the arrival times of the shower particles at the observational level. While the thickness of the electron-photon disc is only a couple of meters at best (< 6 ns spread in time), the heavier particles because of their lower velocity trail behind. The mesons follow the shower front and the nucleon-antinucleons trail behind the mesons.

All these features of the Extensive Air Shower—the composition of the particles, their lateral distributions, their time structure—have been exploited by setting up specially designed arrays composed of detector systems that preferentially respond to the different components. In the last few decades, two other electromagnetic radiations that are secondary to the dominant electron-positron component of air showers have been very effectively utilized in the study of air showers. One is the Cerenkov radiation that is produced predominantly by the electrons and positrons which have such high energies that their velocities in the low density regions of the atmosphere are higher than that of light. The other is the fluorescent radiation that is produced by nitrogen nuclei due to excitation by the high energy electrons and positrons in the core of the shower.

While studies on the nature of the primary spectrum, composition, directional distribution etc., have been done with balloon and satellite borne nuclear emulsion stacks and electronic detector systems up to energies of about a few hundred TeV, such studies become impossible at higher energies because of the steeply falling flux of the primary radiation. Even installations on satellites which can carry much larger area payloads enabling long exposures will not extend the study to much higher energies. This can be seen from the fact that the flux of primaries is of the order of one particle per square meter per steradian per hour at an energy $\geq 10^{14}$ eV and falls by more than a factor of hundred for every decade of energy increase.

Even though the method of Extensive Air Showers is indirect the ingeneous way in which the various parameters of the shower are measured and analyzed and related to the features of the primary radiation have proved eminently successful and new methodologies are continuously evolving with technological advances. In this endeavour the Monte Carlo simulations that have become feasible with the advent of computers have been extremely valuable and have added a new dimension of reliability to this indirect method. Since the air shower is the end product of a large number of high energy collisions, the characteristics of the collisions like the cross-section, secondary multiplicity, composition of secondaries, transverse momentum distribution etc., play a very important role. Though in the early days many of these parameters had to be arrived at from cosmic ray studies themselves, in the last few decades the accelerator experiments have provided the necessary inputs of much higher accuracy.

The primary aim of air shower studies is to determine the spectrum, composition, anisotropy of arrival directions of the primaries up to the highest energies available in the cosmic radiation. And from these studies to locate the possible sources and identify methods of acceleration to such high energies.

In this book, we have endeavoured to present the current status of extensive air shower studies that has evolved out of the last four decades of efforts by a variety of experiments carried out by scientists from different parts of the world. We have discussed the phenomenology of air shower, the detector techniques, the composition of the different arrays at a variety of locations, the types of results that have emerged on the characteristics of interactions beyond the accelerator range, the new developments in the field of TeV and PeV astronomies, the results as they stand in 1996 on the spectrum, composition and anisotropies of the primary radiation up to the highest energies. We have also projected the new excitement that is coming into the field of air shower studies through the prospects of very high energy astronomies and new novel types of installations.

While these are available to some extent in the proceedings of cosmic ray conferences and some review articles, we felt that there is need for compilation of all the information that is distributed in a single comprehensive book with particular emphasis on the experimental methodologies and results.

Contents

List of Figures

List of Tables

Chapter 1
DISCOVERY

It is often said that Nature keeps her secrets. On the contrary Nature keeps on revealing her secrets in the most tantalizing manner and in a way that extracts the best of the hidden intellectual and technological capabilities of man all the time rewarding him with totally unexpected gifts of fundamental knowledge and opening up new avenues for him to pursue with vigor and challenge. This wonderful aspect of Nature is well illustrated by the exhilarating story of cosmic ray research that spans practically the whole of this century.

1.1 Discovery of Cosmic Rays

It all began with a mystery surrounding the continuous and uncontrollable leakage of electrical charge from a well insulated charged gold leaf electroscope. The mystery had remained unexplained almost from the time of Henry Coulomb who had noticed even in 1785, that a charged metal sphere suspended by an insulated silk thread did not retain its charge even though it was isolated.

Near the end of the last century, it was C. T. R. Wilson [1] who connected this phenomenon with the "ionization" of the surrounding air. With the discovery of radioactivity and the finding that the earth itself contained minute traces of radioactive materials all over, it was mistakenly thought that the source of ionization of the air was this radioactive material of the earth. In a classic endeavour Wulf [2] climbed the Eiffel Tower in Paris carrying an ionization measuring device and noticed that the ionization did decrease as he climbed to higher levels. Gockel [3] reached still higher heights by means of balloons. These did cast some doubts on the hypothesis that the radioactivity of the earth was responsible for the mysterious ionization. However it goes to the credit of Victor Hess to have settled the issue unambiguously. On the historic day of August 7, 1912, Victor Hess [4] along with an assistant went up in a Balloon Gondola carrying three electroscopes and floated for several hours at altitudes ranging from 13,000 to 16,000 ft. and established that the radiation responsible for the mysterious ionization was coming from "above" and was possibly of extra-terrestrial origin. Further experiments by a variety of workers all over the world established that this radiation was not only extra-terrestrial, but also extra-solar. The name "Cosmic Radiation" was given to this penetrating radiation from depths of space by Millikan in 1925.

The period 1912 to 1940, though interspersed with two world wars that impeded scientific progress, was characterized by a saga of unique experiments carried out by scientists at widely different locations in the globe, at mountain altitudes, at underground levels and also by deploying instruments to great depths in water and sending instruments on balloons that went up to tens of kilometers.

These slowly unfolded the very complex and multifaceted nature of this radiation. The development of Geiger counter telescopes, counter controlled magnetic and multiplate cloud chambers played a very important role in unraveling the complexities of this radiation.

The experiments carried out at sea level and mountain altitudes showed that the radiation had two components; a soft component that easily got absorbed in a few mm of lead and a penetrating component that passed through several meters of lead and also up to great depths underground and in water. The experiments at different latitudes revealed that the magnetic field of the earth affected the intensity of the radiation, the intensity increasing as one moved from equator to higher latitudes. The absorption characteristics of the penetrating radiation were also baffling. The absorption as measured by the disappearance of particles was more in a rarefied medium like air than in an equal amount of condensed matter like water or lead. The soft component occasionally occurred as a shower of several simultaneous particles with considerable lateral spread.

The only particles that were known till early 30's were the γ rays, the electrons and the protons. The neutral counterpart of the proton, the neutron was discovered in 1932 by Chadwick and this led to a much clearer understanding of the constituents of the atomic nuclei. The neutron was unstable and decayed into a proton and electron with a mean life of 15 minutes. To save the principle of conservation of energy, Pauli postulated the massless neutrino.

A major breakthrough in the understanding of the behavior of the soft component of cosmic rays came through the theoretical developments that were taking place in the field of quantum electrodynamics (Q.E.D.) and more importantly through the discovery of the "positron" the anti-particle of the "electron". The discovery of the "positron" came in 1932 through an ingeneous experiment with a magnetic cloud chamber that had been set up by Anderson [5] at Caltech to study the properties of the cosmic radiation. Experimental evidence soon accumulated on the phenomenon of "pair creation"—the creation of electron-positron pairs by the high energy γ rays passing through matter. The emission of high energy γ rays by the process of "bremsstrahlung" of high energy charged particles through matter and the subsequent creation of electron-positron pairs by these γ rays and the dual process repeating over and over again was the essence of the "cascade theory" that had been developed by Bhabha and Heitler [6] and Carlson and Oppenheimer [7] to explain the "absorption" characteristics of the soft component. According to this elegant theory the soft component was the end product of such cascades in the atmosphere and comprised of electrons, positrons and γ rays.

Where did the high energy electrons that initiated the cascades come from? What are the penetrating particles that penetrated more than a meter of lead? These certainly could not be the electrons since they would, according to the cascade theory, develop into showers and get absorbed in a few cm of lead. Magnetic cloud chamber experiments had established that there were as many penetrating negatively charged particles as positive ones. These ruled out the possibility that the penetrating particles were protons, the only other charged particles known at that time. In a classic paper in 1938 in the Proceedings of the Royal Society, Homi Bhabha [8] after a careful survey of all the experimental results on cosmic radiation and the theoretical formulations of the quantum electrodynamics, came to the conclusion that the anomaly can be best explained if there exist in cosmic radiation a charged particle of mass ~ 100 m_e (m_e is the mass of the electron), which could fill the role of the penetrating particle.

Direct evidence for the existence of particles of mass intermediate between electron and proton came from the cloud chamber experiment of Neddermeyer and Anderson [9] and Street and Stevenson [10] in 1937. The mass of the particle was deduced as 200 m_e and the particle was given the name "Mu-Meson".

This particle was found to be unstable and decayed, in a matter of micro seconds when brought to rest, into an electron and presumably to neutrinos that had been postulated by Pauli in connection with β decay. The anomalous density effect found a natural explanation in terms of this decaying meson, which suffered time elongation in flight expected according to the special theory of relativity.

While several aspects of cosmic ray phenomenology got clarified through these discoveries, the question of the origin of the mu-mesons and high energy electrons still remained a puzzle. Further, a good deal of confusion was caused by the hasty identification of the mu-meson with the particle that had been postulated by Yukawa [11] in connection with the theory of nuclear forces. This particle by the very hypothesis of Yukawa had to have a strong interaction with nuclei and consequently the negatively charged ones had to be absorbed in matter in preference to decay. The negative muons after slowing down preferred to decay spontaneously than get absorbed by the nuclei as revealed in the experiment of Conversi et al. [12]. This anomaly was beautifully resolved by the discovery of the π meson by Lattes et al. [13] in 1947. This discovery was made with highly sensitive nuclear emulsions that had been exposed to cosmic rays at the Swiss high altitude mountain station, Jung-Frau-Joch. Lattes et al. showed that the μ meson was the decay product of the π meson which further experiments showed was just the Yukawa particle with all the requisite properties of interaction. The pion decayed into muon and neutrino in one hundredth of a microsecond.

The next question that had to be answered was—where did the π mesons come from? Marietta Blau and H. Wambacher [14] of Austria had shown as early as 1932, that there was, in cosmic radiation in addition to the soft and penetrating components, yet another component which produced "stars" or nuclear explosions in photographic plates which were sensitive to only highly ionizing tracks of nuclear fragments. Janossy in the early 40's had recorded with shielded counter assemblies events which could be interpreted as production of simultaneous penetrating particles. These were called "penetrating showers". Rossi and Tinlot measured the variation of intensity of these penetrating showers as a function of altitude in the Rocky Mountains of Colorado and also at still higher altitudes with B-29 planes. Their experiments showed that the component responsible for producing the penetrating showers increased in intensity with altitude faster than the single penetrating component namely the muons. The likely candidates were the pions and the protons and the neutrons—the nuclear active component.

With the recognition of different components, their varying altitude dependence, the presence of the short lived components, the pions and the muons, the picture finally emerged that the cosmic radiation comprised two components, the "Primary Component" incident at the top of the atmosphere that gave rise, through nuclear collisions, to the different "Secondary Components", some of which underwent spontaneous decay to give rise to other components. The latitude effect measured by Clay [15] and others and the East-West effect observed by Johnson [16], established that the primary particles were charged. The experiments of Marcel Schein and his collaborators [17] in the early 40's up to altitudes of 70,000 ft had shown that the primary particles are not electrons. Thus among the particles known the best bet was the proton which ultimately proved to be correct.

In 1948 the experiments of Bradt and Peters [18] with nuclear emulsions flown to altitudes of 94,000 ft. revealed that the primary component comprised in addition to protons also a small fraction of completely stripped nuclei—α particles, light ($z = 3 \sim 5$), medium ($z = 6 \sim 9$) and heavy ($z > 10$) nuclei.

Cloud chambers at mountain altitudes and nuclear emulsion stacks at balloon altitudes recorded clear examples of the production of large number of secondary minimum ionizing particles in the collisions of high energy cosmic ray particles. It was soon identified that the secondaries are predominantly π mesons. It was also found that in addition to charged particles, neutral pions are also produced in these collisions and these decay with a life time of $\sim 10^{-16}$ s into pairs of γ rays. Thus by the end of 40's, the different components of the primary and secondary radiations had been identified as well as the generic relations between the different components. The role played by the earth's atmosphere—particularly its exponential nature—in the production, decay and spread of these particles also became clear.

The following picture summarizes the happenings in the atmosphere:

The primary particle of very high energy, a proton or a heavy nucleus that comes from outside the Solar system, on entering the earth's atmosphere collides with an air nucleus (nitrogen, oxygen, etc.) and produces a large number of pions. The charged pions as they travel down either interact with other nuclei and produce more secondaries or decay into muons and neutrinos, which constitute the penetrating charged and neutral components. The neutral pions decay into γ rays and these in turn initiate large electromagnetic cascades and give rise to the soft component observed in the lower atmosphere. A fraction of the soft component also comes from the decay of low energy muons.

As the energy of the primary particle increases not only does the number of particles produced in the first few collisions increases, but also the number of generations that contribute to the nuclear cascade. Because of the exponential fall in the density of the atmosphere as one goes to higher and higher altitudes the secondaries spread out laterally as the longitudinal development of the combined nuclear and electromagnetic cascades progresses downwards. At very high energies ($> 10^{14}$ eV) the cascade development results in the simultaneous incidence of very large number of particles—electrons, positrons, photons, π mesons, nucleons, anti-nucleons, μ meson, neutrinos etc.,—the number amounting to several thousand particles spread over hundreds of square meters at the observational levels of mountain altitude and sea level. At primary energies of 10^{19} eV, the numbers could be a billion particles spread over several square kilometers. These spectacular showers have been given the name "Extensive Air Showers".

Yet another development that took place in the late 40's as a result of investigations in the field of cosmic rays with cloud chamber installations at mountain altitudes and nuclear emulsion stacks flown to balloon altitudes, is the discovery that in addition to pions, several other types of particles whose names were given as "K mesons and Hyperons", are also produced in high energy collisions. The K mesons decayed either into pions and neutrinos or into muons and neutrinos in time intervals of $10^{-8} \sim 10^{-10}$ s. While the K mesons had a mass of ~ 1000 m_e, the hyperons had masses higher than the nucleons. The hyperon Λ^0 decayed into nucleons and pions, Σ^0 into Λ^0 and Ξ^0 into Λ^0 and π^0, with similar decay modes for their charged counterparts. A notable feature was that the K mesons and the hyperons were always produced in pairs—associated production—leading to the introduction of a new quantum member called the Strangeness quantum number.

Table 1.1 gives the properties of the elementary particles discovered in cosmic radiation and Table 1.2 the properties of the strange particles also discovered in cosmic radiation.

1.2 Discovery of Extensive Air Showers

It was Bruno Rossi [19], who as early as 1934, had noticed coincidences between several counters placed in a horizontal plane, far in excess of chance coincidences. He had noted in one of his papers "It would seem that occasionally very extensive groups of particles arrive upon the equipment". The most systematic investigation on these showers were undertaken by Pierre Auger and his collaborators [20]. They recorded coincidences between counters separated horizontally by as far as 75 meters. While the counting rate dropped sharply in going from 10 cms to 10 meters, the rate decreased very slowly at larger distances.

Table 1.1 Properties of elementary particles discovered in cosmic rays.

Particle	Symbol	Mass (MeV)	Mean life (sec)	Decay modes
Photon	γ	0	stable	
Electron	e^-	0.511	stable	
Positron	e^+	0.511	stable	
Mu meson	μ^-	105.66	2.197×10^{-6}	$e^- + \overline{\nu}_e + \nu_\mu$
	μ^+	105.66	2.197×10^{-6}	$e^+ + \nu_e + \overline{\nu}_\mu$
Pi meson	π^+	139.57	2.603×10^{-8}	$\mu^+ + \nu_\mu$
	π^-	139.57	2.603×10^{-8}	$\mu^- + \overline{\nu}_\mu$
	π^0	134.96	0.83×10^{-16}	$\gamma + \gamma$

Table 1.2 Properties of strange particles discovered in cosmic rays.

Particle	Symbol	Mass (MeV)	Mean life (sec)	Decay modes	Branching Ratio(%)
K mesons	$K^\pm$	493.67	1.237×10^{-8}	$\mu^\pm + \nu_\mu$	63.5
				$\pi^\pm + \pi^0$	21.16
				$\pi^\pm + \pi^\pm + \pi^\mp$	5.59
				$\pi^\pm + \pi^0 + \pi^0$	1.73
				$\pi^0 + \mu^\pm + \nu_\mu$	3.20
				$\pi^0 + e^\pm + \nu_e$	4.82
	K_s^0	497.67	0.892×10^{-10}	$\pi^+ + \pi^-$	68.61
				$\pi^0 + \pi^0$	31.39
	K_L^0	497.67	5.183×10^{-8}	$\pi^0 + \pi^0 + \pi^0$	21.5
				$\pi^+ + \pi^- + \pi^0$	12.39
				$\pi^\pm + \mu^\mp + \nu_\mu$	27.1
				$\pi^\pm + e^\mp + \nu_e$	38.7
Hyperons	Λ^0	1115.60	2.632×10^{-10}	$p + \pi^-$	64.2
				$n + \pi^0$	35.8
	Σ^+	1189.36	0.800×10^{-10}	$p + \pi^0$	51.64
				$n + \pi^+$	48.36
	Σ^0	1192.46	5.8×10^{-20}	$\Lambda^0 + \gamma$	100
	Σ^-	1197.34	1.482×10^{-10}	$n + \pi^-$	100
	Ξ^0	1314.9	2.90×10^{-10}	$\Lambda^0 + \pi^0$	100
	Ξ^-	1321.32	1.641×10^{-10}	$\Lambda^0 + \pi^-$	100
	Ω^-	1672.45	0.818×10^{-10}	$\Lambda^0 + K^-$	68.6
				$\Xi^0 + \pi^-$	23.5
				$\Xi^- + \pi^0$	8.0

Air shower studies assumed increasing prominence from mid 40's when it became clear that the primaries responsible for these showers were of very much higher energy than what could be produced at man made accelerators. Large arrays of counter assemblies began to be operated. In the first instance they were GM counter trays. These were replaced by liquid and plastic scintillators in later years. A new dimension was added to air shower research when it was established by Bassi, Clark and Rossi [21] that the thickness of the air shower disc was quite small and all the particles went through the detector in less than a few nanoseconds. This led them to suggest that by recording the arrival times of the shower particles at different scintillators in the array, it would be possible to determine the direction of arrival of the shower particles with an accuracy of $\sim 5°$. The direction determination together with a measurement of the density of particles passing through the different scintillators of the array provided an elegant method of determining the shower parameters—the core position, the shower size and the direction of arrival. Together with other types of detectors in the array that were capable of discriminating between the soft particles, the penetrating particles, the nuclear active particles and the relative time delays between the various components, air shower research became a powerful tool for the study of the characteristics of interactions at super high energies and also for determining the primary spectrum, composition and anisotropies at these very high energies. The particular advantage that air shower studies offered was that the effective area of detection could be orders of magnitude larger than compact assemblies that could be flown to balloon altitudes or carried in satellites. The importance and usefulness of the methods of extensive air showers can be gauged when it is realized that the flux of primary particles at an energy of $> 10^{14}$ eV is about one per square meter per hour.

References

1. C. T. R. Wilson, *Phil. Trans. Royal Soc.* **189** (1897) 265.
2. T. Wulf, *Phys. Z.* **10** (1909) 152;
ibid. **11** (1910) 811.
3. A. Gockel, *Phys. Z.* **11** (1910) 280.
ibid. **12** (1911) 595.
4. Victor Hess, *Phys. Z.* **13** (1912) 1084.
5. C. D. Anderson, *Science* **76** (1932) 238.
ibid. *Phys. Rev.* **43** (1933) 491
ibid. **44** (1933) 406.
6. H. J. Bhabha and W. Heitler, *Proc. Roy. Soc.* **159A** (1937) 432.
7. J. F. Carlson and J. R. Oppenheimer, *Phys. Rev.* **51** (1937) 220.
8. H. J. Bhabha, *Proc. Roy. Soc.* **164A** (1938) 257.
9. S. H. Neddermeyer and C. D. Anderson, *Phys. Rev.* **51** (1937) 884.
10. J. C. Street and E. C. Stevenson, *Phys. Rev* **51** (1937) 1005.
11. H. Yukawa, *Proc. Phys. Math. Soc. Japan* **17** (1935) 48.
12. M. Conversi, E. Pancini and O. Piccioni, *Phys. Rev.* **68** (1945) 232.
13. C. M. G. Lattes, H. Muirhead, Q. P. S. Ochialini and C. F. Powell, *Nature* **159** (1947) 694.
14. Marietta Blau and H. Wambacher, *Nature* **140 (1937) 585;**
ibid. *Proc. Ac. Vienna* **146** (1937) 623.
15. J. Clay, *Proc. K. ned. Acad. Wet.* **30** (1927) 1115.
16. T. H. Johnson,*Phys. Rev.* **43** (1933) 703.
17. M. Schein, W. P. Jesse and E. O. Wollan, *Phys. Rev.* **59** (1941) 615.
18. H. L. Bradt and B. Peters, *Phys. Rev.* **74** (1948) 1828.
19. B. Rossi, *Cosmic Rays* (McGraw Hill 1964).
20. P. Auger, R. Maze and Grivet-Meyer, *C. R. Acad. Sci., Paris* **206** (1938) 1721.
21. P. Bassi, G. Clark and B. Rossi, *Phys. Rev.* **92** (1953) 441.

Chapter 2
PHENOMENOLOGY

When a primary cosmic ray proton or nucleus enters the earth's atmosphere, it interacts strongly with a nucleus in the air (nitrogen, oxygen, etc.). The point at which the particle interacts in the atmosphere is naturally random and decided, on a statistical basis, by its inelastic cross section on the target nucleus. The inelastic cross section, σ_{p-air}, of protons on an "air nucleus" (mean $Z = 7.5$, mean $A = 14.5$) is about 290 mb corresponding to an interaction mean free path of ~ 80 g cm^{-2} at an energy E_0 of $\sim 10^{14}$ eV. So the interaction occurs, on an average, at 80 g cm^{-2} from the top of the atmosphere. In such high energy collisions part of the primary particle energy is lost in the production of secondary particles. These secondary particles consist of mesons, baryons, hyperons etc. The fraction of the energy lost by the primary particle is called the *inelasticity*. While we defer a detailed discussion of the interaction characteristics to Chapter 4, we shall give their broad features here to understand the phenomenology of extensive air showers (EAS).

The energy lost by the primary particle is shared by the secondaries. The number of secondaries, called the total *multiplicity*, slowly increases with the energy of the interaction. The secondaries constitute different types of particles such as pions, kaons, baryons etc. They also acquire *transverse momenta*, p_t, due to which they travel at angle, θ, with respect to the direction of the primary particle, given by $\tan\theta = p_t/p_l$, where p_l are the longitudinal momenta. The p_t distribution is generally of the form

$$f(p_t)\,dp_t \approx \frac{p_t}{p_0^2} e^{-p_t/p_0}\,dp_t \tag{2.1}$$

where p_0 is the average value of p_t, which may depend on the particle type and, to a lesser extent on p_l as well as the energy of the primary particle, E_0. The distribution of p_l, expressed in terms of the Feynmann variable, $x^* = 2p_l^*/\sqrt{s}$, where s is the square of the center of mass (CM) energy and the superscript * denotes quantities measured in the CM system, is independent of E_0 to a first approximation[a], but depends on the particle type.

[a]This independence, known as Feynmann scaling, is violated in the central region of the collision at moderately high energies (few hundred GeV) and perhaps also in the fragmentation region at very high energies.

2.1 The Cascade Process

Majority of the particles produced in the first interaction are pions. Kaons and baryon-antibaryon pairs are also produced, if the energy is high enough, but in smaller numbers. These particles further interact in the air as they travel down. Some of them, such as the pions and kaons, are unstable and so they sometimes decay. If the relativistic time dilated decay mean free path is longer than the interaction mean free path, the interaction process dominates and *vice versa.* The particles produced in these interactions are of the same type as in the first interaction, though of lower energies. These particles, in turn, interact or decay further down in the atmosphere and this *Nuclear Cascade*, thus propagates downwards.

Some of the produced particles in this nuclear cascade are the neutral pions, π^0s. Their proper life time is $\sim 10^{-16}$ s and therefore they almost immediately decay into two γ rays, in spite of the relativistic time dilation, by the following process:

$$\pi^0 \rightarrow \gamma + \gamma \tag{2.2}$$

These γ rays initiate the electromagnetic cascades (see Sec. 2.2). The products of these cascades are photons, electrons and positrons. They constitute the electromagnetic component of air showers.

The charged mesons give rise to muons (μ mesons), when they decay; the dominant decay modes are:

$$\pi^\pm \rightarrow \mu^\pm + \nu \tag{2.3}$$

$$\begin{aligned} K^\pm &\rightarrow \mu^\pm + \nu \\ &\rightarrow \pi^\pm + \pi^0 \end{aligned} \tag{2.4}$$

The branching fractions for the dominant decay modes of the kaons are 63.5% and 21.2% respectively. These muons form the muon component of air showers. The neutrinos, which have only weak interaction, escape and carry away a significant fraction of the primary energy.

The surviving hadrons (all strongly interacting particles such as pions, kaons, baryons etc.) form the hadronic component.

As the shower propagates down in the atmosphere, the hadron and the electromagnetic components increase in size, reach a maximum and then decrease while the muon component does not suffer significant attenuation after reaching maximum because muons lose energy only by ionization and a small fraction of them are lost by decay. This complex *Nuclear-Electromagnetic Cascade* is called the Extensive Air Shower (EAS).

The decay mean free path is usually measured in units of length (m) while the interaction mean free path is measured in the amount of matter traversed, i.e. in g

cm^{-2}. Before we attempt to write the diffusion equations for the various components of the EAS cascade, we need to convert g cm^{-2} to linear distances. The atmospheric depth expressed in g cm^{-2} varies nearly exponentially with the atmospheric height, though the scale height also depends on the atmospheric depth. The standard atmosphere can be represented as

$$\begin{aligned} h\,(x) &= h_0\,(x)\,ln\,(1030/x) \\ h_0\,(x) &= 6,200 + 2.2x \end{aligned} \tag{2.5}$$

where $h(x)$ is the height, above sea level, in meters at a depth x, x is the atmospheric depth in g cm^{-2} and $h_0(x)$ is the variable scale height in meters.

Since the bulk of particles in EAS are photons and electrons, we shall first consider the electromagnetic cascade.

2.2 Electromagnetic Cascade

The electromagnetic cascade is the result of a series of electromagnetic interactions of electrons and photons with the atmospheric nuclei. These electromagnetic processes and the cascade theory are extensively discussed in the classical book by Rossi [1]. Here, we briefly describe them and give the relevant formulae for ready reference.

2.2.1 Interaction of Photons

Photons interact with matter through the following three processes:

- photo-electric effect,
- Compton effect, and
- pair production

2.2.1.1 Photo-electric Effect

Photo-electric effect is the process by which a photon loses its energy completely spending part of its energy in liberating an electron from the atom and the rest is given to the liberated electron as kinetic energy. This process is important only at very low energies comparable to the binding energies of the atomic electrons. At higher energies, the photo-electric cross section is small compared to other processes.

2.2.1.2 Compton Effect

The Compton effect is the process by which the photon interacts with a free electron, or at energies high enough for the orbital electrons to be considered free. In this process, the photon transfers part of its energy and momentum to the free electron initially at rest and emerges with reduced energy. The probability of Compton effect for a photon of energy E is given by

$$\mu_{com} = X_0 \frac{C m_e c^2}{E} \left[\ln \frac{2E}{m_e c^2} + \frac{1}{2} \right] \tag{2.6}$$

where

$$C = \pi N \frac{Z}{A} r_e^2 = 0.15 \frac{Z}{A} \quad \text{g}^{-1}\text{cm}^2, \tag{2.7}$$

m_e is the mass of electron in eV/c^2, c the velocity of light, r_e the classical radius of the electron, N the Avagadro number, Z and A are the charge and mass numbers of the material and X_0 is the radiation length defined later (Eq. 2.10). This process also is important only at low energies, $\sim$ a few tens of MeV. Above these energies, pair production takes over.

Thus, for electromagnetic cascades initiated by high energy photons or electrons, these two processes can be neglected.

2.2.1.3 Pair Production

Pair production is the process by which the photon interacts with the Coulomb field of the nucleus. In this interaction, the photon disappears and an electron-positron pair is created. Therefore, this process can occur only at energies higher than the sum of the rest masses of the electron and the positron. The remaining energy of the photon is almost entirely shared by the two electrons in the form of kinetic energy. The momentum balance is taken care of by the nucleus.

The differential probability, per radiation length of air, that a photon of energy E_0 creates an electron positron pair, one of which has an energy $E = vE_0$ is, under the assumption of complete screening (since we are interested here in high energy particles, see Ref. [1]),

$$\psi_{pair}(v) = 1 - \left(\frac{4}{3} + 2b \right) v (1 - v) \tag{2.8}$$

where

$$b = \frac{1}{18 \ln \left(183 Z^{-1/3} \right)} \tag{2.9}$$

and has a value of 0.0122 for air ($Z = 7.5$). The radiation length, X_0, expressed in units of g cm^{-2}, is defined as

$$\frac{1}{X_0} = 4\alpha \frac{N}{A} Z (Z+1) r_e^2 \ln \left(183 Z^{-1/3} \right) / \left[1 + 0.12 \left(\frac{Z}{82} \right)^2 \right] \tag{2.10}$$

where α ($= 1/137$) is the fine structure constant. The above formula gives a value of 37.7 g cm^{-2} for X_0 in air. Linsley [2] recommends a value of 37.15 g cm^{-2} after taking into account a better representation of the atomic structure.

From Eq. 2.8 it is clear that the pair creation probability is independent of the energy of the photon and depends only on the fraction of energy v taken by one of the pair.

2.2.2 *Interaction of Electrons*

Electrons interact with matter through the following processes:

- ionization, and
- bremsstrahlung

2.2.2.1 Ionization

The energy lost by an electron or a positron due to ionization in a medium is given (see [1]), in units of eV $g^{-1}cm^2$, by

$$\frac{dE}{dx} = -2Cm_e \left[\ln \left(\frac{\pi^2 m_e^2}{(1-\beta^2)^{3/2}\, I^2(Z)} \right) - a \right] \tag{2.11}$$

where $\beta = v/c$, v the velocity of the electron, $I(Z)$ is the average ionization potential in eV of an atom of the medium with atomic number Z. $I(Z)$ for air is 80.5 eV. The constant $a = 2.9$ for electrons and 3.6 for positrons.

2.2.2.2 Bremsstrahlung

Bremsstrahlung is the process by which a high energy electron emits a photon when it interacts with the electromagnetic field of the nucleus in the medium. For high energy electrons, where the formulae for complete screening can be used (see [1]), the differential probability, per radiation length of air, that an electron or a positron of energy E emits a photon of energy vE is

$$\psi_{rad}(v) = v + \left(\frac{4}{3} + 2b \right) \frac{(1-v)}{v} \tag{2.12}$$

where b is given by Eq. 2.9. As in the case of the pair creation probability, the radiation probability also does not depend explicitly on the energy of the electron or the emitted photon, but only on the fractional energy, v, taken by the photon.

Integrating Eq. 2.12, one gets the average fractional energy loss per radiation length as

$$\frac{1}{E}\frac{dE}{dt} = 1 + b \approx 1 \tag{2.13}$$

where t is the thickness of the medium measured in radiation lengths.

We need to introduce another quantity called the critical energy before we proceed to discuss the electromagnetic cascades. Critical energy, ϵ_0, is the energy loss due to collisions by an electron of energy ϵ_0 in one radiation length. It is approximately the energy at which radiation losses and collision loss are equal and has a value of 84.2 MeV in air.

2.2.3 The EM Cascade

When a high energy photon enters, or is created in, the atmosphere it soon interacts with an air nucleus and creates an electron pair. These electrons, which still have high energies, undergo bremsstrahlung process and emit high energy photons. The photons, in turn, materialize and more pairs of electrons are created. Thus, as the particles propagate down in the atmosphere, more and more electrons and photons are created. At each stage, their numbers increase while their energies decrease. The resulting avalanche of electrons and photons is called the electromagnetic cascade or electromagnetic shower. The problem of finding the total numbers and the energy spectra of these particles at various stages of the cascade development solved by many authors under various approximations, is discussed in great detail by Rossi [1]. Here, we briefly mention the approximations and give the final results.

In solving the cascade problem, the asymptotic formulae with complete screening for pair production and bremsstrahlung (Eqs. 2.8 and 2.12) are used. Photo-electric effect and Compton process are neglected. These approximations do not introduce any significant errors since we are interested in cascades generated by high energy particles.

In one of the approximations used, called the "Approximation A", all processes except pair production and bremsstrahlung are neglected. For this reason, the solutions are valid for particles with energy E such that $E_0 \gg E \gg \epsilon_0$, where E_0 is the energy of the primary electron or photon. Furthermore, the method used in obtaining the solutions restricts their validity to $t \gg 1$, where t is the depth measured in radiation lengths from the starting point of the cascade. In "Approximation B", the ionization loss of electrons is also taken into account, but as a constant value equal to ϵ_0 per radiation length. Using this approximation results for particles of all energies have been obtained.

We summarize here some of the results obtained from these calculations (see Rossi [1] and Greisen [3]).

For showers initiated by photons of energy E_0, the average total number of charged particles of all energies, $N_e(E_0, t)$, as a function of depth, t measured in radiation lengths, is given approximately, in the region where the number of particles is large, by

$$N_e(E_0, t) = \frac{0.31}{\beta_0^{1/2}} e^{t\left(1-\frac{3}{2}\ln s\right)} \tag{2.14}$$

where $s \simeq \frac{3t}{t+2\beta_0}$ and $\beta_0 = \ln(E_0/\epsilon_0)$.

The number of particles initially increases with depth, reaches a maximum and then decreases. The parameter s is called the "age" parameter and is a measure of the stage of shower development. Its value at the start of the cascade is 0, < 1 till the shower reaches maximum, $= 1$ at shower maximum and > 1 beyond. At $s > 2$ the number of particles becomes very small.

The variation of the number of particles with depth is approximately given by

$$\frac{\partial \ln N_e}{\partial t} \simeq \frac{1}{2}\left(s - 1 - 3\ln s\right). \tag{2.15}$$

The depth, $t_{\max}$, at which the shower reaches maximum is given by

$$t_{\max}\left(E_0\right) = 1.01\left[\ln\left(\frac{E_0}{\epsilon_0}\right) - \frac{1}{2}\right] \tag{2.16}$$

and the number of particles at maximum by

$$N_{\max}\left(E_0\right) = \frac{0.31}{\left[\ln(E_0/\epsilon_0) - 0.18\right]^{1/2}}\frac{E_0}{\epsilon_0}. \tag{2.17}$$

It is seen that the depth of maximum is approximately proportional to the logarithm of the primary energy and the number of particles to the primary energy.

The average number of particles of energy $> E$ is approximately given by

$$N_e\left(E_0, > E, t\right) = \frac{0.135}{\beta^{1/2}\left(E\right)\left[1 + 0.4\left(s - 1\right)\right]}e^{t\left(1-\frac{3}{2}s\right)} \tag{2.18}$$

where $s = \frac{3t}{t+2\beta(E)}$ and $\beta\left(E\right) = \ln\left(E_0/E\right)$. The depth of maximum and the number of particles at maximum are given by

$$t_{\max}\left(E_0, > E\right) = 1.01\left[\ln\left(\frac{E_0}{E}\right) - \frac{1}{2}\right] \tag{2.19}$$

and

$$N_{\max}\left(E_0, > E\right) = \frac{0.135}{\left[\ln\left(E_0/E\right) - 0.18\right]^{1/2}}\frac{E_0}{E}. \tag{2.20}$$

The integral energy spectrum of shower particles at depth t is proportional to $\left(\frac{E_0}{E}\right)^s$.

The above formulae refer to an average shower of a given primary energy E_0. In individual showers, the number of particles at a given depth fluctuates because of fluctuations in the starting point as well as in the interaction processes along the cascade. The fluctuations are least near the shower maximum and increase on either side. The problem of fluctuations is so complex that analytical solutions, even approximate ones, are not available. It can, however, be studied in detail by Monte Carlo simulations (see Sec. 2.2.4).

2.2.3.1 Lateral Spread

As the shower develops, it also spreads laterally, i.e. perpendicular to the direction of the incident primary particle, due to the multiple Coulomb scattering. The mean square scattering angle, due to multiple Coulomb scattering, of an electron of energy E in traversing a small thickness t of matter is

$$\left\langle \delta\theta^2 \right\rangle = \left(\frac{E_s}{E}\right)^2 \delta t \tag{2.21}$$

where t is measured in radiation lengths and $E_s = m_e c^2 \left(\frac{4\pi}{\alpha}\right)^{1/2} = 21.2$ MeV. The three dimensional cascade problem, including multiple scattering, has been approximately solved by many authors and is discussed in detail by Rossi [1] and Nishimura [4]. The numerical results obtained by Kamata and Nishimura [5] can be well approximated by the function, known as Nishimura-Kamata-Greisen (NKG) function, suggested by Greisen [3]

$$\triangle\left(N_e, r\right) = \frac{N_e}{2\pi r_0^2} \frac{\Gamma\left(4.5 - s\right)}{\Gamma\left(s\right)\Gamma\left(4.5 - 2s\right)} \left(\frac{r}{r_0}\right)^{(s-2)} \left(1 + \frac{r}{r_0}\right)^{(s-4.5)} \tag{2.22}$$

where $\triangle\left(N_e, r\right)$ is the density of particles per m^2 at a distance r in a plane perpendicular to the shower axis in a shower with total number of particles (shower size) N_e, Γ is the gamma function, s is the age parameter and $r_0 = \frac{E_s}{\epsilon_0}\frac{X_0}{\rho}$ is the Molière unit of length or "scattering length" occurring in the scattering theory, and ρ is the density of air at the observation level. Greisen suggests that, because of the exponential nature of the atmosphere, the value of r_0 should be taken at 2 radiation lengths above the observation level. Thus r_0 is 81 m at sea level.

More recently Uchaikin and collaborators [6] solved the three dimensional cascade equations using a different approach and numerical integration methods for photons of energy $10 - 10^5$ GeV. They find that their distribution is steeper than Eq. 2.22. Their function can be represented by

$$\Delta_U\left(N_e, r\right) = m^{-2}\Delta\left(N_e, r/m\right) \tag{2.23}$$

where $m = 0.78 - 0.21s$ for $0.8 \leq s \leq 1.6$. This function agrees better with the results of Monte Carlo simulations.

2.2.4 Monte Carlo Simulations

Recent detailed Monte Carlo calculations by several authors have also shown that the lateral structure function is steeper than given by the NKG function (For a brief description of the Monte Carlo method, see Sec. 2.3.2). Allan et al. [7] find, from their detailed Monte Carlo simulations, that the lateral spread in showers initiated by photons of energy 10 GeV is smaller than the NKG function by a factor of 2. More detailed simulations were carried out by Hillas and Lapikens [8] for photons and electrons of energy 10 and 100 GeV taking careful account of all electromagnetic processes. They also find that the lateral spread is smaller than the NKG function. Their results agree very well with the Uchaikin function given by Eq. 2.23.

2.3 EAS Cascade

The properties of various components of extensive air showers, such as the electron, muon and hadronic components, are theoretically studied, for a given hadron interaction model, by solving a set of coupled diffusion equations which govern the propagation of these components. The longitudinal development can be studied by solving these equations numerically. This numerical method is useful only in studying the average properties of air showers. Even the average lateral structure is difficult to incorporate as the equations become even more complex. Fluctuations and correlations of various components is studied using the Monte Carlo method. The lateral structure can easily be incorporated in this method.

2.3.1 Numerical Method

The longitudinal development of the air shower cascade initiated by a proton of energy E_0 is governed by a set of coupled diffusion equations, which are so complex that an analytical solution is practically impossible. They have to be solved numerically assuming a hadron interaction model. In the following, we shall assume that only pions and baryons are produced in the interactions. Production of other particles like kaons etc. can be taken into account by adding additional terms appropriately.

Let $F_B(E,x), F_\pi(E,x), F_\mu(E,x), F_e(E,x)$ and $F_\gamma(E,x)$ be the number of baryons pions, muons, electrons ($e^\pm$) and photons respectively of energy E in unit energy interval at a depth x g cm^{-2} from the top of the atmosphere. Then, the diffusion equations can be written as follows:

For baryons,

$$\begin{aligned}\frac{\partial F_B(E,x)}{\partial x} &= -\frac{F_B(E,x)}{\lambda_B} + \int_E^{E_0} G_{BB}(E',E)\frac{F_B(E',x)}{\lambda_B}\,dE' \\ &+ \int_E^{E_0} G_{\pi B}(E',E)\frac{F_\pi(E',x)}{\lambda_\pi}\,dE' \end{aligned} \tag{2.24}$$

where

$G_{BB}(E',E), G_{\pi B}(E',E)$ are the number of baryons of energy E in unit energy interval produced in interactions of baryons and charged pions respectively of energy E' and

λ_B, λ_π and are the interaction mean free paths (measured in g cm^{-2}) of baryons and charged pions respectively, which may be energy dependent.

The first term on the right hand side of Eq. 2.24 represents loss of baryons due to interaction. The subsequent terms represent contribution to baryons from interactions of baryons and pions respectively.

For pions,

$$\frac{\partial F_\pi(E,x)}{\partial x} = -F_\pi(E,x)\left(\frac{1}{\lambda_\pi}+\frac{1}{d_\pi}\right)+\int_E^{E_0} G_{B\pi}(E',E)\frac{F_B(E',x)}{\lambda_B}dE' + \int_E^{E_0} G_{\pi\pi}(E',E)\frac{F_\pi(E',x)}{\lambda_\pi}dE' \quad (2.25)$$

where

$G_{B\pi}(E',E), G_{\pi\pi}(E',E)$ are the number of pions of energy E in unit energy interval produced in interactions of baryons and charged pions respectively of energy E' and

d_π is the time dilated decay mean free path of pions measured in g cm^{-2} and is given by $\frac{Ec\tau_\pi}{m_\pi}\rho(x)$ and $\rho(x)=\frac{x}{h_0(x)}$ is the density of air at x. $h_0(x)$ is given by Eq. 2.5, but converted to cm.

The first term on the right hand side of Eq. 2.25 represents loss due to interaction and decay of pions and the subsequent terms represent production of pions in the interactions of baryons and pions respectively as before.

For muons, the equation is

$$\frac{\partial F_\mu(E,x)}{\partial x} = -\frac{F_\mu(E,x)}{d_\mu}+\frac{F_\pi(E_\pi,x)}{d_\pi}+I\frac{\partial F(E,x)}{\partial E} \quad (2.26)$$

where $E_\pi=\frac{m_\pi}{m_\mu}E$ is the energy of the pion required, on an average, to decay into a muon of energy E and $d_\mu=\frac{Ec\tau_\mu}{m_\mu}\rho(x)$ is the decay mean free path of muons.

The first term on the right hand side of Eq. 2.26 represents loss of muons due to decay, the second due to production of muons from decay of charged pions and the last term accounts for loss due to ionization, which is assumed to be a constant of I g^{-1} cm^2 in the same units as E.

The equation for electrons is

$$\frac{\partial F_e(E,x)}{\partial x} = -\frac{F_e(E,x)\mu_e(E)}{X_0}+\int_E^{E_0} G_{ee}(E',E)\frac{F_e(E',x)}{X_0}dE' + \int_E^{E_0} G_{\gamma e}(E',E)\frac{F_\gamma(E',x)}{X_0}dE' \quad (2.27)$$

where

$$\mu_e(E)=\int_0^E \varphi_{rad}(E,E')\,dE' \quad (2.28)$$

and

$$\varphi_{rad}(E, E')\, dE' = \psi_{rad}(v)dv.$$

$G_{ee}(E', E)$ and $G_{\gamma e}(E', E)$ are the probabilities, respectively, of producing an electron of energy E in unit energy interval due to survival of an electron in a radiation process and due to pair production by a photon, of energy E'. X_0 is the radiation length defined in Eq. 2.10.

The first term in the right hand side of Eq. 2.27 represents loss of electrons due to bremsstrahlung process, the second gain due to the same process from higher energy electrons and the third gain due to pair production from higher energy photons.

Similarly the equation for photons is

$$\begin{aligned}\frac{\partial F_\gamma(E,x)}{\partial x} &= -\frac{F_\gamma(E,x)\,\mu_\gamma(E)}{X_0} + \int_E^{E_0} G_{e\gamma}(E',E)\,\frac{F_e(E',x)}{X_0}dE' \\ &+ \int_E^{E_0} G_{B\gamma}(E',E)\,\frac{F_B(E',x)}{\lambda_B}dE' \\ &+ \int_E^{E_0} G_{\pi\gamma}(E',E)\,\frac{F_\pi(E',x)}{\lambda_\pi}dE' \end{aligned} \tag{2.29}$$

where

$$\mu_\gamma(E) = \mu_{pair}(E) + \mu_{com}(E) \tag{2.30}$$

and μ_{pair} and μ_{com} are expressions similar to Eq. 2.28 for pair creation and Compton processes. $G_{e\gamma}(E', E)$ is the probability of producing a photon of energy E in unit energy interval due to bremsstrahlung process by an electron of energy E' and $G_{B\gamma}(E', E)$ and $G_{\pi\gamma}(E', E)$ are similar probabilities due to interactions of baryons and charged pions of energy E', respectively, through production and decay of π^0s.

In Eq. 2.29 the first term represents loss of photons due to pair production and Compton processes, the second term production of photons due to bremsstrahlung process by of higher energy electrons, and the third and fourth terms denote gain due to production and subsequent decay of π^0s in interactions of baryons and charged pions respectively.

These coupled integro-differential equations can be solved only by numerical methods. The boundary condition that the incident particle is a proton leads to the solution of the cascade curves of the various components. On the other hand, if the boundary condition is a spectrum of protons incident on the top of the atmosphere, steady state spectra of the various components at various depths in the atmosphere result.

These equations, however, yield only the average properties of showers. In order to study fluctuations in the various components, one has to take recourse to Monte Carlo simulations.

2.3.2 Monte Carlo Method

The Monte Carlo method involves generation of the spatial and momentum coordinates of the various particles produced in the interactions, governed by their distributions in the assumed interaction model.

The following features of the interactions are specified in the interaction model assumed:

- The inelastic interaction cross sections for the different types of particles such as pions, kaons, baryons etc.,
- The energy dependence of the average value of the inelasticity and its distribution in interactions of different types of particles,
- The energy dependence of the average values of the multiplicities of various types of particles produced in interactions of these different particles and their distributions,
- The distribution of the longitudinal momenta or energies of the different types of particles,
- Their transverse momentum distributions,
- Any other additional features like production of heavy particles etc., which one wants to study.

The value taken by any parameter, like the interaction point or the transverse momentum of one π^+ etc., in a particular instance is decided by a random number generated by a random number generator with a uniform distribution between 0 and 1. A brief description of different methods used to achieve this is described in Appendix A. For example the point of interaction, x, in an individual interaction is given by

$$x = -\lambda \ln R \tag{2.31}$$

where λ is the interaction mean free path and x is the distance in g cm^{-2}, from the point of production (top of the atmosphere if it is the incident primary particle), at which the interaction takes place in that instance and R is the random number. Similarly the values of the various parameters are generated using the appropriate procedures for transforming the uniform distribution of random numbers to the required distributions, as described in Appendix B.

The algorithm of a typical Monte Carlo program consists of two parts; first the generation of individual interactions and the various attributes of the produced particles like the type, energy, momentum etc., and second the propagation of these particles down in the atmosphere through interaction or decay.

First the interaction point of the incident primary particle, say a proton of energy E_0, is decided as discussed above. Then the total number of particles produced in this interaction is decided from the average multiplicity at that energy and the multiplicity distribution, e.g. KNO scaling [9]. The actual numbers of different types of particles produced are decided form their average fractions of the total multiplicity conserving charge, baryon number etc. The transverse and longitudinal momenta and the azimuthal angles of emission are then generated from their distributions conserving energy and momenta simultaneously.

In the second part, each particle in turn is propagated down in the atmosphere. If it is an unstable particle, the interaction point and decay point are generated using the appropriate mean free paths and whichever occurs earlier, is taken as the event that has occurred. If it decays, the decay products and their energy and momenta are generated using the decay kinematics. If it interacts, the interaction is treated as before. If it is a π^0 it is immediately assumed to decay, due to its extremely short life time, and the energy and emission angles of the two photons are generated using the two body decay kinematics. The number of particles at various depths in the atmosphere in the electromagnetic cascades initiated by these photons are then calculated either using the average electromagnetic cascade curve (Eq. 2.14) or by detailed simulation of the EM cascade itself, depending on the detail required. Generally, the former is adequate since the fluctuations in the EM cascade are small and the fluctuations in the electron component are dominated by fluctuations in the hadron cascade. As the cascade propagates, the particles passing through various levels in the atmosphere are accumulated to obtain their cascade curves. The lateral spread of the various components at different altitudes are obtained from the points of production and the opening angles. The arrival time structure of various particles is also calculated by keeping track of the travel time of each particle.

Different techniques and approximations are used by different authors for generating the cascades depending on their needs. A brief description of the algorithm used by the Kolar Gold Fields (KGF) air shower group [10, 11] is given in Appendix B.

2.3.3 EAS Initiated by Primary Nuclei

Cascades initiated by nuclei differ from nucleon initiated cascades essentially in two ways: *i*) the interaction cross section of nuclei is larger than that of nucleons and hence the first few interactions take place higher up in the atmosphere; *ii*) the interaction products in these interactions are, to a first approximation, governed by the energy per nucleon rather than the total energy.

The simplest and the most often used model is the superposition model. In this model, it is assumed that a nucleus of mass number A and total energy E_0 is equivalent to A independent nucleons each of energy E_0/A. Thus the properties of a nucleus initiated shower are obtained by summing up the showers due to these individual nucleons. This leads to two important differences between nucleon and nucleus initiated showers: *i*) nucleus initiated showers develope more rapidly since the energy per nucleon is smaller (as we shall see later) and *ii*) fluctuations in their development are smaller since the nucleus is equivalent to a beam of large number of individual nucleons. In this model, however, the distribution of the points of first interaction is same as that in nucleon initiated showers. In reality, they are different since a nucleon released in a nucleus-nucleus interaction may not interact immediately. The interaction mean free path, therefore, underestimates the extent of fluctuations. It has been shown by Tomaszewski and Wdowczyk [12] that the interaction mean free path adequately describes the average properties of nucleus initiated showers and is therefore useful in studying air shower parameters sensitive to primary mass.

The fragmentation model, which corrects for the deficiencies of the interaction mean free path, is based on low energy cosmic ray data on nucleus-nucleus collisions in nuclear emulsion experiments. A nucleus of mass number A and total energy E interacts with an air nucleus of mass number B with a cross section that can be written as

$$\sigma_{AB} = 10\pi r_0^2 \left(A^{1/3} + B^{1/3} - \delta\right)^2 \left[1 + \frac{2.5}{280} \ln^{1.8}\left(\frac{E/A}{100}\right)\right] \quad \text{mb} \tag{2.32}$$

where $r_0 = 1$ fm is the classical electron radius, $\delta = -0.1$, and E is the total energy of the incident nucleus. The above formula takes into account the increase of cross section with energy. The first interaction, therefore, occurs much higher in the atmosphere than in the case of a primary proton. In this interaction only a few nucleons in the nucleus interact with the air nucleus to produce secondary particles. The mean number of nucleon-nucleon interactions occurring is given by the "wounded nucleon" model. This number, $\langle n \rangle_{AB}$, according to Bialas et al. [13], is given by

$$\langle n \rangle_{AB} = \frac{A\sigma_{pB}}{\sigma_{AB}} + \frac{B\sigma_{pA}}{\sigma_{AB}}. \tag{2.33}$$

Here, σ_{pA} and σ_{pB} are the cross sections of nucleons on target nuclei A and B respectively. The first term in Eq. 2.33 is the number of wounded nucleons in the projectile and the second those in the target. Several other nucleons, a few light nuclei and in general a heavy fragment are also produced. These particles interact further down in the atmosphere and the cascade propagates. The individual interactions are treated as the normal nucleon-nucleus interactions. This model leads to wider fluctuations than the interaction mean free path, but to nearly the same average properties of showers.

2.4 General Features of Air Showers

An excellent account of EAS phenomenology based on early work is given by Hâyakawa [14]. Here we shall summarize the features of air showers obtained from more recent studies.

While the detailed features of air showers depend on the details of the interaction models used in the calculations, some general features of air showers can be summarized as follows:

2.4.1 Electron Component

Based on Monte Carlo simulations using the scaling model for nuclear interactions Gaisser and Hillas [15] (see also [16]) find that the average number of particles, $N_e(E,t)$ at a depth X g cm^{-2}, measured from the point of first interaction, in a shower initiated by a proton of energy E_0 (GeV) can be adequately represented by

$$N_e(E_0,t) = S_0 \frac{E_0}{\epsilon} e^{t_m} \left(\frac{t}{t_m}\right)^{t_m} e^{-t} \tag{2.34}$$

where

$$t_m(E_0) = \frac{X_m(E_0)}{\lambda} = \frac{X_0'}{\lambda} \ln\left(\frac{E_0}{\epsilon}\right) - 1 \tag{2.35}$$

$S_0 = 0.045$, $\epsilon = 0.074$, X_m is the depth of shower maximum in g cm^{-2}, $t = \frac{X}{\lambda}$, $X_o' = 36$ g cm^{-2} and $\lambda = 70$ g cm^{-2}. The depth of shower maximum increases logarithmically with primary energy as in the case of pure electromagnetic showers. The number of particles initially increases with depth, reaches a maximum at $X = X_m$ and then decreases. The cascade curves for protons of energy 10^5 and 10^7 GeV are plotted in Fig. 2.1 to illustrate the shift in the position of shower maximum with primary energy.

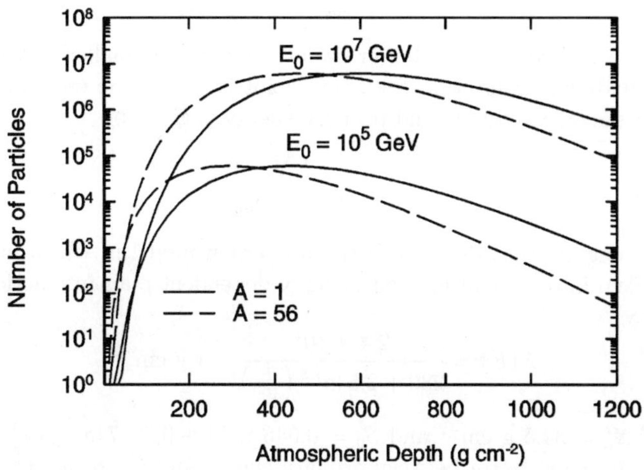

Figure 2.1 Cascade curves for proton and iron primaries from the model of Gaisser & Hillas(1977).

The number of particles at shower maximum is given by

$$N_m(E_0,t_m) = S_0 \frac{E_0}{\epsilon} \tag{2.36}$$

and is proportional to the primary energy E_0.

From the interaction mean free path, the corresponding formulae for a shower initiated by a nucleus of mass number A and total primary energy E_0 are

$$N_e(A, E_0, t) = S_0 \frac{E_0}{\epsilon} e^{t_m} \left(\frac{t}{t_m}\right)^{t_m} e^{-t}, \tag{2.37}$$

where

$$t_m(A, E_0) = \frac{X_m(A, E_0)}{\lambda} = \frac{X_0'}{\lambda} \ln\left(\frac{E_0}{A\epsilon}\right) - 1 \tag{2.38}$$

and

$$N_m(A, E_0, t_m) = S_0 \frac{E_0}{\epsilon}. \tag{2.39}$$

The depth of shower maximum shifts upwards by $0.51\lambda \ln A$, since the energy per nucleon, on which it depends, is smaller by a factor A. The number of particles at shower maximum is independent of A. This is illustrated in Fig. 2.1, where the cascade curves for iron primaries of energy 10^5 and 10^7 GeV are also plotted.

The mean shower size, weighted according to the exponential distribution of the first interaction points, is given by

$$\overline{N}_e(E_0, t) = S_0 \frac{E_0}{\epsilon} \frac{t_m}{t_m + 1} e^{t_m} \left(\frac{t}{t_m}\right)^{t_m+1} e^{-t}. \tag{2.40}$$

Note that the weighted mean of shower size at maximum is smaller by the factor $\frac{t}{t_m+1}$ than the individual showers whose starting point is fixed. The relation between the mean shower size at maximum and primary energy is given by

$$E_0 = \epsilon \frac{t_m + 1}{t_m} \frac{\overline{N}_m(t_m)}{S_0}. \tag{2.41}$$

Gaisser [16] suggests that, for a realistic interaction model in which nuclear target effects are taken into account and with energy dependent p-air interaction mean free path given by

$$\lambda(E) = \frac{2.4 \times 10^4}{280 + 2.5 \ln^{1.8}\left(\frac{E}{100}\right)} \quad \text{g cm}^{-2}, \tag{2.42}$$

the value of $X_o' = 34.5$ g cm^{-2} and $S_0 = 0.045 \times \left\{1 + 0.0217 \ln\left(\frac{E}{10^5}\right)\right\}$. With these values, the conversion factor between primary energy and mean shower size at maximum varies from 1.7 GeV/particle at $E_0 = 10^6$ GeV to 1.5 GeV/particle at $E_0 = 10^8$ GeV. Fluctuations due to shower development even for a fixed starting point are not taken into account in this treatment.

When air showers are observed at some depth in the lower third of the atmosphere, where they are beyond the shower maxima, showers of a given size arise (i) from different starting points for a given primary mass number and energy, (ii) different primary energies for the given mass number of the primary and (iii) different primary mass numbers. The realistic shower size spectrum, $G(N_e, x)\, dN_e$ at depth x is, therefore, given by

$$G(N_e, x)\, dN_e = dN_e \sum_A \int p(A, E_0, N_e, x)\, F_A(E_0)\, dE_0 \tag{2.43}$$

where $p(A, E_0, N_e, x)\, dN_e$ is the probability that a primary of mass number A and total energy E_0 gives rise to a shower of size N_e in a differential interval dN_e at depth x, which has to be obtained by Monte Carlo simulations using some interaction model, and $F_A(E_0)\, dE_0$ is the energy spectrum of primaries of mass number A. It is clear that the procedure of deducing the primary energy spectrum from the average primary energy to shower size relation discussed above, is only an approximation. If, in the region of interest, the mass composition of the primaries is changing with energy, this procedure can lead to erroneous results. In this context, it is interesting to see the relative contributions to a given shower size, from primaries of different energies and mass numbers, at a fixed atmospheric depth. Fig. 2.2 shows the relative contributions from different primary energies of protons and iron nuclei, with an energy spectrum $\propto E_0^{-2.6} dE_0$, obtained by Acharya et al. [10]using scaling model of nuclear interactions for showers of size $10^4 - 2 \times 10^4$ at an atmospheric depth of 920 g cm^{-2}. It can be seen that protons of energy over a factor of 8 and iron nuclei over

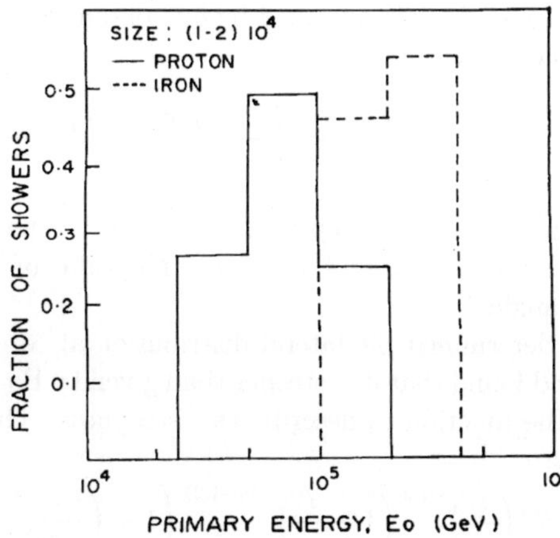

Figure 2.2 Relative contribution to showers of size $10^4 - 2 \times 10^4$ at 920 g/cm^2 from protons and iron nuclei of different energies with a primary energy spectrum $\propto E_0^{-2.6} dE_0$ for the scaling model of hadron interactions.

a factor of 4 contribute to showers in this size range. The relative contributions from primaries of different mass numbers depends on their relative fluxes at the top of the atmosphere, i.e. the primary mass composition.

2.4.1.1 Lateral Structure

Even though the EAS cascade is a superposition of many electromagnetic cascades initiated by γ rays of various energies arising as decay products of π^os produced in nuclear interactions at various depths in the atmosphere, the NKG function, Eq. 2.22, is found to be a reasonable approximation to the lateral distribution of charged particles and has been very useful in the analysis of experimental data. However, careful experimentation in the last two decades has revealed that the experimentally observed distribution can not be described by a single age parameter in Eq. 2.22 at all distances from the core. Because of this and also since different experiments measure the distribution in different distance ranges depending on the array configuration, it is difficult to compare the results of different experiments. Capdevielle and Gawin [17] find that the experimental values of the age parameter estimated assuming NKG function agree better with their own Monte Carlo simulations using the Uchaikin function, Eq. 2.23. They also find that the sizes of showers and hence the shower size spectrum obtained from these experiments could also be effected by the use of a single age parameter over all distances.

In the Volcano Ranch experiment, where the lateral distribution is measured for large sizes in the distance range 70 to 1,000 m, Linsley et al. [18] use a different function whose logarithmic slope is not fixed by one parameter only as in the case of the NKG function. Their function is

$$\Delta\left(N_e, r\right) = \frac{N_e}{2\pi r_0^2}\frac{\Gamma\left(\eta-\alpha\right)}{\Gamma\left(2-\alpha\right)\Gamma\left(\eta-2\right)}\left(\frac{r}{r_0}\right)^{-\alpha}\left(1+\frac{r}{r_0}\right)^{-(\eta-\alpha)} \tag{2.44}$$

with $r_0 = 100m$ at the atmospheric depth of 820 g cm^{-2}. The logarithmic slope of this function is $-\alpha$ for $r \ll r_0$ and $-\eta$ for $r \gg r_0$. The value of η depends on shower size as well as zenith angle.

The Akeno group determined the lateral distribution at core distances beyond 1 km in large showers and found that it is steeper than given by Eq. 2.44. An additional term is necessary in the function to describe this steepness. Their function is given by

$$\Delta\left(r\right) = C\left(\frac{r}{r_0}\right)^{-1.2}\left(1+\frac{r}{r_0}\right)^{-(\eta-1.2)}\left(1+\left(\frac{r}{1000}\right)^2\right)^{-0.6} \tag{2.45}$$

where C is the normalization constant.

2.4.2 Hadron Component

The hadron component is the back bone of the air shower. It constantly feeds energy into the electron component via the decay of π^0 mesons produced in the nuclear interactions of the hadronic cascade. It constitutes about 1% of the total number of particles in the shower. It consists of pions, kaons and baryons. Because of the transverse momenta acquired by these particles at production, the hadron component has a lateral spread. The characteristic spread, r_h (m), of hadrons of momentum p is given, to a first approximation, by

$$r_h = l\frac{p_0}{p} \tag{2.46}$$

where l is the interaction mean free path, in meters, of the parents of these particles and p_0 the average transverse momentum, which is typically ~ 0.4 GeV/c. Thus hadrons of energy ~ 40 GeV have a characteristic spread of ~ 8 m at sea level where l is ~ 800 m. The lateral distribution is, in general, of the form

$$\rho_h(E_h, r) = N_h(E_h)\, r^{-\alpha_h} e^{-r/r_h} \tag{2.47}$$

where $\rho_h(E_h, r)$ is the density of hadrons of energy E_h at a distance r from the core and $N_h(E_h)$ the total number of hadrons of energy E_h at the observation level. The total number of hadrons is nearly proportional to the primary energy of the shower. The integral energy spectrum of hadrons can be expressed as a power law. The spectrum is flat at lower energies and gradually becomes steeper as the hadron energy increases.

The hadron component carries important information on the characteristics of high energy interactions.

2.4.2.1 Temporal Structure of Hadrons

Hadrons, being more massive than electrons, lag behind the electron shower front. The time lag depends on the mass of the hadron and its energy. The time lag, δt, of a hadron of mass m and energy E_h with respect to the shower front, which travels almost with the velocity of light, in traversing a distance l is

$$\delta t \propto \gamma^{-2} \tag{2.48}$$

where $\gamma = E_h/m$ is the Lorentz factor. Thus measurement of the time lag distribution of hadrons of a given energy yields information of production cross section of heavier hadrons such as kaons, baryons and any new heavy mass particles.

2.4.2.2 Charged to Neutral (C/N) Ratio

If pions were the only particles produced in nuclear interactions, the only neutral hadron expected is the surviving primary nucleon which, starting off as a proton, has a probability of 0.5 of being a neutron after several interactions. Thus the C/N ratio would be very high. In general, if kaons and nucleon-antinucleon pairs are also produced, the ratio would be

$$C/N = \frac{n_{\pi^\pm} + n_{K^\pm} + 0.5 n_{N\bar{N}} + 0.5}{0.5 n_{N\bar{N}} + n_{K_L^0 K_S^0} + 0.5} \tag{2.49}$$

where n_i is the number of particles of type i at the observation level. The C/N ratio for pions, kaons and nucleons is ∞, 1.5 (since K_S^0s are short lived and not many of them survive) and 1 respectively. By comparing the total number of hadrons and the C/N ratio with Monte Carlo simulations, the production cross sections of different types of particles can be obtained.

2.4.3 Muon Component

The muon component arises from the decay of pions and kaons. At very high energies, where the production

of heavy flavor particles such as charm and beauty mesons and baryons, can be substantial, their decay also can contribute to the muon component, particularly to high energy muons. The muon component constitutes about 10% of the total number of particles in the shower. Since muons do not interact strongly, and the only losses are due to ionization and decay, they travel almost in straight lines down to the observation level. The lateral spread of the muon component is essentially due to the transverse momenta acquired by the parent pions, as in the case of hadrons, but it is much wider since muons travel larger distances compared to hadrons. Multiple scattering and deflections in the geomagnetic field also contribute to the lateral spread, but to a lesser extent; the latter also introduces asymmetric azimuthal distribution. High energy muons and also low energy muons at large distances from the core are produced at very high altitudes and hence carry important information about highest energy interactions in the cascade as well as the nature of the primary particle.

The lateral spread of the muons clearly depends on their energy since the opening angles of the parent pion are smaller for higher energy particles as the transverse momentum distribution is nearly independent of energy. The higher the muon energy the steeper the distribution. Based on some early investigations, Greisen [19] suggested the following function to describe the lateral spread of muons of energy $>$ 1 GeV at sea level:

$$\rho_\mu(N_e, r) = 18\left(\frac{N_e}{10^6}\right)^{3/4} r^{-3/4}\left(1+\frac{r}{320}\right)^{-2.5} \qquad m^{-2} \tag{2.50}$$

where $\rho_\mu(N_e, r)$ is the density of muons of energy > 1 GeV at a core distance r in showers of size N_e. Since the characteristic spread of muons is $\sim$ 300 m as compared to the Molière unit ($\sim$ 100 m) for the electron component, muons are spread out to much larger distances than electrons. At very large distances, larger than a km, muons contribute appreciably to the particle densities.

The energy spectrum of muons can be represented by a power law, as in the case of hadrons. It is flat at low energies because of increasing losses due to ionization and decay with decreasing muon energy, and becomes steeper at higher energies. The relation between the number of muons of energy $> E_\mu$ and the shower size can be expressed as

$$N_\mu(> E_\mu) \propto N_e^{\alpha_\mu(E_\mu)}. \tag{2.51}$$

The exponent $\alpha_\mu(E_\mu)$ decreases with E_μ presumably because the decay probability of higher energy pions decreases as their energy increases. The value of α_μ is 0.8 – 0.9 at $E_\mu \sim 1$ GeV and decreases to about 0.7 for $E_\mu \sim 200$ GeV. This feature makes the muon component sensitive to the primary mass number. In the interaction mean free path, the number of muons of energy $> E_\mu$ in a shower of primary mass A is given by

$$N_\mu(A) \quad \propto \quad A\left(\frac{N_e}{A}\right)^{\alpha_\mu} \quad \propto \quad A^{1-\alpha_\mu}. \tag{2.52}$$

Thus showers generated by heavy primaries have larger number of muons. The higher the energy of the muons, the more sensitive they are to primary mass number.

2.4.4 Cerenkov Light from EAS

Cerenkov light is emitted by a charged particle when it travels in a dielectric medium with a velocity greater than that of light in that medium. The light is emitted at an angle θ, with respect to the direction of motion of the particle, given by

$$\cos\theta = \frac{1}{\beta n} \tag{2.53}$$

where $\beta = v/c$, v and c are the velocities of the particle and light respectively and n the refractive index of the medium. Particles above $\beta = 1/n$ only can emit Cerenkov light, thus setting a threshold energy. Light is emitted along the surface of a cone of half angle θ around the direction of the particle.

For highly relativistic particles $\beta = 1$, and the angle of emission is maximum given by $\theta_{\max} = \cos^{-1}(1/n)$. The value of n is 1.000292 for air at sea level at N.T.P. and $\theta_{\max} = 1.3°$. Since the refractive index in air is close to 1, it is convenient to write $n = 1+\eta$. The refractive index in air varies with atmospheric depth and temperature and can be written as (see Ref. [20])

$$\eta = 0.000292\left(\frac{x}{1030}\right)\left(\frac{273.2}{T}\right) \tag{2.54}$$

where $T = 204 + 0.091x$ is the temperature at depth x (g cm^{-2}) in °K. In this approximation,

$$\theta_{\max} = \sqrt{2\eta} \qquad \text{rad.} \tag{2.55}$$

With these values the threshold energy for electrons at sea level is 21 MeV. The number of Cerenkov photons, between wave lengths λ_1 and λ_2 per cm of path length, $\frac{dN}{dl}$, emitted by a particle in air is given by

$$\frac{dN}{dl} = 2\pi\alpha\left(\frac{1}{\lambda_2} - \frac{1}{\lambda_1}\right)\sin^2\theta \tag{2.56}$$

where $\alpha = \frac{1}{137}$ is the fine structure constant. For highly relativistic particles the above formula can be written, with the approximation $\sin\theta \approx \theta$, as

$$\frac{dN}{dl} = 4\pi\alpha\eta \left(\frac{1}{\lambda_2} - \frac{1}{\lambda_1}\right). \tag{2.57}$$

Thus, the number of photons emitted per cm of air at sea level between wavelengths 3500 Å and 5000 Å is 0.23, which corresponds to 230 per g cm^{-2}.

Most of the electrons in an air shower are relativistic and therefore emit Cerenkov radiation. Because of the weak absorption of light in the atmosphere, most of it reaches the observation level. The number of Cerenkov photons in a shower is, therefore, proportional to the track length integral above the observation level and is a very good measure of the primary energy. Fluctuations in the amount of light for showers of a given primary are much less than those in shower size, thus making it a more accurate primary energy estimator.

An important feature of Cerenkov light, in contrast with other EAS components, is that it carries vital information about the longitudinal development of the shower. Calculations show that the lateral distribution as well as the temporal structure of the Cerenkov pulse at large distances from the core depend on the position of the shower maximum. This dependence is independent of the model of nuclear interactions and primary mass (see Ref. [21] and references therein). The lateral distribution can be approximated as

$$\rho(r) \sim (r+a)^{-\delta} \tag{2.58}$$

where

$$\delta = b + cX_m \tag{2.59}$$

and X_m is the depth of shower maximum in g cm^{-2} and a, b and c are constants.

The full width at half maximum (FWHM), $\tau(r)$, of the Cerenkov pulse at a core distance r is also related to the position of shower maximum as follows

$$\tau(r) \propto r^{d+eH_m} \tag{2.60}$$

where H_m is the height of shower maximum in linear dimensions and d and e are constants. Thus Cerenkov radiation is very useful for studies of primary charge composition.

The drawback of Cerenkov radiation is that observations can be made only on dark moonless, cloudless nights. Because of this the duty cycle is small and only 5 to 10% of the time is available for observations depending on the local conditions. Also, special techniques have to be used to detect the Cerenkov radiation against the night sky background.

2.4.5 Fluorescent Light

When charged particles in the air shower pass through the atmosphere, they ionize and excite the air molecules. The excited nitrogen molecules emit fluorescent light.

Most of the light comes from the 2P band system of molecular nitrogen ($\sim 80\%$) and the 1N band system of the N_2^+ molecular ion ($\sim 20\%$). As the light is emitted isotropically, it can be detected at large distances from the axis and thus can be distinguished from Cerenkov light which is emitted in the forward direction and therefore confined mostly to small distances. The light yield is inversely proportional and the ionization loss per unit path length is directly proportional to the atmospheric density and so the light yield per unit path length (in meters) is nearly constant, independent of altitude. The number of photons in the blue and ultraviolet region, after taking the atmospheric absorption into account, is nearly 1 per meter of track length. Because the light yield is small, the method can be used only for large showers. However, the effective area for recording showers is very large compared to conventional particle detector arrays and compensates the low duty cycle resulting from observations having to be made only on clear dark moonless cloudless nights. The effective aperture for a shower of energy 10^{20} eV is about 10^3 km^2 sr. As in the case of Cerenkov light, special techniques have to be used to detect it against the night sky background and other terrestrial sources such as lightning, auroras, airplanes etc.

The detector, as we shall see later, is essentially an assembly of large number of photomultipliers, each looking at a designated region of the sky. Each of them collects light from a particular stage of longitudinal development of the shower. Thus the longitudinal development of the shower can be measured, unlike conventional arrays. The primary energy is estimated integrating the shower development curve. Since the energy loss over a large part of the shower development is measured, fluctuations are much less and the primary energy estimate is more accurate compared to other techniques. The fluorescence technique is very effectively used by the Fly' Eye group [22].

References

1. Bruno Rossi, *High-Energy Particles* (Prentice-Hall, Inc., Englewood Cliffs, N. J., 1952).
2. J. Linsley, *Proc. 19th International Cosmic Ray Conference, La Jolla* **7** (1985) 163.
3. K. Greisen, *Progress in Cosmic Ray Physics* **3** (1956) 1.
4. J. Nishimura, *Handbook der Physik,* **XLVI/2** (1967) 1.
5. K. Kamata and J. Nishimura, *Suppl. Prog. Theor. Phys.* **6** (1958) 93.
6. A. A. Lagutin et al., *Proc. 16th International Cosmic Ray Conference, Kyoto* **7** (1979) 18.
7. H. R. Allan et al., *Proc. 14th International Cosmic Ray Conference, Munich* **8** (1975) 3071.
8. A. M. Hillas and J. Lapikens, *Proc. 15th International Cosmic Ray Conference, Plovdiv* **8** (1977) 460.
9. Z. Koba, H. B. Nielsen and P. Olesen, *Nucl. Phys.* **B40** (1972) 317.
10. B. S. Acharya et al., *Proc. 19th International Cosmic Ray Conference, Kyoto* **9** (1979) 109.
11. Srikantha Rao, Ph. D. thesis (unpublished), University of Bombay (1981).
12. A. Tomaszewski and J. Wdowczyk, *Proc. 14th International Cosmic Ray Conference, Munich* **8** (1975) 2899.
13. A. Bialas, M. Bleszynski and W. Czyz, *Nuclear Pysics* **B111** (1976) 461.
14. S. Hayakawa, *Cosmic Ray Physics* (Wiley-Interscience, 1969).
15. T. K. Gaisser and A. M. Hillas, *Proc. 15th International Cosmic Ray Conference, Plovdiv* **8** (1977) 353.
16. Thomas K. Gaisser, *Cosmic Rays and Particle Physics* (Cambridge University Press, Cambridge, 1990).
17. J. N. Capdevielle and J. Gawin, *J. Phys. G: Nucl. Phys.* **8** (1982) 1317.
18. J. Linsley, L. Scarci and B. Rossi, *J. Phys. Soc. Japan* **17 Suppl A-III** (1962) 91.
19. K. Greisen, *Ann. Rev. Nucl. Sci.* **10** (1960) 63.
20. A. M. Hillas, *J. Phys. G: Nucl. Phys.* **8** (1982) 1475.
21. M. V. S. Rao, *Proc. Int. Workshop on Very High Energy Gamma ray Astronomy, Ootacamund, India*, ed. P. V. Ramana Murthy and T. C. Weekes (1982) 197.
22. R. M. Baltrusaitis et al., *Nuclear Instruments and Methods in Physics Research* **A240** (1985) 410.

Chapter 3
TECHNIQUES

We have seen in the previous chapter that the most abundant component in air showers is the electron component. Because of this, the electron component is most generally used to trigger and record air showers. The electrons in air shower are confined to a thin disk of about a meter in thickness and travel with almost the velocity of light. The disk is several tens to hundreds of meters in radius depending on the size of the shower. These properties are made use of in triggering air showers against the background of small local showers generated by the low energy hadrons essentially originating from primaries of much lower energies.

In this chapter, we discuss how the showers are selected, how the various components are recorded and their intensities measured and how the various parameters such as shower size, age, arrival direction etc. are estimated. We shall first describe the different types of detectors used to study the various components in typical air shower arrays. We shall also describe a few multicomponent air shower arrays, past and present.

3.1 Detectors

Several types of detectors, ranging from simple Geiger-Muller (GM) counters to highly sophisticated total absorption calorimeters, spark chambers and time track detectors have been used depending on the component being measured.

3.1.1 Electron Detectors

The number of electrons passing through each of the array of detectors is measured for the purpose of estimating the total number of electrons in the shower. Two types of detectors are used for these measurements: those whose response is independent of the number of particles traversing and hence give information only on whether or not one or more particles have passed through them (YES/NO information); GM counters and neon flash tubes are of this type; the other type are those whose response is proportional to the energy deposited such as ionization chambers, scintillation detectors etc. In the former case, the number of particles is obtained from the number of counters discharged. Hence the accuracy depends on the number of counters in each location. In the second case, since most of the electrons are relativistic, and therefore minimum ionizing, the energy loss in the detector is nearly proportional to the total number of electrons passing through the detector. By calibrating the detector for the mean pulse height for a minimum ionizing particle, one obtains the average number of particles passing through the detector.

3.1.1.1 YES/NO Type of Detectors

Historically, the first counters to be used in EAS studies for measuring electrons, are the GM counters. They are cheap, fairly stable over long periods of operation and can be easily fabricated in various sizes. In order to obtain the density of particles, expressed in particles/m^2, trays with a large number of counters are used. The density of particles, Δ_e, can, then, be obtained from

$$\Delta_e = \frac{1}{S} \ln \frac{m}{m-n} \tag{3.1}$$

where m and n are the total number of counters, each of area S m^2, in the tray and the number discharged respectively. To cover a large dynamic range of density, sets of counters of different areas are used. Details of early work with GM counters are described by Galbraith [1].

Another YES/NO type of detector used, particularly for electron core structure studies, is the Neon Flash Tube. These are typically about 1 cm in diameter and filled with 99% pure Neon. Several thousands of them are arranged in a closely packed array in a horizontal plane forming a hodoscope. The hodoscope is placed between two electrodes across which high voltage is applied on an EAS trigger. The tubes traversed by charged particles then flash up and the image is photographed. This type of hodoscope was first used by the Tokyo air shower group [2] and subsequently by the Kiel group [3]. A cross section of the neon hodoscope used by the Kiel group is shown in Fig. 3.1 and a photograph of the electron and hadron components in the

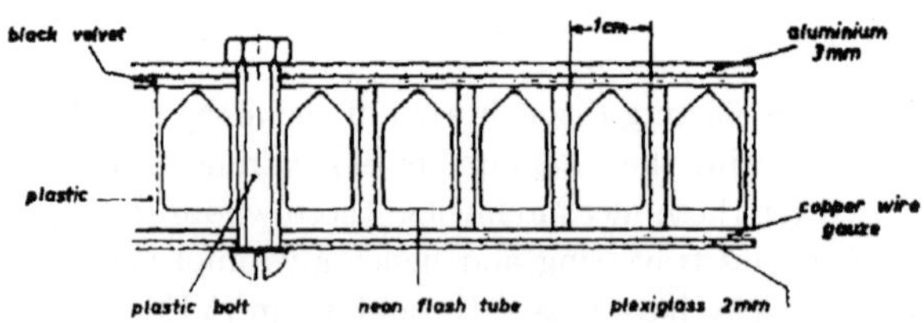

Figure 3.1 Cross section of the neon hodoscope in the Kiel experiment (Samorski 1973).

core of an air shower are shown in Fig. 3.2. The mean density of particles is given by

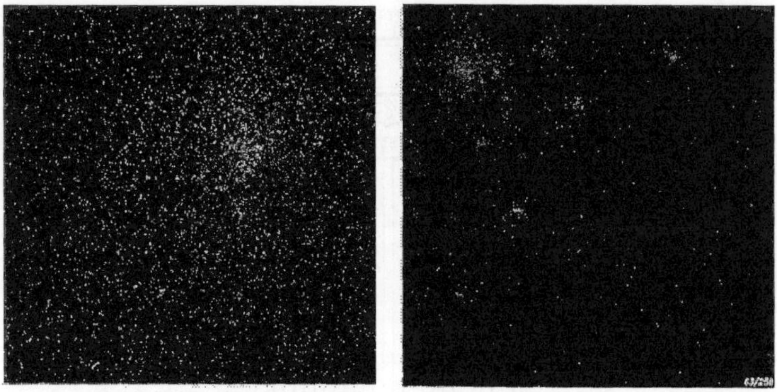

Figure 3.2 Neon hodoscope photograph of the electron component of a core of a shower of size 1.8×10^5 particles (left) and several bursts induced by the hadron component of a shower of size 2×10^6 particles (right) from the Kiel air shower experiment (Samorski 1973).

$$\Delta_e = \frac{1}{f_t e_t} \ln \left(1 - \frac{n_f}{\rho_t S}\right)^{-1} \tag{3.2}$$

where f_t and e_t are the effective area and the internal efficiency of the flash tubes, n_f is the number of tubes flashed over a subarea S and ρ_t the density of the tubes in the array (see Ref. [3] for details).

Yet another type of detectors used for core structure studies are visual detectors like spark chambers and cloud chambers.

The Tokyo air shower group [4] is one of the first to use spark chambers also for core structure studies. A schematic diagram of the chamber used by them is shown in Fig. 3.3. The chamber is made of a sealed glass box of size 50 cm × 50 cm × 2 cm filled with 99.9% pure neon gas. One of the electrodes is a duraluminum plate and the other a wire grid, both covered by the non-conducting glass plates. This feature increases the efficiency for multiple tracks dramatically. About 20 kV high voltage pulse of duration about 100 ns is applied on EAS trigger and the tracks are photographed through the grid electrode using a mirror system. Particle densities up to $1,500\ \mathrm{m}^{-2}$ are measured without any loss of efficiency with these chambers. An array of 66 such chambers covering an area of $20\ \mathrm{m}^2$ was used in the INS air shower experiment [5].

An array of 35 spark chambers, each of area $1\ \mathrm{m}^2$, has been operated by the Leeds group [6] also for core structure studies. Kino et al. [7] also have operated a $54\ \mathrm{m}^2$ array of 160 spark chambers, each of dimensions 50 cm × 50 cm × 1.7 cm in an air shower array at Mt. Norikura in Japan.

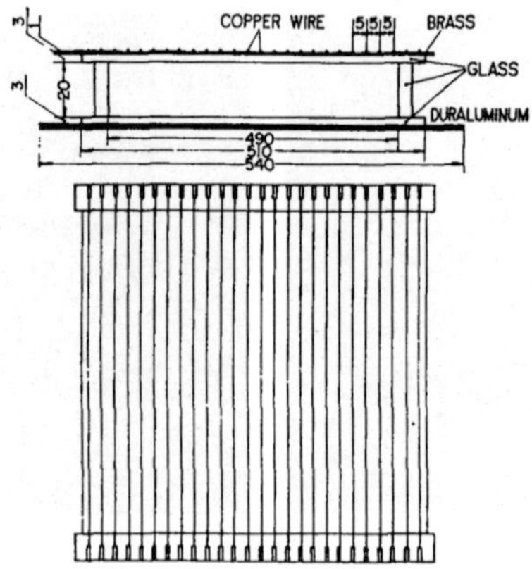

Figure 3.3 Schematic diagram of the spark chamber used by the Tokyo group (Nagano and Shibata, 1965).

Spark chambers have also been used to determine the arrival direction of the showers by mounting them in a vertical plane.

While cloud chambers have also been used for core structure studies, particularly by the Sydney group [8], the advantage being that one can cover a much larger range of densities, large areas are not possible as in the case of spark chambers.

3.1.1.2 Proportional Detectors

The earliest proportional detectors used are the ionization chambers by Williams [9], Campbell and Prescott [10] and others, but they soon gave way to the now popular and extensively used scintillation detectors. Experiments using ionization chambers are discussed in detail by Galbraith [1].

Scintillation detector essentially consists of a material which emits light on the passage of a charged particle and a photomultiplier which detects the light and converts it into a voltage pulse. The M.I.T. group [11] pioneered the development of large area scintillators suitable for air shower studies. At first liquid scintillators were used. The liquid scintillator essentially consists of a solvent—usually organic liquids like Toulene, Benzene etc.—which fluoresces on the passage of a charged particle traversing it. Since the fluorescence light is in the ultraviolet region, which is beyond the sensitive region of the photomultiplier, a converter like terphenyl or paraterphenyl is added. This converter absorbs and emits the light in a lower frequency region. A wave length shifter, such as Popop is generally also added to further shift the light into the blue region, where the photomultiplier is most sensitive and the absorption of light in the medium is small. These liquids have the disadvantage of being toxic and inflammable and messy to handle. The liquid scintillators in use now, particularly in Russia, have mineral oil as solvent, which is less toxic and has a higher flash point, but still messy to handle.

The M.I.T. group has pioneered also in developing plastic scintillators [12]. The base material is polystyrene or acrylic; a converter and a wavelength shifter are added as in the case of liquid scintillators. The plastic scintillators can easily be cast into large area blocks of any shape and size as required and are highly stable and easy to handle. The scintillator blocks are typically of 0.25 m^2 to 1 m^2 in area and 5–10 cm thick. Each detector generally contains 1–9 blocks covering an area of 0.25–2.25 m^2, depending on whether the array is designed for large or small showers, and is looked at generally by one photomultiplier. Though the light is emitted isotropically, because of the large size of the detector care has to be taken in designing the geometry of the detector so that approximately the same amount of light reaches the photomultiplier irrespective of where the particle traverses in the scintillator block.

The M.I.T. design is shown in Fig. 3.4. The scintillator is a circular disk of

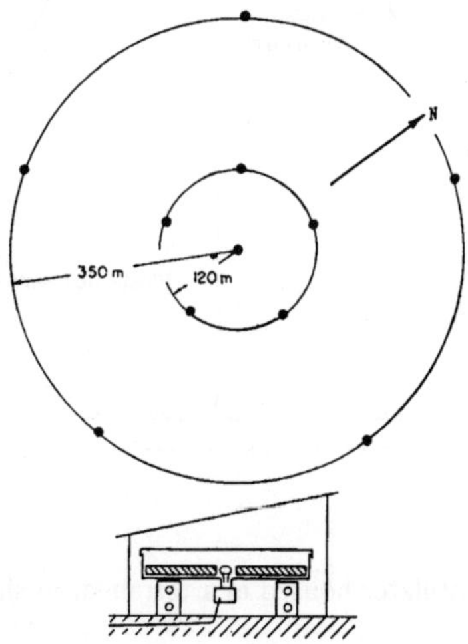

Figure 3.4 The disk shaped plastic scintillator(bottom) and the El Alto air shower array (top) of the M.I.T. group (Clark et al., 1957).

diameter 1 m and thickness 10 cm. It is housed in a galvanized iron container whose inner surface is white-painted to diffuse the light and make it nearly uniform. A 5″ diameter Dumont 6364 photomultiplier is placed in a well at the center of the disk to collect the light. The detector is calibrated with minimum ionizing cosmic ray muons. The response of the detector varied by less than a factor of two from center to edge.

The other design, which is used in most of the experiments nowadays, has the container in the shape of a tetrahedron. Schematic of a typical detector used by the Tata group [13] is shown in Fig. 3.5. The scintillator blocks are placed at the

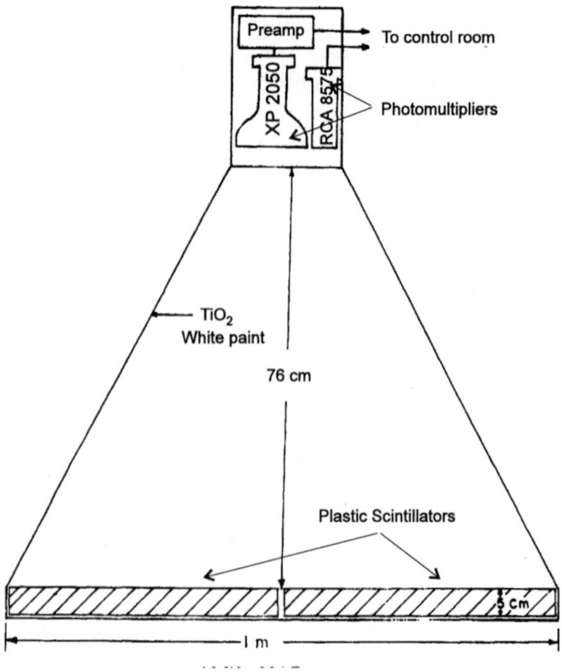

Figure 3.5 The palstic scintillator housed in a tetrahedron shaped container used by the Tata group (Sinha, 1987).

bottom and the inner walls are painted with white TiO_2 paint. The photomultiplier is housed at the top looking downwards. This design gives better uniformity of response over different parts of the scintillator. A substantial part of the light seen by the photomultiplier reaches it after multiple reflections in the walls of the container, thus making the response more uniform. Since the photocathode area forms only a small fraction of the inner surface, the number of photons reaching it is smaller than in the earlier design. Use of higher quantum efficiency photomultipliers ($\sim$ 20%), like Phillips XP2050, compensates this loss. The detector is calibrated for the average pulse height for a minimum ionizing muon traversing vertically. The response is found to be uniform to within 15%.

In the very early experiments, pulses from the detectors were pre-amplified, brought to a central recording station where they were suitably amplified and displayed on an oscilloscope. Pulses from several detectors were displayed on the same oscilloscope by introducing suitable delays between them for identification. When digitization techniques became available, they were digitized and recorded on paper tape. With the advent of modern electronic techniques, minicomputers and microprocessors, the pulse height information from the various detectors are stored on analog storage devices, such as Sample/Hold devices, digitized sequentially through a digital multiplexer, read by the processor and recorded on magnetic tapes.

In the proportional detectors, the pulse height for particles arriving at a zenith angle θ is larger by $\frac{1}{\cos\theta}$ since it is proportional to the track length of the particle. Thus the number of particles incident at an angle θ is given by $\frac{h\cos\theta}{h_0}$, where h is the pulse height recorded and h_0 is the average pulse height for a vertically incident muon. However, the area offered by the detector at this angle is $S\cos\theta$, and therefore, the density is given by

$$\Delta_e = \frac{1}{S\cos\theta}\frac{h\cos\theta}{h_0} = \frac{1}{S}\frac{h}{h_0}. \tag{3.3}$$

Thus the density of particles is independent of θ.

One of the factors to be taken into consideration in measuring densities with scintillation detectors is that the electron component consists of electrons and photons over a wide range of energies. These particles can undergo multiplication and absorption processes in the detectors. Chatterjee et al. [14] measured the ratio of particle density in liquid scintillator of thickness varying between 1–8 cm to that in a plastic scintillator of thickness 4.5 cm integrated over core distances larger than 5 m in showers of all sizes and find that the density decreases with increasing thickness with a mean free path of 17–20 g cm^{-2}. Their results essentially refer to small shower sizes. This variation is known as transition effect. Nagano et al. [15] carried out more detailed investigations by comparing the density measured in plastic scintillators of thickness 5 cm and 0.3 cm as a function of shower size and core distance. Their results are shown in Fig. 3.6. The data points are from the Akeno group, the solid line is the ratio of the density in 5 cm thick scintillator to that in a spark chamber measured by Shibata et al. [16] and the dotted line is the transition effect in a 1 mm thick iron plate measured by the Kobe group referred to by Nagano et al. It can be seen that the densities measured by 5 cm thick scintillator are overestimated at distances less than 5 m while at larger distances the effect is negligible. A 5 cm thick plastic scintillator offers only about 1/8 of a radiation length and hence this transition effect is not serious except within a few meters from the core where energetic particles are present.

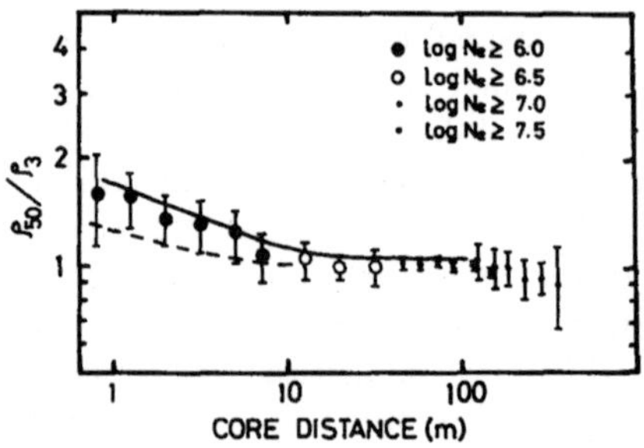

Figure 3.6 The average ratio of the density measured by a 5 cm thick plastic scintillator to that measured by 3 mm thick scintillator, obtained by Nagano et al (1984).

3.1.1.3 Fast Timing

The arrival directions of the showers are obtained from the relative times of arrival of the shower front at different detectors (see Sec. 3.3.1). The output of the phototube is fed to a discriminator, whose leading edge is used to measure the relative time of arrival. The typical delays are in the range of a few to a few tens of nanoseconds. The scintillation detectors have a response time of a nanosecond or less and thus ideally suited for the fast timing measurements. The photomultipliers used for the density measurements typically have rise times of about 20 ns and hence are not suited for fast timing measurements. A fast tube with high gain, like the RCA8575 (2″ diameter), is used for this purpose. The fast tubes generally have small photocathode area (to reduce the dispersion of transit time from different parts of the cathode) and so collect less light. In the early experiments deep ($\sim$ 20 cm) liquid scintillators were used to increase the light yield. With the availability of discriminators which can be set to a few tens of millivolts bias, plastic scintillators of thickness 5 cm are also now being used for fast timing measurements also. Fig. 3.5 shows the dual assembly of a fast photomultiplier tube mounted next to the 5″ tube.

3.1.2 Muon Detectors

GM counters, proportional counters, scintillators, spark chambers etc. have been used for measuring low energy muons in air showers. The detectors are kept under an absorber, whose thickness determines the minimum energy required by the muon to reach the detector. The energy loss in the overburden is essentially due to the continuous process of ionization. The detectors are either buried underground at shallow depths (as in the case of Akeno, CASA-MIA and Ooty experiments—see Sec. 3.2) or placed in underground tunnels (as in Tien Shan and Moscow State University experiments) or under concrete structures erected over the counters (Akeno and Kolar Gold Fields experiments). A typical muon detector of threshold energy 1 GeV under concrete absorber in the Kolar Gold Fields (KGF) air shower array is shown in Fig. 3.7. These detector assemblies provide information about muons above

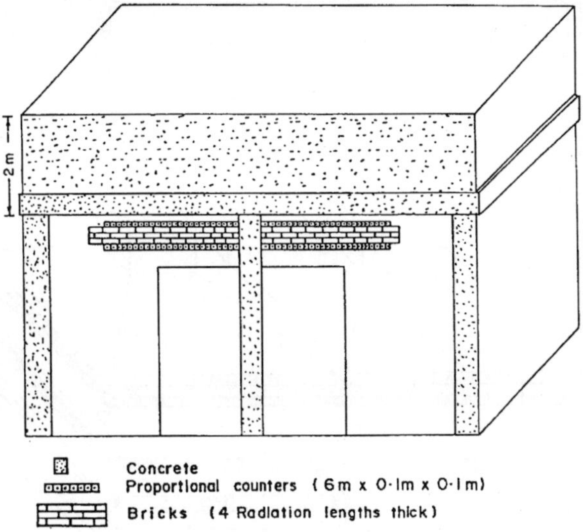

Figure 3.7 The muon detector with proportional counters placed under concrete absorber used in the KGF air shower experiment.

certain energy threshold. Muons of threshold energy 0.2–10 GeV have been measured in such arrangements.

Very high energy (> 100 GeV) muon densities are much smaller and so even larger area detectors can be used in the YES/NO mode. Two techniques are used to study high energy muons; one is to deploy detectors at deep underground levels, the overburden of earth determining the threshold energy of the muons and the other is to use magnet spectrographs to measure the muon momenta. In the former case, not only the collision loss but also other losses, in particular the losses due to direct pair production, bremsstrahlung and nuclear interaction processes, have to be taken into account. While the collision loss is approximately proportional to the logarithm of the muon energy, the losses due to the other processes are proportional to the muon energy itself and therefore become important at high energies. The bremsstrahlung and nuclear interaction processes are not continuous, but lead to catastrophic losses and thus to wide fluctuations in the energy loss. The threshold energy at deep underground levels is, therefore, not unique. But an average minimum energy can still be assigned at these deep levels, taking into account the muon energy spectrum

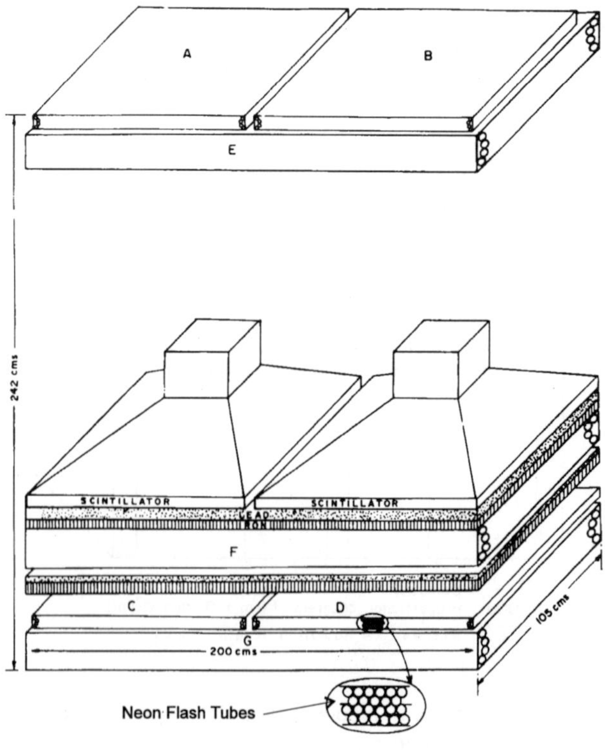

Figure 3.8 The Neon Flash Tube telescope operated at 270 m underground with the second EAS array at KGF (Acharya et al, 1988).

(see Menon and Ramana Murthy [17]).

The first deep underground detector for the study of high energy muons in EAS was by the Cornell group [18]. They operated GM counter trays at a depth of 1574 m.w.e. (meters water equivalent) corresponding to a muon threshold energy of 600 GeV. Subsequently the KGF group used scintillators, water Cerenkov detectors and a neon flash tube telescopes at various depths in the KGF mines to study high energy muons in detail (see Sec. 3.2.1.2). A schematic diagram of the neon flash tube telescope used by the KGF group [19] to study the lateral distribution of muons of energy ≥ 220 GeV, is shown in Fig. 3.8.

Currently two large area detectors are being operated at the Gran Sasso underground laboratory with an overburden of 1 km of rock in conjunction with the EAS-TOP air shower array to study high energy muons. One of them is the MACRO detector [20], which uses streamer tubes of total area 24×12 m^2 to track the muons with a positional accuracy of 1 cm. The other is the LVD detector [21], which is a compact liquid scintillator and comprises of 1520 units, each of volume 1.5 m^3, making up a total volume (including space between units) of $40 \times 13 \times 12$ m^3. The threshold energy of the muons is 1.4 TeV in both cases.

The first magnetic spectrograph was used by Bennet and Greisen [22] in their pioneering experiment to study high energy muons. They used an air gap magnet with a field of 12 kG and GM counters for tracking the muons. The spectrograph had a maximum detectable momentum (mdm) of about 20 GeV/c. Subsequently detailed studies of high energy muons were carried out by the Durham group [23], who operated a solid iron magnet with a field of 14.6 kG and neon flash tubes to track the muons in the Haverah Park EAS array. The mdm in their case was 60 GeV/c. Their spectrograph is shown in Fig. 3.9. The top of the spectrograph was covered by

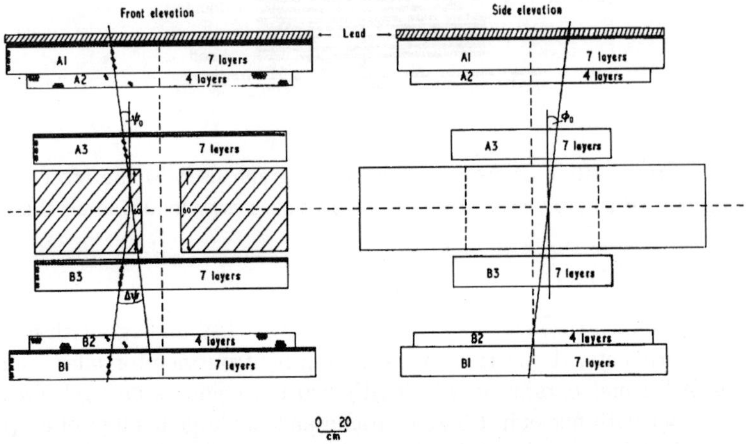

Figure 3.9 The magnetic spspectrograph used by the Durham group (Earnshaw et al, 1967) in the Haverah Park air shower array to study high energy muons.

Pb to filter out the electron component.

In recent years, the Moscow State University [24] operated a solid iron magnetic spectrograph at a shallow underground level of 40 m.w.e. in their Moscow air shower array. They used wide gap spark chambers to track the muons. The overburden of the tunnel filtered out other components. Fig. 3.10. shows the diagram of their spectrograph. The mdm in their case is 640 GeV/c.

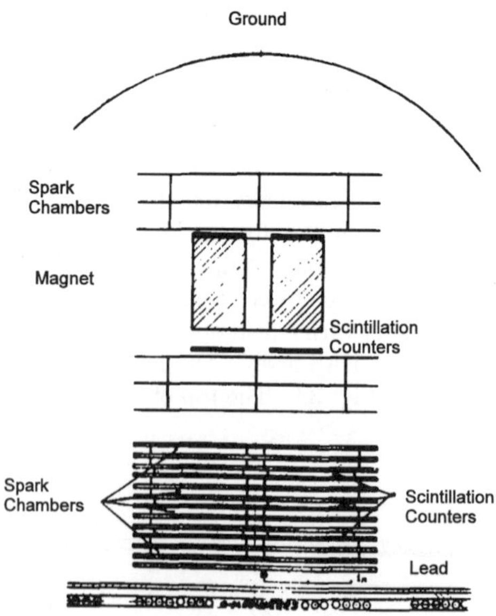

Figure 3.10 Cross section of the magnetic spectrograph used by the Moscow State University group (Bazhutov et al., 1977) to study high energy muons in air showers.

3.1.3 Hadron Detectors

Hadrons are detected through the strong interactions they undergo with matter. In the early experiments, they were detected using neutron monitor piles. These are assemblies of Pb and paraffin in which BF_3 neutron counters are embedded. The hadrons interact with nuclei in Pb and paraffin and the large number of evaporation neutrons produced are moderated to thermal energies due to interactions with light nuclei in paraffin and detected in the BF_3 counters. A typical detector of this type used by Chatterjee et al. [25] is shown in Fig. 3.11. The efficiency of these detectors is an increasing function of the energy of the hadron and is about 5% at about 2 GeV for the detector shown in Fig. 3.11. Nevertheless, because of the steeply falling energy spectrum of the hadrons, the mean energy of the hadrons detected is small—a few GeV.

The other type of hadron detectors, particularly to study hadrons of high energies, are the hadron calorimeters. In these detectors the hadron is allowed to interact, develop the nuclear-electromagnetic cascade and dissipate its energy in the calorimeter. The energy dissipated at several stages in the development is sampled by proportional type of detectors and the total energy estimated, after applying corrections to various

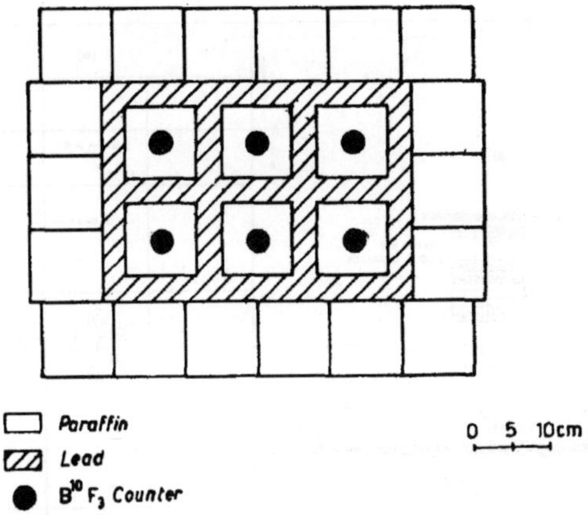

Figure 3.11 Neutron monitor pile used as a low energy hadron detector in the Tata air shower experiment(Chatterjee et al, 1960).

losses.

In the early experiments, a crude calorimeter, called the burst detector was used. In these detectors the cascade is sampled at only one stage of the development. The detector typically consisted of a layer of 4–5 radiation lengths of Pb to let the electromagnetic component develop, a layer of detectors to measure the electromagnetic component, a further layer of Pb of about 5–10 radiation lengths to completely absorb the electromagnetic component. Below this is a layer of light material like Carbon of thickness 1–2 interaction mean free paths to let the hadrons interact, and then a layer of 4–5 radiation lengths thick Pb for the π^0s produced in the interaction to develop the soft cascade and finally a layer of counters to measure the energy of the π^0s. The detectors used in the BASJE (Bolivian Air Shower Joint Experiment) experiment [26] and by a Soviet group [27] are shown in Figs. 3.12 and 3.13. It was assumed that the cascades are sampled at their maximum development and a constant conversion factor between the number of particles detected and the hadron energy was used. The conversion factor includes the assumed values for the inelasticity and the fraction of energy given to π^0s. These values fluctuate widely from event to event. Thus the energy estimated from this type of detectors for individual hadrons is not accurate.

The earliest total absorption calorimeter to study high energy hadrons in EAS was the one used by the Tata group [28]. Their calorimeter, called TASS (Total Absorption Scintillation Spectrometer) is shown in Fig. 3.14. It consists of 750 g cm^{-2} of iron, shielded from the soft component by $6\frac{1}{2}$ radiation lengths of lead from the top up to zenith angles of 25°, and on the sides by 5 radiation lengths of

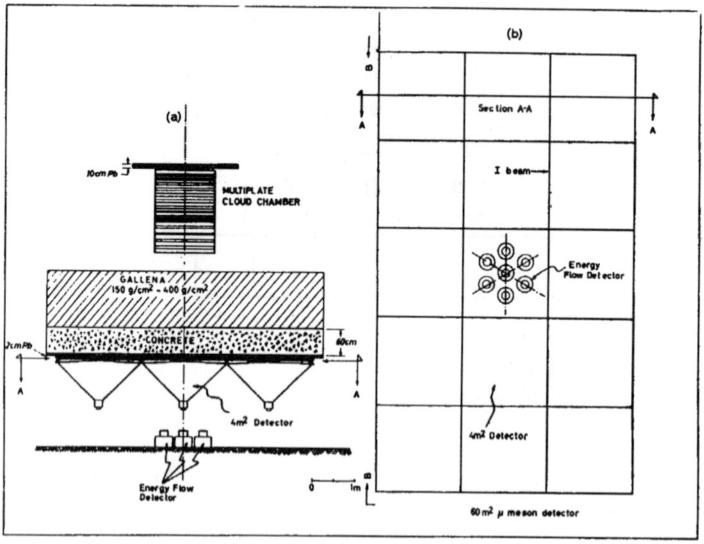

Figure 3.12 The burst detector used in the BASJE experiment to study high energy hadrons (Suga et al, 1962).

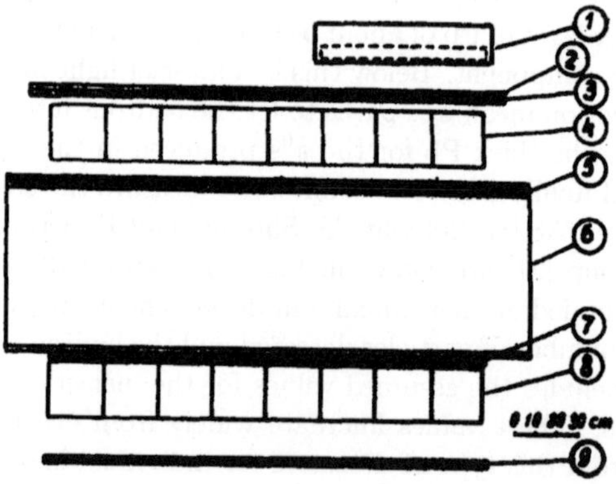

Figure 3.13 The burst detector used in the early Russian experiment (Goryunov et al., 1960) to study high energy hadrons: 1,4,8 — ionisation chambers; 2,3,5,7,9 — lead plates; 6 — carbon block.

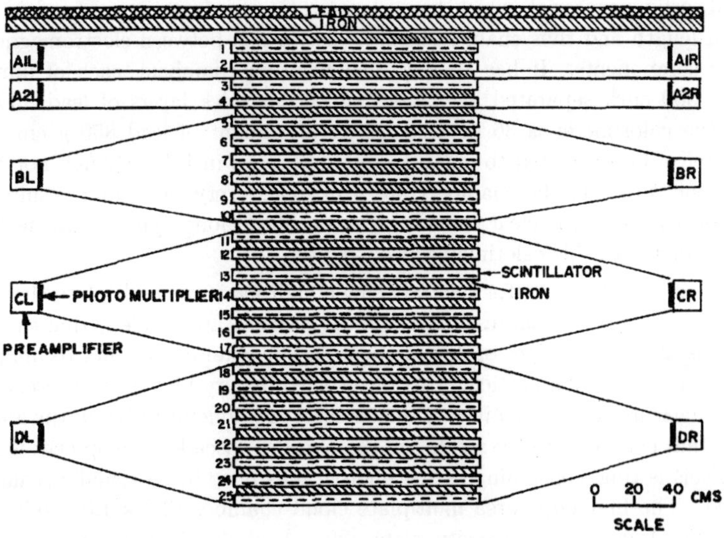

Figure 3.14 The Total Absorption Scintillation Spectrometer used as a high energy hadron detector in the Tata air shower experiment.

lead. The spectrometer has two identical sections, covering an overall area of 1.4 m^2. Each section consists of a sandwich of alternate layers of iron plates (2 r.l.) and liquid scintillators. There are 25 iron plates and 25 scintillators in each section. The 750 g cm^{-2} of iron corresponds to 6 interaction mean free paths for hadron collisions. Each section is subdivided vertically into 5 channels containing 2, 2, 6, 7 and 8 layers reckoning from the top. Each channel is viewed by one 5″ Dumont 6364 photomultiplier. The top two channels detect the soft component and the remaining 3 measure the hadronic energy. Pulses from the photomultipliers were amplified and displayed on oscilloscopes. The scintillators are calibrated for the pulse height due to the passage of a single minimum ionizing particle using cosmic ray muons. The total energy is estimated by considering the energy loss in the iron plates by minimum ionizing particles corresponding to this pulse height. The total energy was corrected for losses due to heavily ionizing particles produced in interactions in iron using Monte Carlo simulations. Because of the large segmentation used in this calorimeter (114 cm × 64 cm), the probability of more than one hadron going through one segment is not small enough for individual hadrons to be measured. Energy flow in the hadron component can, however, be measured.

A large area calorimeter was used by the Tien Shan group with their air shower array at Mt. Tien Shan. A distinct feature of this calorimeter is that it has Pb as the absorbing material, which has a large ratio of interaction mean free path to radiation length and thus enables them to distinguish between electromagnetic and hadronic components. It has 19 rows of ionization chambers, each of dimensions $5 \times 22 \times 300$ cm^3, separated by 2.5 cm and 5 cm thick layers of lead. The total area of the calorimeter is 36 m^2 and the total amount of lead 850 g cm^{-2}. This calorimeter is better suited to measure single hadrons in EAS, though not the best since in one direction the chambers are 3 m long. Very important results on the absorption features of nuclear cascades, the so called "long flying component", have been obtained with this calorimeter (see Chapter 4).

The most suitable detector to study single hadrons is the multiplate cloud chamber. It has very good spatial resolution and the cascades are visible so that the number of particles at various stages of development can be counted or other techniques can be used when individual particles are not resolved in very high energy cascades. The disadvantages are that large areas of the detector are difficult to build, the amount of material that can be installed inside the chamber for cascade development is limited, and analysis is time consuming as the photographs have to be visually scanned and measured. The first large area multiplate cloud chamber ($2.0 \times 1.3 \times 0.9$ m^3) was used by the Osaka City University group [29]. A similar chamber was used by the Tata group [30] also. Their chamber, shown in Fig. 3.15, has an area of 2m $\times$ 1m. It has 21 iron plates of 2 cm thickness comprising of $\sim$ 2.2 interaction mean free paths. The front and side walls are of toughened glass, so that it can be illuminated from the sides and photographed from the front. The plates are tilted so that they do not cast any shadow over the entire depth. The chamber is filled with Argon at a pressure of $\sim$ 1.3 atmospheres. The liquid used for condensation is a mixture of ethyl alcohol, isopropyl alcohol and water in the ratio 4.5 : 4.5 : 1. The gas is normally is saturated with the vapor of the liquid mixture. On an external trigger, due to the passage of an air shower, the chamber is expanded by moving a magnetically operated rubber diaphragm at the back, thereby cooling the chamber by a few degrees because of which the vapor becomes supersaturated. The vapor then condenses on the argon ions formed along the track of the charged particle. These droplets are then illuminated by 4 Xenon filled flash tubes mounted on the sides and photographed. The chamber is photographed by 4 cameras focussed at different depths in the chamber. The camera arrangement is shown in Fig. 3.16. A photograph of a hadron of estimated energy 2500 GeV at a core distance of 0.6 m in a shower of size 9×10^4 is shown in Fig. 3.17. Notice the electron component in the top section of the chamber and several lower energy cascades ($\sim$ 25 GeV) initiated by γ rays or electrons. Fig. 3.18 shows a photograph of an event with multiple hadrons. Thus it is clear that a detector with high spatial resolution is required to study high energy hadrons.

Another detector with high spatial resolution is the X-ray emulsion chamber. This detector consists of several layers of X-ray films and/or nuclear emulsions and Pb stacked alternately. The electrons in the core of the hadron cascades leave a dark spot

Figure 3.15 Multiplate cloud chamber used by the Tata group in their air shower experiment.

in the X-ray films. The intensity and the size of the spots is a measure of the energy deposit. The spatial resolution of X-ray films is better than a millimeter. In the case of nuclear emulsions, the number of electrons can be counted under a microscope with a resolution of a micron. The threshold energy of the cascades that can be recorded in these detectors is usually about a TeV for X-ray films and a hundred GeV for nuclear emulsions. These sensitive layers are changed every few months so that not many cascades are recorded. A family of cascades which are parallel to each other belong to the same event. When operated along with an air shower array, the correspondence between the families and the air shower is established by matching the arrival direction of the cascades with that of the air shower, which is independently determined. This type of detectors are used first by the Japan-Brasil collaboration and also by the Russians in the HADRON experiment. Details of this detector are

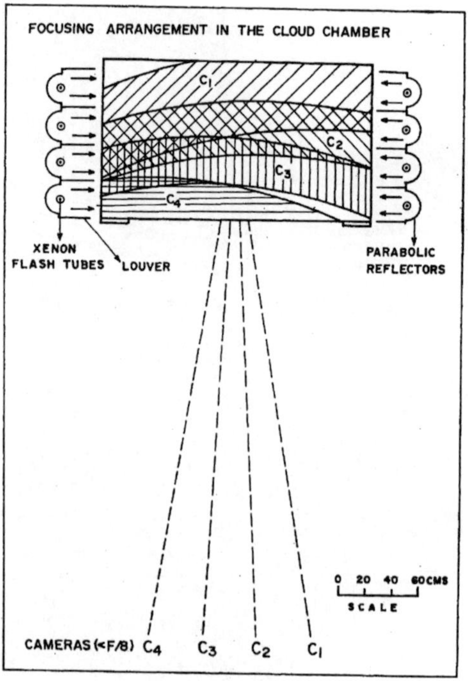

Figure 3.16 Arrangement of the cameras for photographing the large multiplate cloud chamber used by the Tata group.

discussed in Chapter 4.

3.2 Multicomponent Techniques of EAS Studies

In order to derive information regarding nuclear interactions or mass composition of primary cosmic rays it is essential to study, apart from the electron component, many of the EAS components such as hadrons, muons, Cerenkov radiation, etc. We shall now describe some of the multicomponent air shower arrays that have been operated.

3.2.1 Surface and Deep Underground Installations Operated in Coincidence

One of the EAS components that gives rather direct information on high energy interactions and primary mass composition is the high energy muon component since these muons are produced early in the cascade—first few interactions if the energy of the muons is high enough. The energy of the muons is measured using magnet

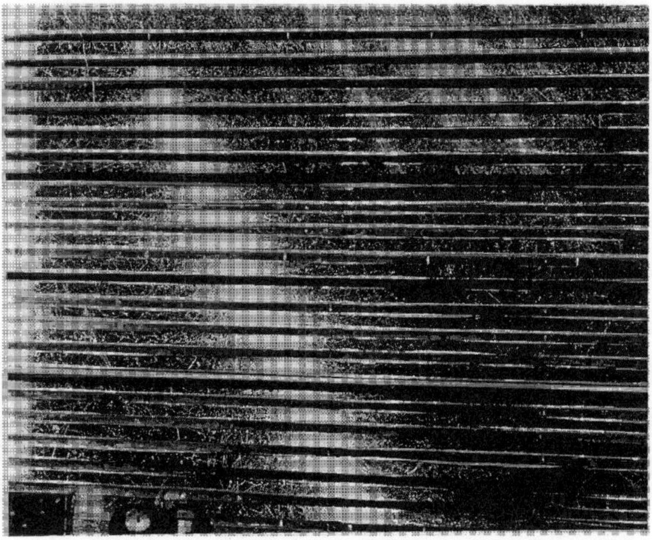

Figure 3.17 Cloud chamber photograph of a hadron of energy 2500 GeV in a shower of size 9×10^4 at a core distance of 0.6 m recorded by the Tata group.

spectrographs or by locating the muon detectors deep underground so that muons reaching the underground level will have a minimum energy corresponding to their energy loss in the overburden. We shall now describe the details of some of the experiments which used this technique.

3.2.1.1 The Cornell Experiment

Barret et al. [18] were the first to operate an extensive air shower array of Geiger counters at the surface with a muon detector again of Geiger counters at a depth of 1574 m.w.e. (meters water equivalent). The muons had to have an energy ≥ 600 GeV for penetrating the over burden and the mean energy of the muons recorded was greater than 10^{12} eV. The number spectrum of muons underground was determined from a comparison of single, two fold and three fold coincidences and from the size distribution of showers at the surface. It was deduced that the number of muons of such high energy varied with shower size as $N_\mu \propto N_e^{0.6}$. In EAS of size 10^6, the number of muons of energy greater than 600 GeV worked out to be about 60.

Figure 3.18 Cloud Chamber picture of multiple high energy hadrons in the core of an extensive air shower. There are at least 3 hadrons, one starting a cascade in the second plate from the top and the other two in the third. The second cascade from the left could also be a hadron starting its cascade perhaps in the air.

3.2.1.2 The KGF Experiment

The pioneering experiments just with Geiger counter arrays and Geiger counter telescopes of Barret et al., were followed by very elaborate and sophisticated series of experiments at the Kolar Gold Fields (KGF) starting from 1965 and lasting for three decades. The experimental arrangement [31] in the first of the series of experiments carried out at KGF, is shown in Fig. 3.19. Shower size was measured with an array of 20 plastic scintillators, each of area 1 m^2, on the surface of the mines and muons of threshold energies 220 GeV and 640 GeV were measured with 4 plastic scintillators, each of area 1.44 m^2, located at a depth of 270 m underground and 5 water Cerenkov detectors, each of area 2 m^2, at 600 m depth respectively. Each of the underground detectors was viewed by two 5″ photomultipliers (DUMONT 6364), in coincidence, to reduce photomultiplier noise since the counting rate of muons underground is very low. The electronics used provides only "yes" or "no" information from these detectors. Number of muons above threshold energy in showers of a given size was obtained from the fraction of showers in which the respective underground detectors responded. The variation of the muon number with the two threshold energies with shower size is found to be $N_\mu \propto N_e^{0.41}$, somewhat slower than the Cornel result.

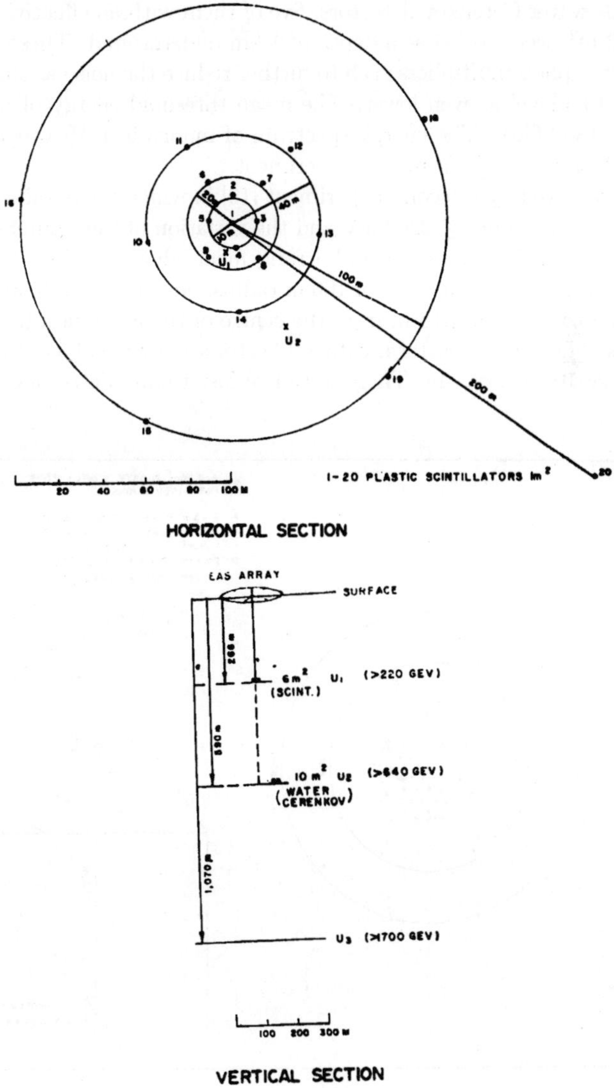

Figure 3.19 The first EAS array at KGF with detectors at two underground levels corresponding to muon energies of 220 and 640 GeV.

This array was later expanded (Naranan et al. 1973 [32]) to 31 detectors within 100 m with 21 of them within 20 m from the centre to facilitate measurement of small showers. Nine water Cerenkov detectors, five of them with an effective area of 1.6 m^2 and four of 2 m^2, were added at a depth of 1 km underground. These detectors were viewed by three photomultipliers each to further reduce the noise as the counting rate of muons at this level is even lower. The mean threshold energy of muons to reach this depth is 1800 GeV. The energy spectrum of muons in EAS was determined for the first time up to 2 TeV from this experiment.

This was followed by a second experiment [19] in which the detailed lateral distribution of muons of energy $\geq$ 220 GeV and the variation of their number with shower size were measured. The experimental arrangement, shown in Fig. 3.20 consisted of 70 scintillators spread over a circle of 100 m radius, out of which 37 were of area 2.25 m^2 each and located within 15 m from the centre of the array in a highly symmetric densely packed hexagonal pattern, with a detector separation of 5 m to measure small showers of size 10^4 accurately. There were five fast timing detectors to measure the

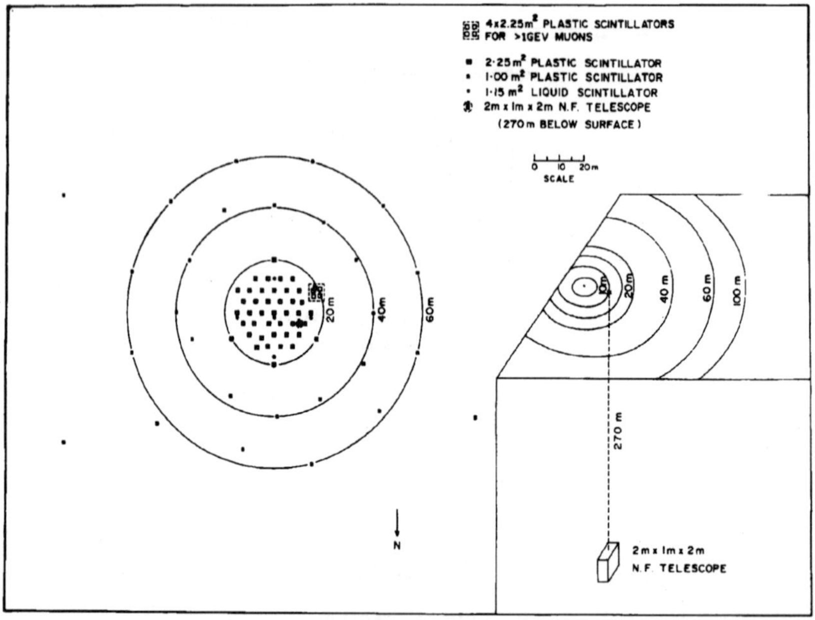

Figure 3.20 The second EAS array at KGF with densely packed central area and a Neon Flash Tube telescope at 270 m underground.

arrival directions and a muon detector of area 9 m^2 to measure low energy muons (>

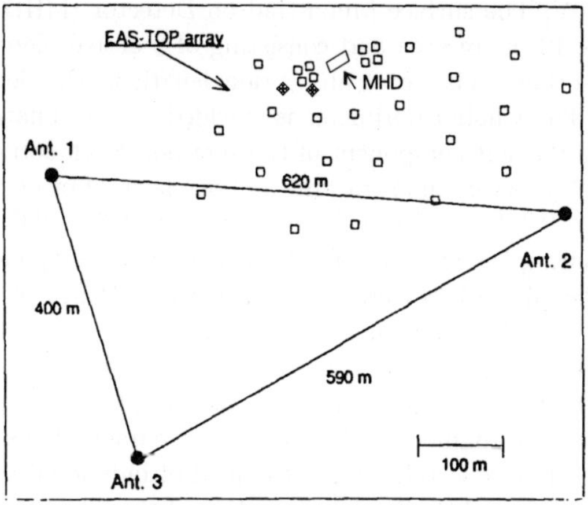

Figure 3.21 The EAS-TOP air shower array:; □, scintillation detectors; ✠, Cerenkov mirrors; •, radio antennas.MHD, Muon-Hadron Detector.

1 GeV). Vertically below the array, at a depth of 270 m underground, 4 scintillators, each of area 1 m^2, were operated to measure muons of threshold energy 220 GeV. Two of the detectors were surrounded by a neon flash tube telescope, shown in Fig. 3.8, to locate the muon track. The spatial angle of the muon could be determined to an accuracy of better than 15′. The overall accuracy of the distance of high energy muons from the shower core was 2.5 m and the worst case accuracy of shower size estimation was $\sim 20\%$.

The lateral distribution of muons of energy ≥ 220 GeV has been determined with good accuracy for the first time in this experiment. It is found to be flatter than expected, particularly at large distances from the core. More important, while the variation of $N_\mu\,(E_\mu \geq 220GeV)$ with N_e in the shower size range $10^4 - 5 \times 10^5$ at 920 g cm^{-2} is consistent with predictions for the primary charge composition directly measured at lower energies, it flattens above this range and recovers back to the expected variation for protons above a shower size of 2×10^6. This suggests that the composition changes from a mixed one to protons in this shower size range.

3.2.1.3 The EAS-TOP Array

In recent years an elaborate multicomponent array has started operating over the Grand Sasso Laboratory in Italy. The EAS array [33] which is laid out along a mountain slope at a mean altitude of 2005 m above sea level has 35 modules of scintillators each of area 10 m^2 and spaced 17 m apart at the centre and 80 m apart at the edges. The array is suitable for detailed analysis of showers in the energy range $10^{13} - 10^{17}$ eV. The surface Muon-Hadron Detector (MHD) comprises of two modules each 12×12 m^2 in area and consisting of 9 active planes interleaved with 13 cm of iron absorber. The total absorption length in the detector is 7 nuclear mean free paths. The whole calorimeter is shielded by 15 radiation lengths of iron and lead to absorb the soft component of the showers. Each active plane consists of two streamer tube layers for muon tracking and one layer of proportional counters for hadron calorimetry. While the threshold for Muon detection is 2 GeV, the calorimeter is capable of measuring the energies of the nuclear active component up to a TeV.

The Imaging Cerenkov Light Detector in the EAS-TOP array consists of 3 parabolic mirrors each 90 cm in diameter and 65 cm focal length, supported on alt-azimuth mount. One of the mirrors is equipped with a 64 pixel multianode photomultiplier used as an imaging device and the other mirrors have photomultipliers with cathode areas restricted to 2 cm diameter. The pointing accuracy of the telescope is 0.05°. The field of view is 1.76×1.76 deg^2 and the field of individual pixels is 0.22×0.22 deg^2. The arrangement enables the determination of the longitudinal development of showers and the height of production of Cerenkov light. The final plan is to have 8 such telescopes.

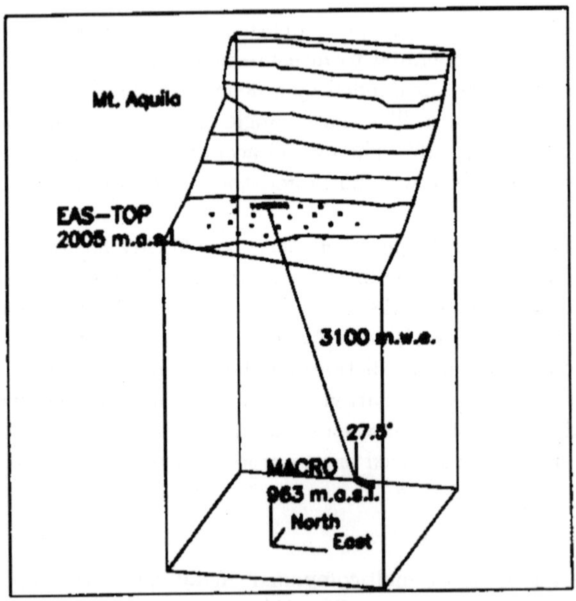

Figure 3.22 Relative locations of the EASTOP array and the MACRO detector.

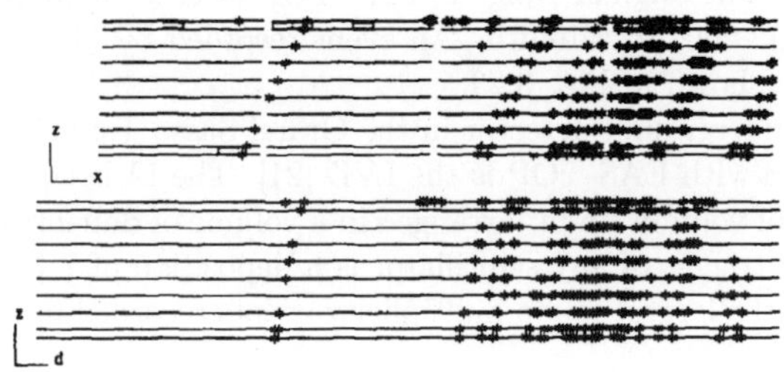

Figure 3.23 An event of muon bundle (23 muons) recorded by MACRO in coincidence with the EASTOP array.

The EASRADIO [34], shown in Fig. 3.21, is a detector of LF and VLF radio pulses that accompany air showers, and consists of 3 stations located at the corners of a triangle of sides 400 m, 590 m and 620 m. At each station there is 15 m vertical antenna connected to multichannel radio receiver tuned to 300–500 kHz and 2.3–3.5 MHz. The installation is operated in conjunction with EAS-TOP which gives information on the associated shower size, core distance, angle of arrival. Measurements have been made of the intensity of the radio pulse as a function of distance and also the radio spectrum. The data has shown [34] that the energy in the radio pulse I is related to the primary energy E_0 by the relation $I \propto E_0^{0.5}$. The coincidences between the radio pulses of the 3 antennae themselves provide triggers of EAS and the rate is found to increase with the gate width of coincidence. The maximum gate width required is about 2 micro seconds which is, however, small enough to eliminate background.

In the underground Grand Sasso tunnel there are two important detectors the MACRO and the LVD operated in conjunction with EAS-TOP. The MACRO detector [20], shown in Fig. 3.22, has a modular structure whose basic block is a "supermodule" of dimensions $24 \times 12 \times 4.8$ m^3. Each supermodule is equipped with streamer tube chambers, liquid scintillators and track-etch plastics. Muon tracking is provided by 10 horizontal layers of streamer tubes. Each tube is 12 m long and 3×3 cm^2 in cross section, and utilizes a 100 μm anode wire and a graphite cathode. A two-dimensional readout is performed using the signals from the anode wires and external 26.5° stereo angle strips. Each projected view has a spatial resolution of about 1.1 cm. The intrinsic angular resolution is 0.2° for muons crossing ten horizontal planes, considerably less than the mean multiple Coulomb scattering angle of muons in the rock (0.6°). The location of the EAS-TOP is such that the surface and underground detectors are separated by 1000 m. in altitude and 500 m on the horizontal projection. The minimum and average rock overburden above MACRO are 3,200 and 3,700 m.w.e. respectively. The Muons recorded by MACRO have a threshold energy of 1.4 TeV at the surface for the minimum MACRO depth. The rock overburden varies from

3100–3500 m.w.e. The angular range of EAS-TOP as seen by MACRO is 25° to 37° in Zenith and 160°–200° in azimuth. An event recorded by MACRO in coincidence with EAS-TOP is shown in Fig. 3.23.

Yet another underground detector in the Grand Sasso Tunnel that is being operated in association with EAS-TOP is the LVD [21]. The LVD consists of 1520 liquid scintillators each of volume 1.5 m^3 arranged in a volume of dimensions $40\times13\times12$ m^3. The threshold for trigger of any scintillator is a deposition of 7 MeV energy. Figure 3.24 shows one of the largest muon multiplicity events recorded in LVD in association with EAS-TOP.

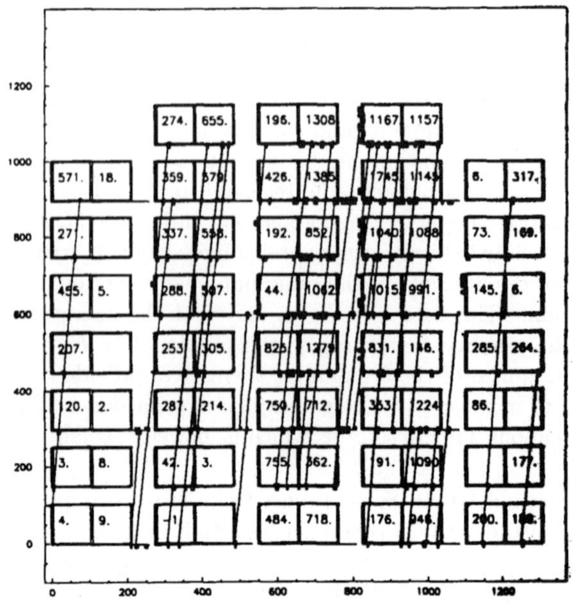

Figure 3.24 Event with the largest muon multiciplicity recorded by LVD in coincidence with the EASTOP array. The LVD view from tracking, and the total energy loss in the scintillators are shown.

3.2.2 *The MILAGRO*

The MILAGRO [35], a new type of air shower detector designed for operation over a wide range of energy, much below the normal range of air shower arrays, effective in the energy range 1–100 TeV, is being set up in Los Alamos, U.S.A., at an altitude of 2600 m as an extension of the currently operating CYGNUS array. The MILAGRO is a water Cerenkov detector (Fig. 3.25) built in an existing pond of dimensions 60 m × 80 m and 8 m deep. The capacity of the tank is 5 million gallons, and will be

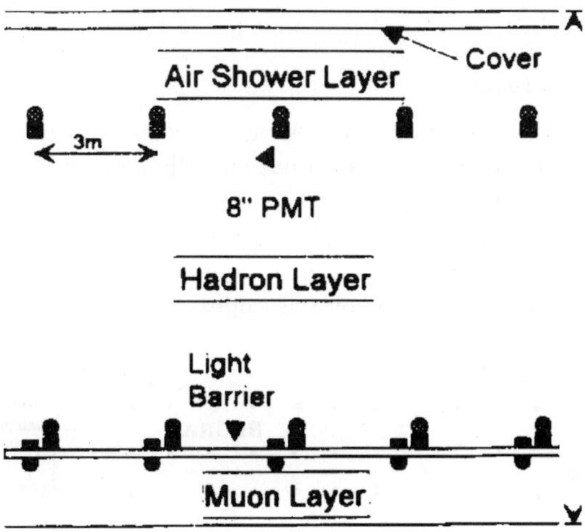

Figure 3.25 Schematic diagram of a section of the MILAGRO detector.

filled with purified water. The tank is fitted with a light tight cover.

There are several layers of 8″ diameter phototubes fitted inside the tank at different depths. The first layer of 450 tubes is at a depth of 2 m and is essentially for the measurement of the arrival times of the shower front particles and to determine the angle of the shower in zenith and azimuth. A second layer of 170 tubes facing upwards are located at a depth of 6.5 m and these measure the energy deposited by the shower particles—the soft component, the penetrating component and also the hadron component which produces penetrating showers and leads to nuclear cascades in the water. A third layer of 170 phototubes at a depth of 8 m facing downwards and optically isolated from the light from above, view 1500 m^2 area of the bottom 1 m depth of water and this layer is sensitive to the muon component.

The CYGNUS array (see Fig. 8.9) with its 204 scintillation counters covers an effective area of 40,000 m^2, surrounds the MILAGRO pool and provides the basic parameters of the showers like the shower size, the core distance and angle of arrival etc. for showers of energy greater than ~100 TeV. Monte Carlo simulations of the showers have shown that the threshold for detection by MILAGRO is about 1 TeV and the angular resolution ~0.5°. The muon detection efficiency is high so that by demanding low muon content (above a certain threshold energy of the shower) gamma ray induced showers can be preferentially separated out. The hope is that the MILAGRO will have a considerably lower threshold for gamma ray astronomy, lower than that of the atmospheric Cerenkov detectors. The MILAGRO's large sky coverage and low threshold may make it possible to look for anisotropies in the TeV range and may lead to a better understanding of the source of galactic TeV Cosmic rays.

3.2.3 The HEGRA Detector

The HEGRA (High Energy Gamma Ray Array) is located at Canary Island, La Palma at an altitude of 2240 m and is a multi-nation collaboration [36]. It is designed to cover the energy range 0.5–10 PeV.

The array, shown in Fig. 3.26, has 257 scintillators each of area 1 m^2 and 20 muon detectors of total area 320 m^2 and an array of 49 wide angle air Cerenkov counters (AIROBICC) and 5 imaging Cerenkov telescopes.

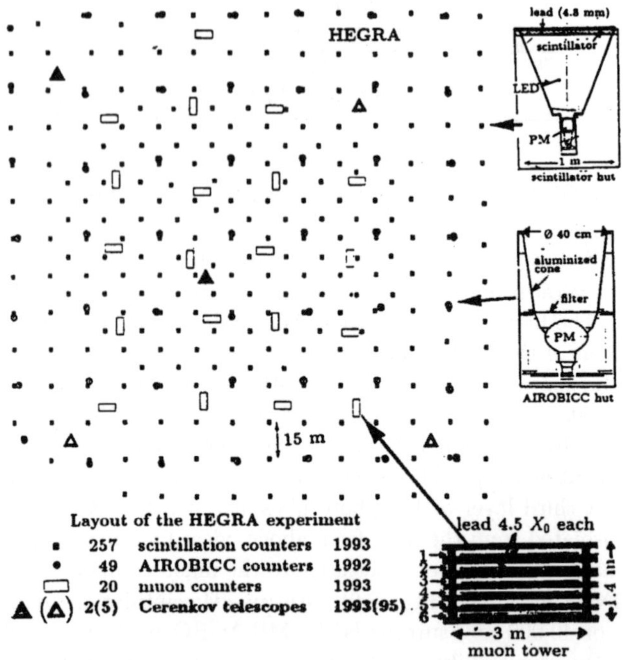

Figure 3.26 Layout of the HEGRA air shower experiment.

Each wide angle AIROBICC consists of a hemispherical photomultiplier of diameter 20 cm that views the night sky over 1 steradian with a light collection cone of diameter 40 cm, through a blue filter. The final number of such detectors in the array will be 80. Discrimination in favor of γ rays is expected to be made on the basis of the lateral distribution of Cerenkov light. Even though the duty cycle is only 15%, because of the wide angle design the possibility of searching for γ rays from different sources within the field of view enhances its capability compared to the narrow angle Cerenkov detectors. The information on the longitudinal shower development is also obtained by measuring the arrival time distributions with fast flash ADCs in counters placed outside the array.

Out of the array of 5 imaging Cerenkov telescopes, a prototype with a mirror of area 5 m^2 and a camera of 37 pixels has been installed by 1992. The final design is for mirrors of 8.5 m^2 area and cameras of 271 pixels which will give an expected threshold of 300 GeV and field of view 4°–6°.

3.2.3.1 Muon Towers in HEGRA

There are in all 20 muon towers in the HEGRA distributed as shown in Fig. 3.26.

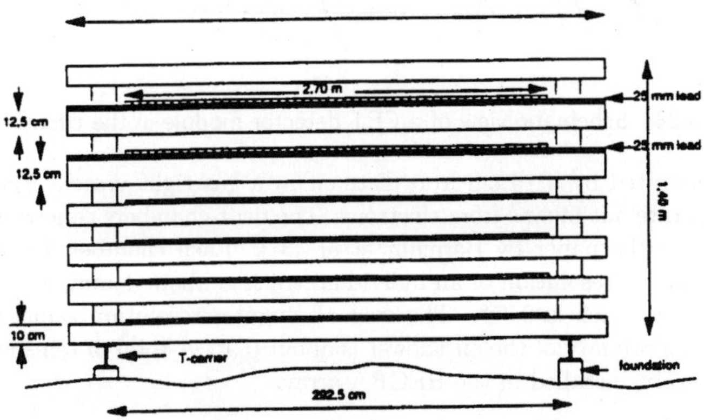

Figure 3.27 Cross sectional view of one muon tower in the HEGRA array.

The cross sectional view of a Tower is shown in Fig. 3.27. It consists of 6 active layers separated by spacers of light concrete and two shielding layers of lead 4.5 radiation lengths, below the top two layers. Each active layer consists of 160 individual tubes with a cross section of 1.5 × 1.5 cm^2 operating in the Geiger mode. The tubes are read out by threshold electronics.

The preliminary runs of the HEGRA array with just one tower have shown that with 20 muon towers, it will be possible to measure muon content of showers up to few hundred TeV.

3.2.3.2 Tracking Detectors in HEGRA

The CRT (Cosmic Ray Tracking) detectors have added a new dimension to air shower studies. The chief virtue of these is that the directions of individual shower particles (electrons or muons) can be recorded with high accuracy and connected very well to the direction of arrival of the primary. The schematic view of the tracking detector module that is operating in La Palma is shown in Fig. 3.28. Two large volume drift

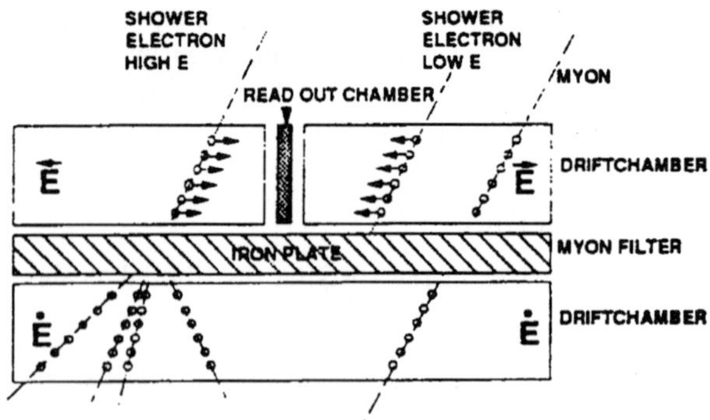

Figure 3.28 Schematic view of a CRT detector module in the HEGRA array.

chambers separated by a 10 cm iron plate sit in a gas tight vessel. The iron plate serves to separate out muons from electrons. The drift chambers construction details are available in the paper by Berniöhr et al. [37]. Each chamber has 6 wires and the average spatial resolution of an individual track is about 1 mm, and the angular resolution of the track is 0.35°. The over all angular resolution achievable for the direction of the primary of the air shower is about 0.9°. A total of ten such detectors are planned to be installed in the HEGRA array.

3.2.4 Time Track Detectors in Air Shower Arrays

At the International Cosmic Ray Conference held at Bangalore in 1983, Linsley [38] suggested that the arrival time distributions of particles in the shower fronts of air showers contained information on the longitudinal profiles of the showers analogous to the Fly's Eye arrangement (see Sec. 7.1.1) based on the measurement of the air scintillation light from different heights. This new methodology is attempted in the GREX/COVER PLASTEX experiment [39] that is being carried out at Haverah Park, U.K. The GREX (Gamma Ray astronomy EXperiment) is an array of 36 scintillators each 1 m^2 in area with a spacing of 30 m in the central part which responds to showers of > 400 TeV and spacing of 50 m at the edges responding to showers > 600 TeV. The array has an angular resolution of 1°. In the array, shown in Fig.

3.29, there are 5 COVER PLASTEX hodoscopes. There are layers of Resistive Plate

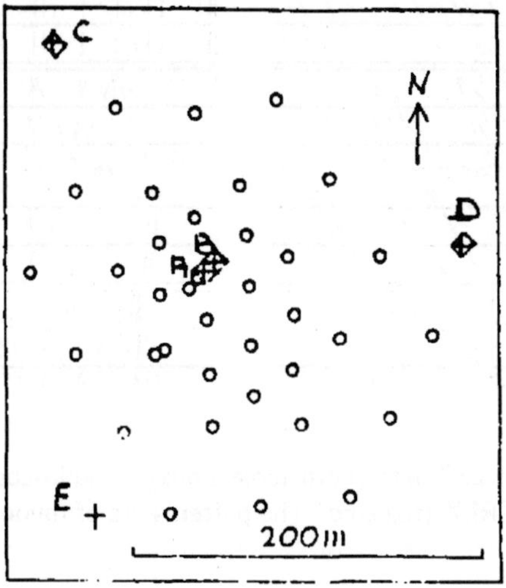

Figure 3.29 GREX/COVER PLASTEX array of the Haverah Park group: o – scintillators; × – PLASTEX hodoscopes; ◇ RPC covers.

Counters (RPC) and Limited Streamer Tubes (LST). The details are available in the paper of Agnetta et al. [39]. Data from the LST hodoscopes are subjected to pattern recognition software which enables the reconstruction of the tracks (see Fig. 3.30).

The fast timing detector is the Resistive Plate Counters which are highly segmented and make possible the discerning of the individual tracks and their time of arrival. Clearly this is a very sophisticated and intricate arrangement which if it succeeds, as the preliminary trials do indicate, will open tremendous new possibilities in the analysis of air shower data.

3.2.5 The CASA-MIA Array

The Chicago Air Shower Array (CASA) in combination with Michigan Muon Array (MIA) and the Hi-Res Fly's Eye [40] is another powerful combination for multiparameter studies of air showers. The CASA (see Fig. 8.10) consists of 1089 scintillators each 1.5 m^2 in area capped with one radiation length of lead sheet, arranged in a square grid of 15 m spacing. The MIA consists of 1024 scintillators buried 3 m below CASA. The counters in MIA are distributed in 16 widely spaced clusters and the active area is 2500 m^2. The Hi-Res prototype detector is so constructed that its 14 mirrors overlook the CASA-MIA array 3.4 km away (see Fig. 3.31) and is capable of

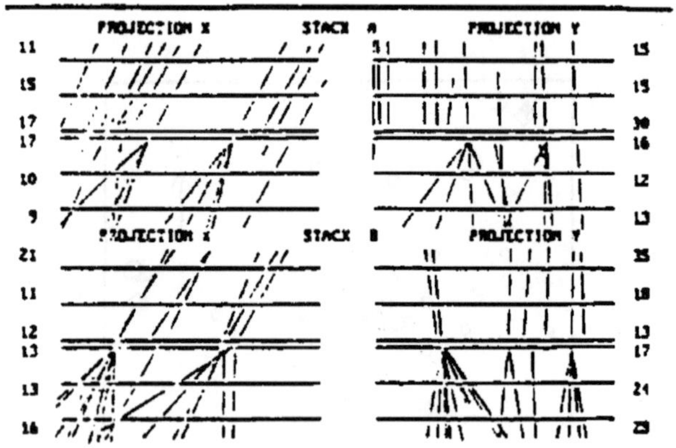

Figure 3.30 Display of LST hits (both projections) in hodoscopes A (above) and B (below). The event is GREX-triggered. The patterns are of muons, high- and low-energy electrons, and photons.

observing showers of energy above 5×10^{16} eV.

This unique combination thus provides the opportunity to observe showers both from the ground based array and the atmospheric fluorescence techniques just above the "knee" region of the primary spectrum (see Chapter 5 for knee) and enable cross-calibration on the energy estimation.

The Hi-Res will give the depth of shower maximum of incident shower and the ground based array will give the arrival direction and core location and the Muon array will give the muon content. The distribution of the events can then be compared with Monte Carlo simulation for different composition models.

3.2.6 The KASCADE Array

The KASCADE Array [41] coming up at the Karlsruhe Nuclear Research Centre in Germany (Fig. 3.32) is organized in 16 clusters of 16 detector stations. Each cluster contains 4 circular liquid scintillators of 1 m diameter for detection of electrons and photons, a 1.8×1.8 m^2 plastic scintillator under 10 cm of lead and 4 cm of iron for detection of muons ($>$ 200 MeV). The electron detectors are covered by 1 radiation length of lead.

The central detector is a large iron calorimeter and two layers of MWPCs below the calorimeter. The calorimeter has 8 active layers of liquid ionization chambers filled with tetramethylsilane, 0.8 cm thick. The chambers are of size 0.5×0.5 m with 4 readout pads 25×25 cm wide. These are interspersed with iron stacks of thickness ranging from 12 cm at the top to 36 cm at the bottom. The plastic scintillator layer

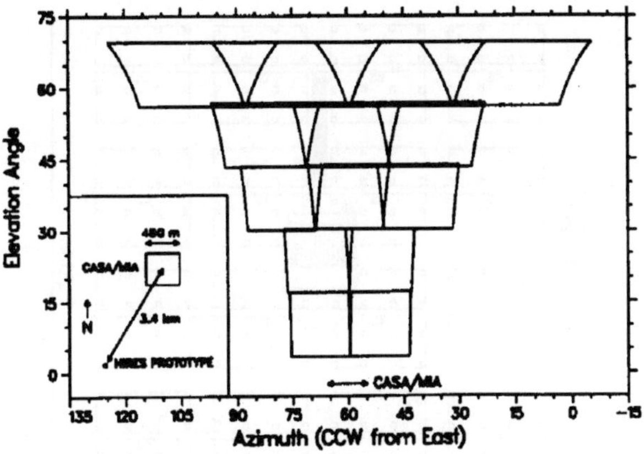

Figure 3.31 Geometrical arrangement of the HiRes prototype and CASA-MIA experiments at Dugway, Utah. The HiRes prototype is 140 m higher in altitude than the array complex.

covers two thirds of the effective area of the calorimeter. The bottom MWPCs in two layers serve as the muon detectors, the threshold for muon detection is 2 GeV. The total area of the hadron detector at the centre will be 16 m × 20 m.

The long tunnel north of the central building will contain a muon detector of total area 150 m^2 (see Fig. 3.33). Each muon tower comprises of 3 layers of streamer tubes 4 m in length, and is under 15 radiation lengths of shielding.

The KASCADE array is expected to go into operation in the second half of 1995, and is designed to obtain multi-parameter inputs for analysis of showers in the range 10^{14}–10^{17} eV.

3.2.7 GRAPES III Array

A new 500 detector array is coming up at Ooty (altitude: 2.2 km) with muon detectors of area 544 m^2 in 16 modules of 34 m^2 each. The surface detectors are 332 plastic scintillators and the muon detectors are proportional counters 10 cm × 10 cm × 6 m. RCC blocks of size 60 cm x 60 cm x 15 cm are stacked to a height of 3 m above the proportional counter arrays. The array is designed for studies of air showers produced by primaries in the energy range 10^{13}–10^{17} eV.

3.2.8 SPASE Array

The SPASE (South Pole Air Shower Experiment), shown in Fig. 3.34, is an air shower array that is operating since 1987 at the South Pole and has twenty four scintillators

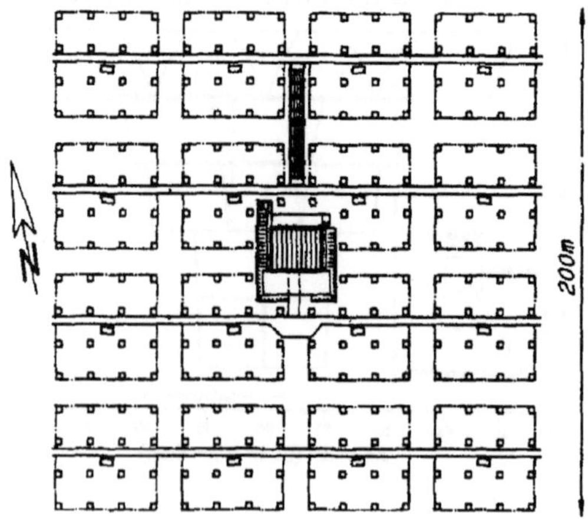

Figure 3.32 Layout of the KASKADE experiment. The array is divided into 16 clusters as "independent" experiments. The central detector building contains the calorimeter, the muon MWPCs, and the central data acquisition. The long tunnel north of the center houses streamer tube muon detectors.

each 1 m^2 in area and is effective for recording showers in the energy range $\geq 10^{14}$ eV.

The array has been operated in conjunction with a muon telescope at a depth of 800 meters in ice. The muon telescope is essentially a string of four 7.5 cm upward looking photomultipliers lowered to a depth of 800 m into the ice. The figure also shows the core locations in 117 showers determined by SPASE which had muon coincidences (> 200 GeV). An interesting aspect of this experiment is that the Monte Carlo simulations have shown that the fraction of events which give coincidences for all species of primary cosmic rays is about 20% and that heavy nuclei contribute significantly to the coincidence fraction suggesting that 200 GeV Muons could be quite sensitive to the composition of the primaries [42].

An exotic detector, The POOL, is also being tested out in the SPASE array [43]. The POOL is an Ice detector 7 m in diameter and 1.5 m deep. Twenty upward facing photomultiplier tubes arranged at the bottom of the pool filled with clean water and allowed to freeze make the POOL detection system.

SPACE-2 experiment [44] is being constructed at Amundsen-Scott South Pole Station and is optimized to work with the AMANDA assembly of muon and neutrino detectors at an ice depth ranging from 810–1000 m [45]. The proposed array is shown in Fig. 3.35. The array is designed to study primary cosmic ray composition in the

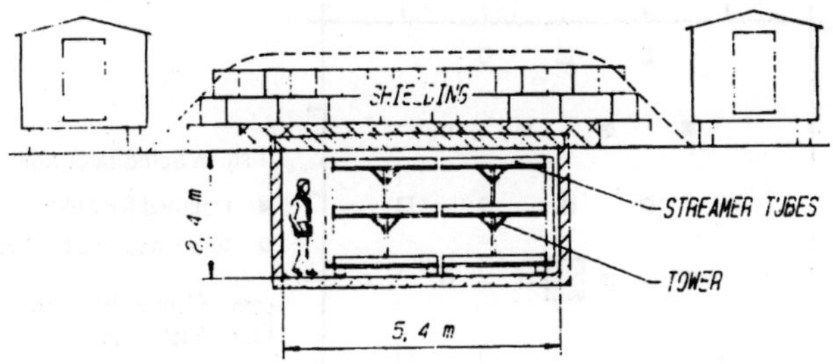

Figure 3.33 Muon detector tunnel located between two detector stations in the KASKADE array.

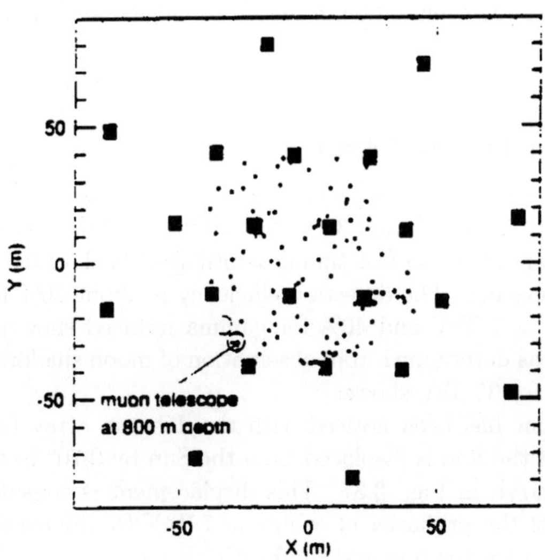

Figure 3.34 Location of the muon telescope and the detectors of the SPASE array. Also shown are the core locations of 117 showers which yield muon coincidences.

knee region and sources of UHE gamma rays.

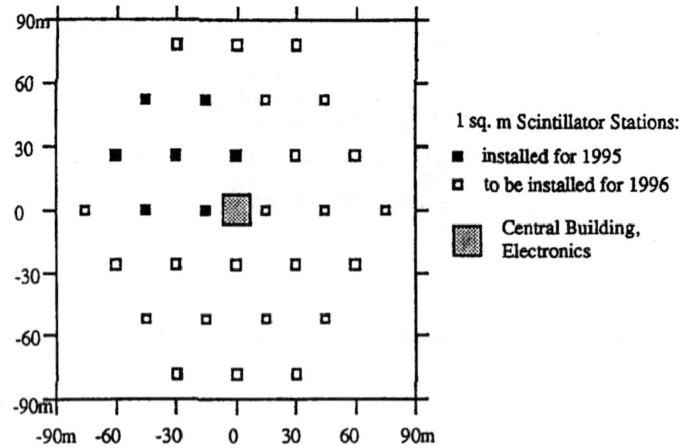

Figure 3.35 The proposed SPACE-2 array of Gaisser at al. (1995) at the South Pole to study primary cosmic ray composition in the "knee" region and sources of UHE gamma rays.

3.2.9 The Tibetan Air Shower Array

The Tibetan air shower array [46], shown in Fig. 3.36, is set up at an altitude of 4,300 m (606 g cm^{-2}, 90.5° E and 30.1° N) in Tibet and has been operational since 1990. The array consists of 49 fast timing scintillators each of 0.5 m^2 area placed in a lattice of 15 m spacing. The detection efficiency is about 20% for proton induced showers of energy > 7 TeV and 40% for gamma induced showers of 10 TeV. The angular resolution as determined from observation of moon shadow is 0.88° for 7 TeV showers and 0.52° for 35 TeV showers.

An anomaly that has been noticed with the Tibetan array for the first time is that the shadow of the Sun is displaced from the Sun by 0.70° to the west and 0.40° to the South as shown in Fig. 3.37. This displacement is considered to be caused by the deflection of the primaries of energy > 7 TeV by the combined effect of the magnetic fields between the Sun and the Earth.

The array is planned to be extended to cover an effective area of 3.7 x 10^4 m^2 by increasing the number of scintillation detectors to 221 with the same lattice spacing, as shown on Fig. 3.38. The array will also have 100 m^2 of Emulsion Chamber which will enable the study of the primary composition in the knee region of the primary spectrum.

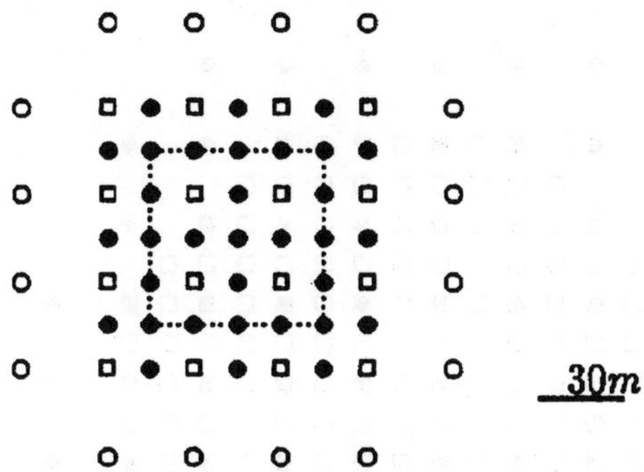

Figure 3.36 Layout of the Tibetan air shower array. Closed circles and open circles except the detectors at the four corners are equipped with fast photomultipliers tubes. Open circles and open squarés are equipped with wide range tubes.

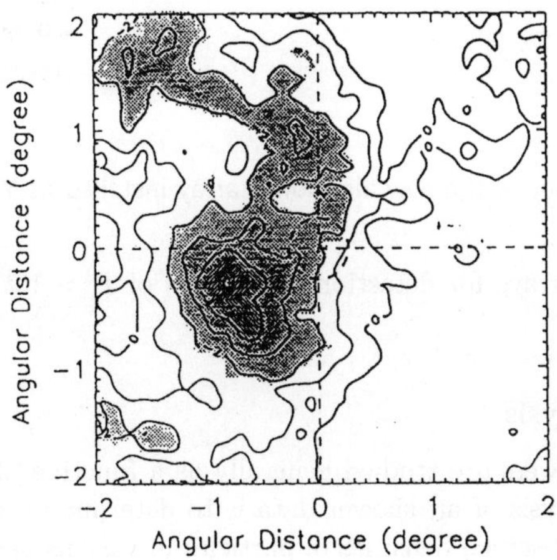

Figure 3.37 Arrival directions of all events in the $4^\circ \times 4^\circ$ area centered at the Sun (intersection of the dashed lines) recorded by the Tibetan Array.

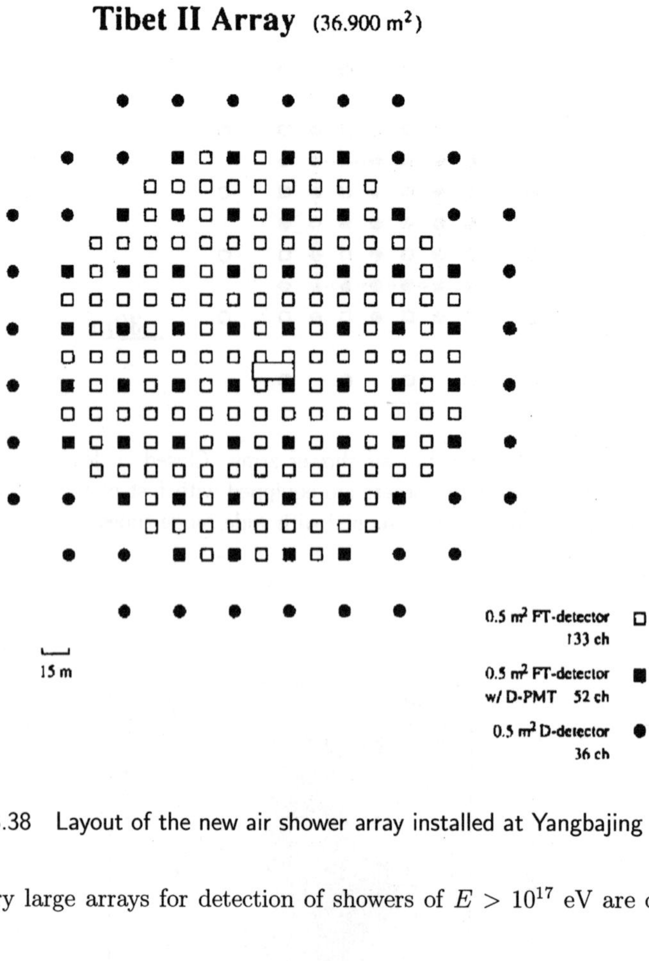

Figure 3.38 Layout of the new air shower array installed at Yangbajing in Tibet.

The very large arrays for detection of showers of $E > 10^{17}$ eV are described in Chapter 7.

3.3 Shower Analysis

Properties of air showers are studied generally as a function of shower size. So, the first task in the analysis of air shower data is to determine the shower size. This is done by fitting the observed densities of particles in various detectors to an assumed lateral distribution function—e.g. the NKG function (Eq. 2.22)—in the shower plane. The shower plane, which is perpendicular to the direction of the primary particle (also called the shower axis) is determined from the relative arrival times of the particles recorded in various detectors. The shower axis is characterized by the zenith and azimuth angles in the local coordinate system.

3.3.1 Determination of Zenith Angle

The plane front is represented by the equation

$$lx + my + nz - ct = 0 \tag{3.4}$$

where $l = \sin\theta\cos\theta\cos\varphi$, $m = \sin\theta\cos\theta\sin\varphi$, and $n = \cos\theta$ are the direction cosines of the shower axis, θ and φ the zenith and azimuth angles, x, y and z are the Cartesian coordinates, c the velocity of light and t is the time of arrival of the front at that point relative to some arbitrary time. The fitting is carried out by minimizing the quantity ψ^2 defined as

$$\psi^2 = \sum_i \omega_i \{lx_i + my_i + nz_i - c(t_i - t_0)\}^2 \tag{3.5}$$

with the constraint

$$l^2 + m^2 + n^2 = 1 \tag{3.6}$$

where t_i is the arrival time at the ith detector with coordinates x_i, y_i and z_i, $\omega_i = 1/\sigma_i^2$ is the weight factor, σ_i the error in the arrival time and t_0 is some arbitrary reference time. The summation is carried out over all the timing detectors which have responded. The error in the arrival time includes the instrumental uncertainty σ_t as well as the spread in the arrival time due to the finite thickness of the shower disc σ_d and is given by

$$\sigma^2 = \sigma_t^2 + \sigma_d^2 \tag{3.7}$$

σ_t is usually about 1 ns. Linsley [47] found empirically that

$$\sigma_d = 1.6\left(\frac{r}{30} + 1\right)^{1.5} \quad \text{ns} \tag{3.8}$$

with the core distance r measured in meters. Thus, while instrumental errors are important near the shower core σ_d becomes dominant at larger distances. The free parameters l, m and t_0 are obtained by solving the equations

$$\frac{\partial\psi^2}{\partial l} = \frac{\partial\psi^2}{\partial m} = \frac{\partial\psi^2}{\partial t_0} = 0. \tag{3.9}$$

The accuracy achieved for the zenith angle with this plane front approximation is usually around 3°.

The plane approximation is valid only near the shower core—say up to about 20 m. The shower front has a curvature because of which the arrival time lags behind that expected from the plane front approximation by an amount which increases rapidly with core distance. This leads to a systematic error in the determination of the zenith angle if arrival times measured away from the shower core are used. Measurements show that this deviation from plane front is a linear function of core distance [48, 49]. Acharya et al. [49] find that the mean deviation $\langle\delta t\rangle$ is given by

$$\langle\delta t\rangle = b(\theta, N)\,r + c \quad \text{ns} \tag{3.10}$$

$$b(\theta, N) = 0.72N^{-0.45} \exp(-0.7885 \sec\theta) \quad \text{ns m}^{-1}. \tag{3.11}$$

where N is the number of particles recorded in the detector. Note that the coefficient of linearity is a function of θ and N. The shower front, thus, is an inverted cone moving along the shower axis. The deviation from the plane front has to be taken into account, particularly when arrival time measurements at large core distances are used, if better angular accuracy is required as in the case of search for point sources of ultra high energy gamma rays against the cosmic ray background to improve the gamma ray signal (see Chapter 8). This is done by correcting the arrival times for this deviation before a plane front is fitted. Eq. 3.5 is then rewritten as

$$\psi^2 = \sum \omega_i \left\{ lx_i + my_i + nz_i - c\left(t_i - t_0 - br_i/\sqrt{N_i}\right)\right\}^2. \tag{3.12}$$

First the zenith and azimuth angles are determined using the plane front approximation. The approximate x and y coordinates of the shower core is obtained by taking the density weighted average of the x and y coordinates of the detectors. The distances of the fast timing detectors in the shower plane are then calculated and Eqs. 3.9 solved for θ and φ. Using these values of θ and φ the core location is determined more accurately (see Sec. 3.3.2) and the angles are determined again. This iterative procedure is repeated a few times till the required accuracy is achieved. Generally two or three iterations are adequate. The error in the zenith angle is approximately proportional to $1/\sqrt{\nu}$ where ν is the number of timing detectors used. The angular accuracy obtained is typically about 1° or better.

3.3.2 Determination of Shower Size

Shower size, N_e, is determined by fitting the observed particle densities to an assumed lateral distribution function, e.g. Eq. 2.22) in the shower plane using least square procedure. In this case the quantity χ^2 is defined as

$$\chi^2 = \sum \omega_i \left(\Delta_i^o - \Delta_i^e\right)^2 \tag{3.13}$$

where Δ_i^o and Δ_i^e are the observed and the expected (according to the lateral distribution function) densities in the ith detector and $\omega_i = 1/\sigma_i^2$ is the weight factor. The error σ_i includes the statistical as well as instrumental errors as before. The free parameters N_e, the shower size, s, the age parameter and the coordinates X_0, Y_0 of the shower core in the plane of the array are determined by solving the equations.

$$\frac{\partial\chi^2}{\partial X_0} = \frac{\partial\chi^2}{\partial Y_0} = \frac{\partial\chi^2}{\partial s} = \frac{\partial\chi^2}{\partial N_e} = 0. \tag{3.14}$$

These equations are highly non-linear and are solved by an iterative procedure using the method of steepest descent. First an initial estimate of the parameters is made.

For example the initial values X_1, Y_1 of the shower core are estimated by taking the density weighted average of the detector coordinates that responded. Thus,

$$X_1 = \frac{\sum \Delta_i^o X_i}{\sum \Delta_i^o}, \qquad (3.15)$$
$$Y_1 = \frac{\sum \Delta_i^o Y_i}{\sum \Delta_i^o}$$

Using these values, weighted averages of s and N_e are made using the lateral distribution function. The iteration starts with these values. The values of the parameters in the $(k+1)^{th}$ iteration are obtained from those in the kth iteration as

$$\xi_{k+1}^j = \xi_k^j - \lambda_k \frac{\left(\frac{\partial\chi^2}{\partial\xi^j}\right)_k}{\sqrt{\sum_j \left(\frac{\partial\chi^2}{\partial\xi^j}\right)_k^2}} \qquad (3.16)$$

where ξ^j stands for any of the parameters and λ_k is the step length, which depends on the array configuration and is adjusted by trial and error. The iterative procedure is stopped when a predetermined criteria is satisfied, which ensures that the minimum value of χ^2 is reached, e.g. the difference between successive values of χ^2 may be required to be less than a predetermined value and the gradients should be nearly equal to zero.

Errors on the estimated parameters are obtained generally by analyzing artificially generated showers. For each shower with a given set of values of the parameters, particle densities at all the detector positions are calculated using the lateral distribution function, errors imposed using Monte Carlo technique and analyzed using the same procedure. The errors are estimated from the distribution of the parameter for a large sample of showers.

References

1. W. Galbraith, *Extensive Air Showers*, (Butterworths Scientific Publications, London, 1958).
2. S. Fukui, *J. Phys. Soc. Japan.* **16** (1961) 604.
3. M. Samorski, *Nucl. Instr. Meth.* **108** (1973) 285.
4. M. Nagano and S. Shibata, *J. Phys. Soc. Japan* **20** (1965) 685.
5. T. Matano et al., *Can. J. Phys.* **46** (1968) S56.
6. A. L. Hodson et al., *Proc. 15th International Cosmic Ray Conference, Plovdiv* **12** (1977) 160.
7. S. Kino et al., *Proc. 15th International Cosmic Ray Conference, Plovdiv* **8** (1977) 68.
8. R. J. Reid et al., *Proc. Phys. Soc.* **78** (1961) 103.
9. R. W. Williams, *Phys. Rev.* **74** (1948) 1689.
10. I. Campbell and J. R. Prescott, *Proc. Phys. Soc.* **A65** (1952).
11. G. Clark et al., *Nature* **180** (1957) 353, 406.
12. G. Clark, F. Schreb and W. B. Smith, *Rev. Sci. Instr.* **28** (1957) 433.
13. S. Sinha, *Ph. D. Thesis* University of Bombay (Unpublished) 1987.
14. B. K. Chatterjee et al., *J. Phys. Soc. Japan* **17 Suppl A-III** (1962) 254.
15. M. Nagano et al., *J. Phys. Soc. Japan* **53** (1984) 1667.
16. S. Shibata et al., *Proc. 9th International Cosmic Ray Conference, London* **2** (1965) 672.
17. M. G. K. Menon and P. V. Ramana Murthy, *Prog. Cosmic Ray Phys.* **9** (1967) 163.
18. P. H. Barrett et al., *Phys. Rev.* **95** (1954) 1571.
    ibid. *Rev. Mod. Phys.* **24** (1952) 133.
19. B. S. Acharya et al., *Nucl. Instr. and Meth.* **A270** (1988) 556.
20. The MACRO-EAS-TOP Collaboration, *Proc. 22nd International Cosmic Ray Conference, Dublin* **4** (1991) 563.
21. EAS-TOP and LVD Collaboration, *Proc. 23rd International Cosmic Ray Conference, Calgary* **4** (1993) 696.
22. S. Bennet and K. Greisen, *Phys. Rev.* **124** (1961) 1982.
23. J. C. Earnshaw et al., *Proc. Phys. Soc.* **90** (1967) 91.
24. Yu. N. Bazhutov et al., *Proc. 15th International Cosmic Ray Conference, Plovdiv* **6** (1977) 167.
25. B. K. Chatterjee et al., *Il Nuovo Cimento* **18** (1960) 1148.
26. K. Suga et al., *J. Phys. Soc. Japan* **17 Suppl A-III** (1962) 128.
27. N. N. Goryunov et al., *Proc. Moscow Cosmic Ray Conference* **II** (1960) 70.
28. B. K. Chatterjee et al., *Proc. Int. Conf. Cosmic Rays, London* **2** (1965) 802.

29. S. Miyake et al., *Proc. Moscow Cosmic Ray Conference* **II** (1960) 67.
30. B. V. Sreekantan, *Proc. Ind. Nat. Sci. Acad.* **44** (1978) 125.
31. B. K. Chatterjee et al., *Proc.* 9th *International Cosmic Ray Conference, London* **2** (1965) 627,
32. S. Naranan et al., *Proc.* 13th *International Cosmic Ray Conference, Denver* ***3*** (1973) 1872.
33. M. Aglietta et al., *Nucl. Instr. Meth.* **A277** (1989) 23.
34. C. Castagnoli et al., *Proc.* 23rd *International Cosmic Ray Conference, Calgary* **4** (1993) 258.
35. S. Barwick et al., *Proc.* 23rd *International Cosmic Ray Conference, Calgary* **1** (1993) 436.
    G. E. Allan et al., *Proc.* 24th *International Cosmic Ray Conference, Roma* **1** (1995) 942.
36. F. Aharonian et al., *Proc.* 23rd *International Cosmic Ray Conference, Calgary* **4** (1993) 291.
    ibid. *Proc.* 24th *International Cosmic Ray Conference, Roma* **1** (1995) 474.
    M. Panter, *Proc.* 24th *International Cosmic Ray Conference, Roma* **1** (1995) 958.
37. K. Berniöhr et al., *Proc.* 23rd *International Cosmic Ray Conference, Calgary* **4** (1993) 199.
38. J. Linsely, *Proc.* 18th *International Cosmic Ray Conference, Bangalore* **12** (1983) 135.
39. G. Agnetta et al., *Proc.* 23rd *International Cosmic Ray Conference, Calgary* **4** (1993) 270.
40. D. J. Bird et al., *Proc.* 23rd *International Cosmic Ray Conference, Calgary* **4** (1993) 283.
41. K. Bekk et al., *Proc.* 23rd *International Cosmic Ray Conference, Calgary* **4** (1993) 674.
    H. O. Klages, *Proc.* 24th *International Cosmic Ray Conference, Roma* **1** (1995) 946.
42. S. Tilav et al., *Proc.* 23rd *International Cosmic Ray Conference, Calgary* **4** (1993) 565.
43. S. Barwick and J. Beaman, *Proc.* 23rd *International Cosmic Ray Conference, Calgary* **4** (1993) 683.
44. T. K. Gaisser et al., *Proc.* 24th *International Cosmic Ray Conference, Roma* **1** (1995) 938.
45. B. Erlandsson et al., *Proc.* 24th *International Cosmic Ray Conference, Roma* **1** (1995) 1039.
46. M. Amenomori et al., *Proc.* 24th *International Cosmic Ray Conference, Roma* 2 (1995) 736.
47. J. Linsley, *Proc.* 19th *International Cosmic Ray Conference, La Jolla* 3 (1985) 461.
48. P. J. V. Eames et al., *Proc.* 20th *International Cosmic Ray Conference, Moscow* 2 (1987) 449.

49. B. S. Acharya et al., *J. Phys. G: Nucl. Part. Phys.* **19** (1993) 1053.

Chapter 4
HIGH ENERGY INTERACTIONS

The Extensive Air Shower, as we have seen earlier, observed at mountain altitude or sea level is the end product of the nuclear and electromagnetic cascades initiated at the top levels of the atmosphere by the arrival of just one high energy cosmic ray primary which could be a proton, α particle or a heavy nucleus. The primary, being a hadron, undergoes repeated nuclear collisions with air nuclei as it traverses down losing a fraction of its energy to the creation of a large number of secondaries in each collision. The secondaries are predominantly pions, kaons and nucleon-antinucleon pairs and in a small measure other exotic particles. These secondaries themselves being hadrons collide further with air nuclei and give rise to succeeding generations of nuclear interacting particles. Since the density of the atmosphere is low at high altitudes and the particles traverse considerable distances before encountering with other air nuclei, some of them like the charged pions and kaons decay and give rise to mu-mesons, electrons and neutrinos. The neutral pions decay into γ rays which initiate electromagnetic cascades. The nuclear cascade thus feeds continuously into the electromagnetic cascade. The large distances between the nuclear encounters also results in the lateral spread of the particles to tens of meters to hundreds of meters even though the particles, generated in the high energy collisions are fairly collimated in the forward direction. Because of the different velocities of the particles of different mass and the momentum distribution, there is a spread in the arrival times of the particles at the observational level. The photons which traverse with the velocity of light will arrive first, then the electrons, then the muons and then the nucleons and antinucleons and the other heavy particles. The spread in the arrival time distribution is also controlled by the extent of scattering the particles undergo and the distance from the core at which the times are measured.

The details of the lateral distributions of the different components, the time structure of the different components, the number and the energy spectra of the different components naturally depend on the collision characteristics at different energies for the different types of hadrons, as well as on the decay characteristics of the unstable particles. The collision characteristics that are relevant are:

1. cross section for proton-air nucleus collision,
2. cross section for pion-air nucleus collision,

3. cross section for nucleus-air nucleus collision,
4. inelasticity in the collisions,
5. multiplicity of secondaries,
6. composition of secondaries, and
7. transverse momentum distribution.

Historically, before the advent of very high energy accelerators, the cosmic ray scientists had to depend on the extrapolation of the characteristics of collisions obtained from mountain altitude cloud chamber experiments (up to a TeV or so) and jet studies with nuclear emulsion stacks exposed at stratospheric balloon altitudes.

With the advent of accelerators, the information on the collision characteristics at least on proton-proton collisions became much more reliable and quantitative almost up to $\sim 10^{15}$ eV. In the case of nucleus–nucleus, and pion–nucleus collisions, reliable results are available from accelerators up to 10^{13} eV and in the case of nucleus-nucleus collisions up to 2×10^{11} eV per nucleon [1–3]. Considerable amount of information on the special characteristics at very high energies have also come from the emulsion chamber experiments at mountain altitudes and hybrid installations of emulsion chambers and air shower arrays which we shall consider separately.

A most important development in the efforts to derive some of the characteristics of very high energy interactions and the composition of the primaries, is the advent of large computer systems which have enabled very detailed and sophisticated Monte Carlo simulations that can be compared with experimental observations. Let us first consider some historical examples of discernment of special features of high energy collisions from cosmic ray air shower studies.

4.1 Inelasticity of High Energy Collisions

As early as 1953, the Soviet Group reported at the Oxford Conference on Extensive Air Showers the presence of at least one very high energy nuclear active particle on the average in the core regions of extensive air showers. This gave rise to the idea that the very high energy collision was not totally inelastic and a leading particle that carried away a considerable amount of the colliding particle energy persisted in the core of the air shower even after tens of collisions. This unique feature has been later confirmed in cosmic ray experiments as well as in accelerator experiments and has played a crucial role in the interpretation of air shower data.

4.2 Variation of Interaction Mean Free Path with Energy

A small fraction of the protons incident at the top of the atmosphere as primary cosmic rays will leak into the observation levels at mountain altitude, where air shower

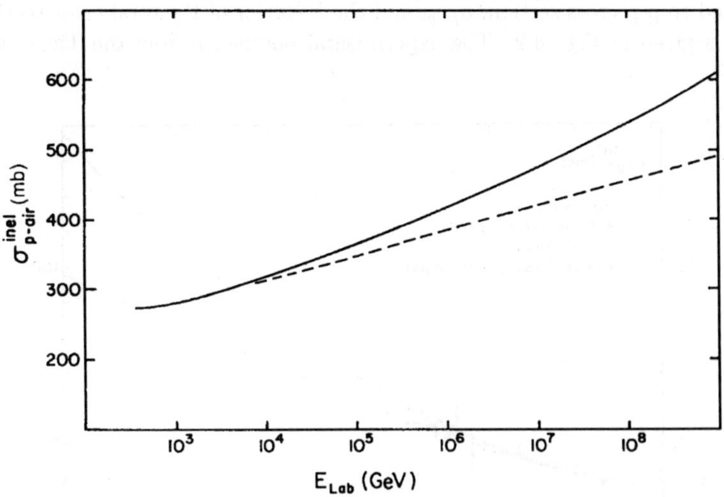

Figure 4.1 The lower bound on p-air inelastic cross section obtained by Yodh et al (1972).

installations are normally set up, without undergoing any interaction in the intervening atmosphere. If the flux of such single unaccompanied protons are determined at mountain altitude and compared with the primary flux then one can arrive at the interaction mean free path through the simple relation

$$J_{mountain}(E) = J_{primary}(E)e^{-X/\lambda(E)}$$

where X is the depth of the observation level in g cm^{-2} and $\lambda(E)$ is the interaction mean free path in g cm^{-2} and E is the energy of the proton. The two main difficulties in this method are:

1. knowledge of the primary flux of protons is essential and
2. the experimental difficulty of establishing that the protons recorded at the mountain level are unaccompanied.

Nevertheless it has been shown by Yodh et al. [4, 5] that the method can lead to a lower bound on the p-air cross section at various energies (Fig. 4.1).

As the energy increases, the efficiency of establishing the unaccompanied nature of the proton increases. Also, the complication that arises because of the mixed nature of the primary composition is not important at least up to 10^{14} eV where protons dominate.

Using Glauber Theory [6,7] the $\sigma_{p-air}^{inelastic}$ for proton-air nucleus collisions have been converted to p-p cross section, σ_{p-p}, and the behavior of the total cross section with energy is given in Fig. 4.2. The experimental points are from the Tienshan group

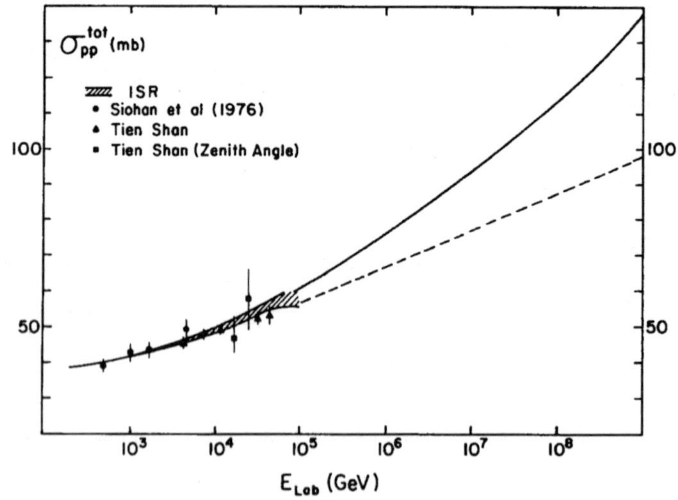

Figure 4.2 The total p-p cross section derived from the p-air cross section using Glauber theory by Yodh et al (1972) (see Yodh et al, 1972 for references).

who are really pioneers in this field. The behavior of the σ_{p-p} over a much wider energy range deduced from a number of experiments, given by Tonwar [8], is shown in Fig. 4.3.

At energies above 10^{14} eV the measurement of σ_{p-air} is based on a comparison of the flux of air showers having the same number of muons and electrons in the vertical and inclined directions.

Tonwar has emphasized that systematic effects due to fluctuations in shower development, the steepness of the primary spectrum, the uncertainty of primary composition may result in deducing the behavior of the cross section variation with energy less dependable. One can summarize the results as showing that the cross section does increase as $\ln^2 s$ up to 10^{14} eV and at higher energies there is uncertainty.

A summary of the more recent results from the Akeno, Yakutsk and Fly's Eye experiments on the variation of σ_{p-air} with energy in the range beyond 10^{16} eV, quoted by Nagano et al. [9], is shown in Fig. 4.4.

While the Akeno measurements are based on the intensities at various zenith angles for the same N_e and N_μ, the Fly's Eye and Yakutsk results are based on the distribution of the shower maxima in the atmosphere.

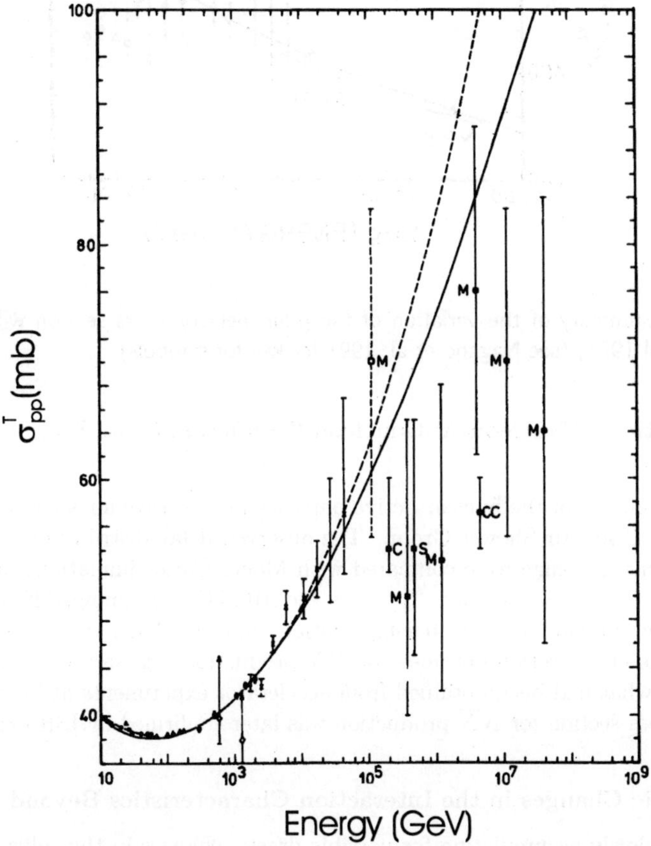

Figure 4.3 Variation of the p-p total cross section with energy (Tonwar 1979) (see Tonwar, 1979 for key for symbols).

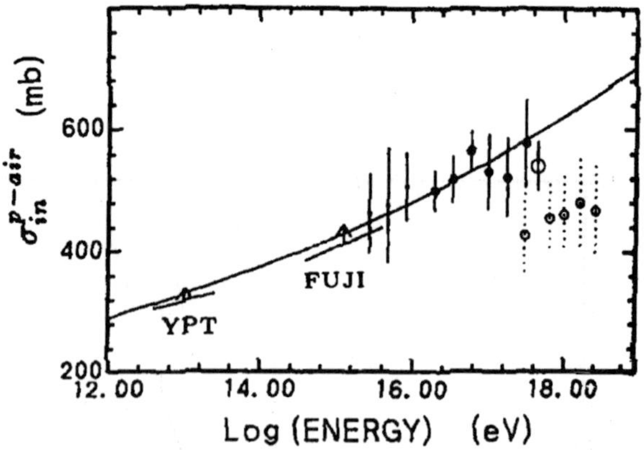

Figure 4.4 Summary of the variation of the p-air inelastic cross section with energy (Nagano et al 1991) (see Nagano et al, 1991 for key for symbols).

4.3 Variation of Nucleon-Antinucleon Production Cross Section with Energy

The time structure of the nuclear active particles in extensive air showers was measured by the Ooty Air Shower Group. The observed delay distributions of particles of different energy range were compared with Monte Carlo simulations for particles of different masses—pions, kaons, protons etc. [10, 11], assuming different models of interaction "fireball" and "isobar-pionization" models. Their results are shown in Fig. 4.5. The cross section obtained for $N\overline{N}$ production was an order of magnitude higher than what had been obtained from accelerator experiments at lower energies. The high cross section for $N\overline{N}$ production was later confirmed by ISR experiments.

4.4 Drastic Changes in the Interaction Characteristics Beyond 10^{15} eV ?

Evidence is slowly accumulating for possible drastic changes in the collision characteristics beyond a few times 10^{15} eV (just a little beyond the present day accelerator experiments with colliders) from Japanese and Russian Emulsion Chamber experiments at high mountain altitudes and air shower experiments some of which are operated in conjunction with emulsion chambers and Electron Hadron Calorimeters. In this section we will consider the type of evidence that is coming from hybrid air shower experiments.

The Lebedev Institute Group [12] have been operating the large X-ray emulsion chamber (162 m^2 in area) installation HADRON, shown in Fig. 4.6, in conjunction

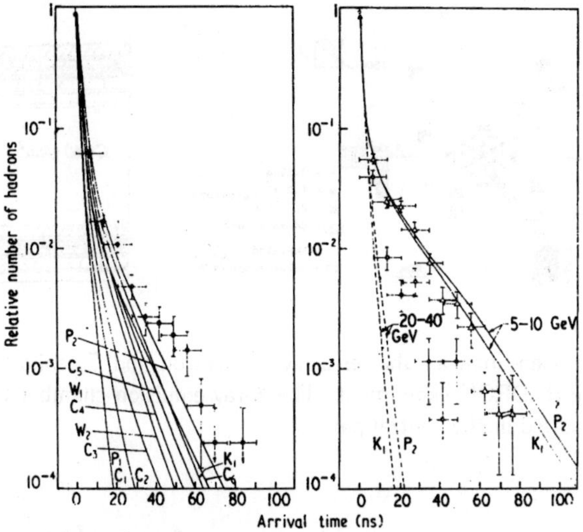

Figure 4.5 Comparison Monte Carlo simulations of the time structure of hadrons at core distances $\leq$ 20 m in showers of size $6.7 \times 10^4 - 1.8 \times 10^6$ with experimental data by Tonwar and Sreekantan (1971) (see .Tonwar and Sreekantan, 1971 for details of the interaction models used).

with the Tienshan extensive air shower array at an altitude of 690 g cm^2 since 1984. The HADRON is being used for measuring the energy spectra of electrons, gamma quanta and hadrons in the cores of air showers with primary energy greater than 10^{15} eV. The extensive air shower array has more than 80 scintillation detectors for measurement of density of electrons at various distances. The most interesting results from this experiment are shown in the Figs. 4.7, 4.8 and 4.9.

The energy spectra of gamma quanta and electrons in showers of energy grater than 2×10^{16} eV are distinctly very steep compared to the spectra at lower energies. This drastic change in the behavior of the electron gamma quanta spectrum is interpreted by Nikolsky as evidence for the disappearance of the fragmentation region in the collision that normally feeds to the electromagnetic component. Nikolsky is of the opinion that in the collisions at these energies, there may be the onset of the generation of Higgs Bosons (with mass greater than 400 GeV c^{-2}) that decay directly into the W and Z bosons, muons and neutrinos. An interesting aspect of this experimental result is that the change over takes place in an extremely narrow range of energy of 5×10^{15} eV–2×10^{16} eV. This is evident from Fig. 4.8 which shows the dependence of the energy flux in the EAS core with primary energy.

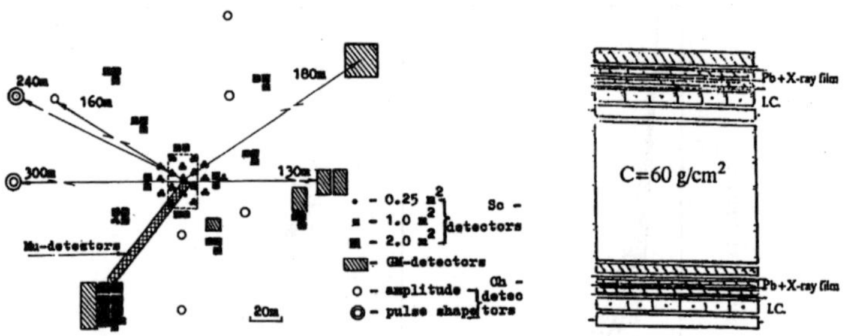

Figure 4.6 The Tien-Shan air shower array of electron and Cerenkov light detectors (left) used in the HADRON experiment. The X-ray emulsion chamber (right), shown as a rectangle, is located in the central part.

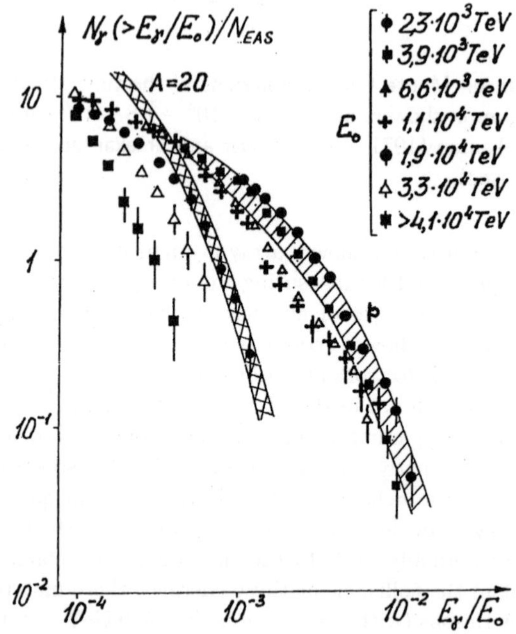

Figure 4.7 Energy spectrum of gamma-quanta and electrons in cores of EAS of different primary energies (Nilolsky 1993)

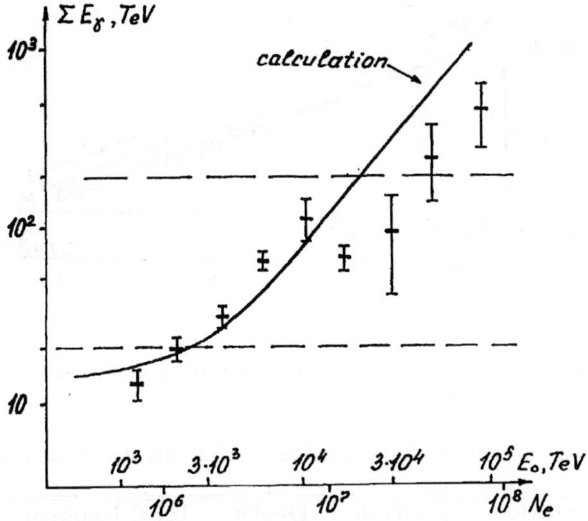

Figure 4.8 Dependence of energy flux on the energy of the primary particle (Nikolsky 1993)

Clearly these trends, which are important for the interpretation of air shower observations, need to be confirmed by further experiments by other groups as well. They have a direct bearing on the results on primary composition at these energies.

4.5 Emulsion Chambers

During the period 1945–65, studies on cosmic ray interactions with magnetic and multiplate cloud chambers at mountain altitudes and specially designed nuclear emulsion sandwiches flown to high balloon altitudes, revealed many of the intricate aspects of high energy nuclear collisions resulting in meson production and also opened up the field of elementary particle physics. This development naturally encouraged the construction of higher and higher energy accelerators and the field of high energy and elementary particle physics was gradually taken over by the accelerator physicists. The cosmic ray physicists had however the advantage that they could shift their attention to still higher energies and also study nucleus-nucleus collisions which could not be done with the colliders. This race resulted in the advent of the Emulsion Chamber Technique for the study of interactions at energies above 10^{14} eV, where the flux of particles in cosmic rays had fallen to such low values that the area of the chambers had to be at least a hundred square meters to get any meaningful results

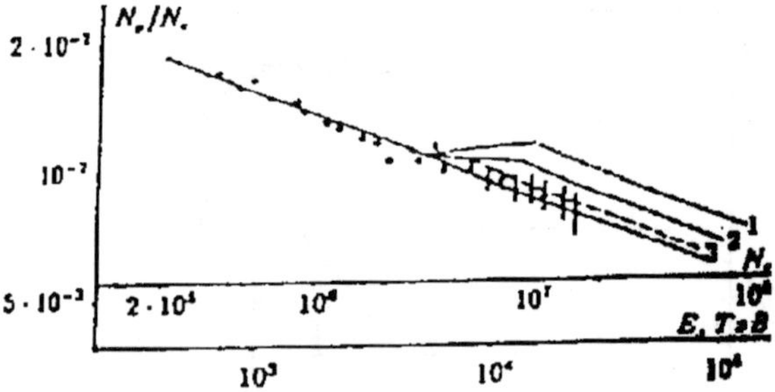

Figure 4.9 Dependence of number of muons on primary energy from the HADRON experiment at Tien-Shan (Nikolsky. 1993).

Table 4.1 Details of emulsion chamber installations at mountain altitudes.

Station	Altitude (m)	Depth (g cm^{-2})	Total Exposure till 1990 (m^2 yr)
Chacaltaya	5,200	540	1,000
Pamir	4,370	600	10,000
Mt. Fuji	3,750	650	1,100
Kanbala	5,500	520	1,000

in a reasonable time.

The Emulsion Chamber is essentially designed from the point of view of developing and recording the energy and spacings of individual very high energy electromagnetic cascades (10^{12} eV). The installations which are fairly large in area (~ 100 m^2) consist of alternate layers of lead (1–2 cm thick) and either X-ray emulsions (of the type RT6-1) or regular high sensitivity nuclear emulsions. The intensity and diameter of the dark spots left by the electrons in the core of cascades are a measure of the energy of the cascade. In the case of nuclear emulsions the number of electrons can be counted under a microscope with a spatial resolution of about a micron. The installations are at very high mountain altitudes (see Table 4.1).

The emulsion chambers are designed to

1. record the secondaries from very high energy collisions of individual hadrons with the nuclei of air a few tens to a few hundred meters above the installation,
2. record families of such cascades which result from several collisions at different heights of the particles which are essentially part an extensive air shower

initiated by a cosmic ray primary in the atmosphere.

The families are identified by their parallelism in the cascade directions. There are several designs of the chamber. In the older versions, there used to be a top section and a bottom section with an absorber like carbon in between. The chamber used by the Pamir group is shown in Fig. 4.10. The top section responded to the pure

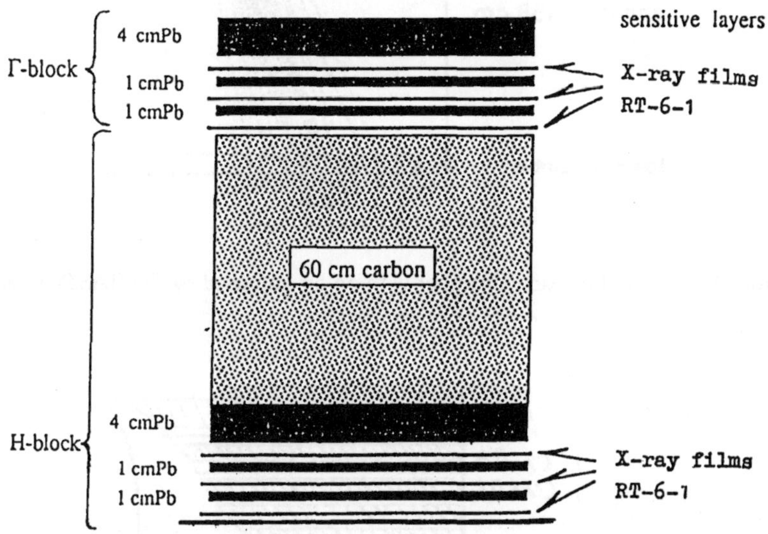

Figure 4.10 Lay-out of carbon X-ray emulsion chamber of the Pamir group.

electromagnetic component from the decay of very high energy π^0 mesons produced in the collisions above. The bottom section responded to the high energy hadrons that interacted in the carbon block and to the π^0s that were produced in these interactions. Another variant has been to have a thin carbon block and a large air gap below it, the hadron interactions taking place in the lead plates themselves or in the carbon block. The air gap provides for separation of the cascades. Such a chamber used by the BASJE group is shown in Fig. 4.11. A third design has been to have a "deep lead chamber" of a little over a meter of lead in between the emulsion plates. The chamber used by the Pamir group is shown in Fig. 4.12. In yet another version the emulsion chamber

is operated in conjunction with an extensive air shower array. The air shower array enables the determination of the direction of the shower to an accuracy of a few degrees. This is good enough to relate with the emulsion chamber jets since the emulsion plates are changed after an exposure of a few months.

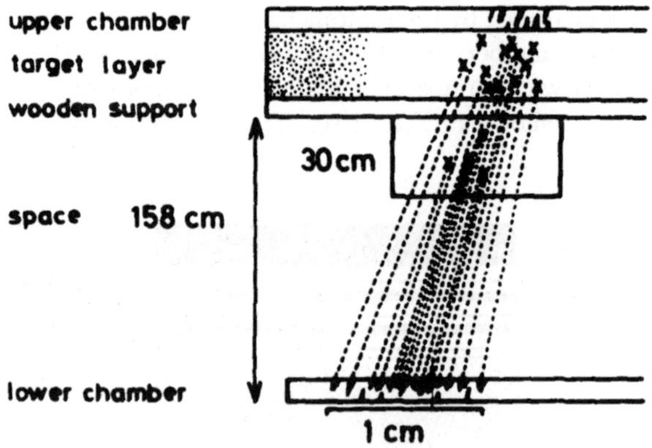

Figure 4.11 The Emulsion Chamber with an air gap used by the BASJE group.

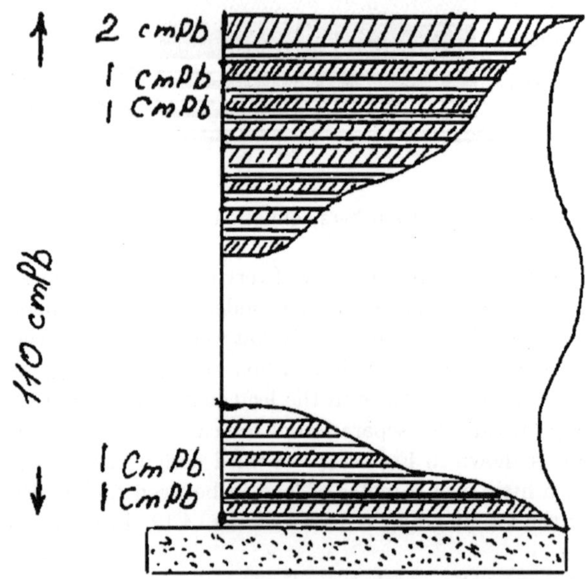

Figure 4.12 The deep lead Emulsion Chamber used by the Pamir group.

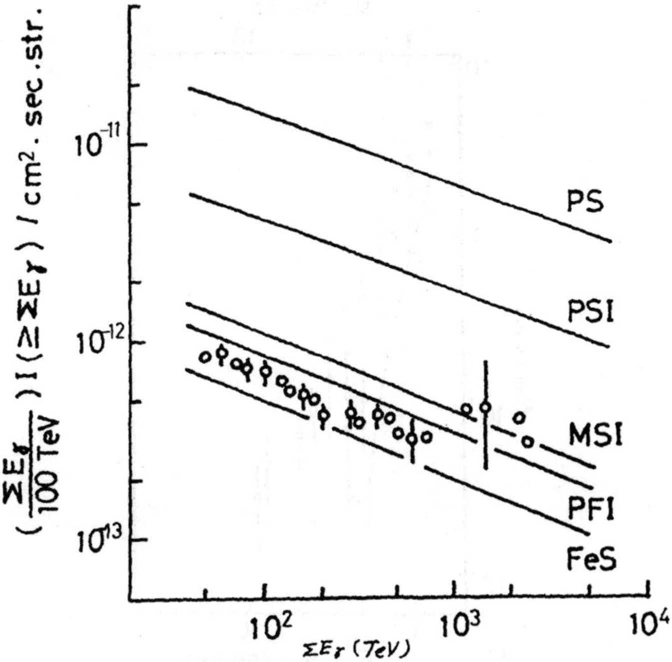

Figure 4.13 Energy spectrum of gamma-ray families ($E_{thres} = 2$ TeV) observed in the Mt. Fuji experiment. Simulation results, shown as straight lines, are by Kasahara, Tori and Yuda for different models of high energy interactions and primary mass composition.

4.5.1 Results from Emulsion Chamber Experiments

We shall now discuss the main results from the Emulsion Chamber installations: (see Holynski [13] and Wlodarczyk [14]).

4.5.1.1 The Energy Spectrum of Gamma Families: Scaling Violation

The results from the Mt. Fuji experiment are given in Fig. 4.13. It is seen that the measured intensity is at least an order of magnitude lower than what is calculated on the basis of proton dominant composition and validity of Feynmann scaling in the fragmentation region. The agreement is better for the following models of composition and scaling hypotheses:

1. Mixed primary composition + scaling with increasing σ_{inel}, the inelastic cross section.

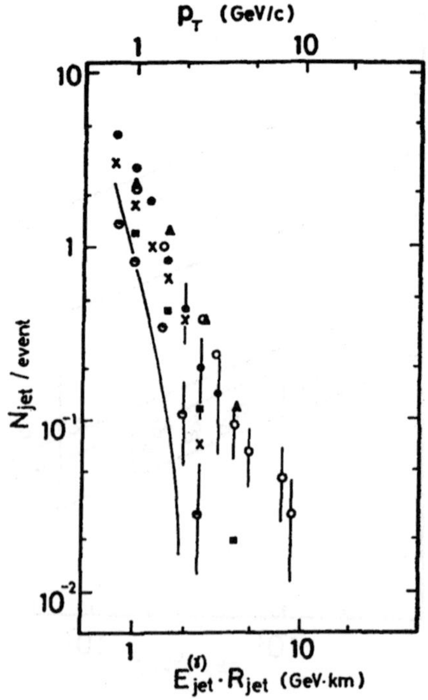

Figure 4.14 Integral energy spectrum of $E_{jet}^{(\gamma)} \cdot R_{jet}$ of jets in families with $\sum E_\gamma > 100$ TeV. The data arae from the Pamir, Fuji and Chacaltaya experiments.

2. Iron primary with scaling

3. Proton with $E_0^{1/4}$ law for multiplicity increase. Feynmann scaling assumes that the number of high energy secondaries with fractional energy $X_F = E/E_0 \geq 0.1$, produced in p-p interactions is independent of the primary energy.

Yet another result which highlights the departure from scaling and the production of high p_t secondaries with momenta up to 10 GeV/c is shown in Fig. 4.14. The data is from Pamir and Fuji experiments.

4.5.1.2 Attenuation of Cosmic Ray Hadrons in Lead

The Japan-Russian Collaboration with Uniform-type deep lead chambers has given rather surprising results on the differences in attenuation lengths of (a) single hadrons and (b) hadrons which are members of families. These results are shown in Fig. 4.15.

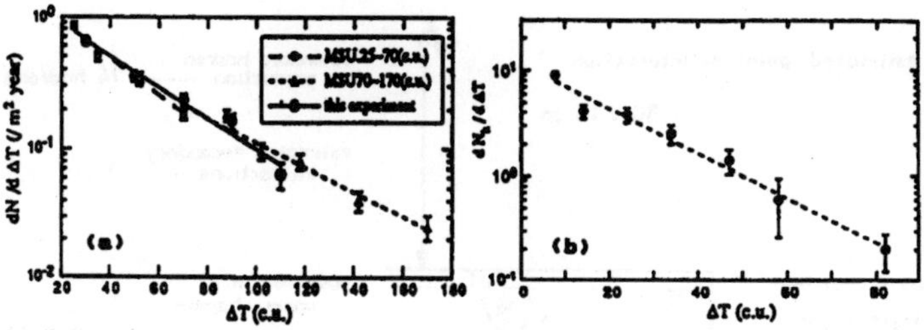

Figure 4.15 Distribution of starting points of cascades of single hadrons of (a) $\sum E^h_\gamma \geq 6$ TeV, and (b) $\sum E^h_\gamma \geq 10$ TeV in high energy families of $\sum E > 700$ TeV.

For single hadron with $E_h \geq 6$ TeV, The $\Lambda_{Pb} = 208 \pm 15$ g cm^{-2} and is in agreement with the Moscow State University result of 208 ± 17 g cm^{-2} in the cascade unit depth range 20–78. At larger depths (78–92) the $\Lambda_{Pb} = 310 \pm 36$ g/cm^2. The number of observed cascades at depths larger than 20 c.u. is quite high. To account for this excess, role of decay of charmed particles has been proposed by some of the Russian scientists. However, the cross section for charm production has to be as large as $(1.8 \pm 1.0)A$ millibarns, which is inconsistent with the extrapolation from accelerator data by a large factor.

For individual hadrons of energy ≥ 10 TeV in families with ≥ 700 TeV, the $\Lambda_{Pb} = 136$ g cm^{-2} which is much smaller than that for single hadrons.

4.5.1.3 Centauro Events

The Chacaltaya and Pamir Collaboration groups came up with evidence for special type of families of energy 500 TeV, in which the number of hadrons were in far excess of gamma quanta. Normally, family events at this atmospheric depth have more gamma rays than hadrons and therefore more jets are recorded in the upper chamber than in the lower one. Thus, looking at the lower chamber, one can't predict the contents of the upper chamber. Hence these events were given the name Centauro[a].

The Centauro type of event was first reported by the Chacaltaya group (Brazil-Japan Collaboration) at the Denver Cosmic Ray Conference in 1973 [15]. In the Emulsion Chamber at Chacaltaya (540 g cm^{-2}) a big family event, shown in Fig. 4.16, was recorded with 43 hadrons in the lower chamber, 6 hadrons in the upper chamber and just ONE gamma cascade, all of them confined within about a centimeter from the center of the event. The visible energy of the event was estimated as 2.3×10^{14} eV and the total energy $\sim 1.5 \times 10^{15}$ eV. Being the first of such events observed, they named this event Centauro-I. The interpretation was that, in an interaction at

[a]Centauro is a Greek mythical beast with human head and the body of an animal.

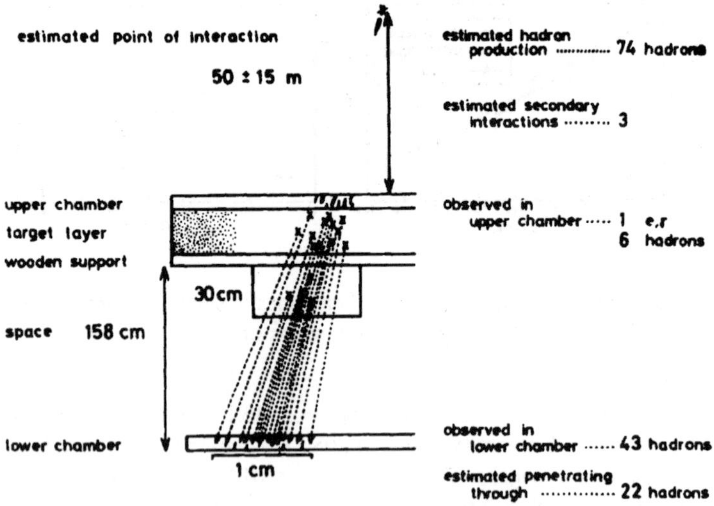

Figure 4.16 The first hadron-rich family, named Centauro, recorded by the Brasil-Japan group in their emulsion chamber detector exposed to cosmic rays at Mt. Chacaltaya (540 g cm^{-2}).

$\sim$ 50 m above the detector, a fireball of mass 200 GeV/c^2 had been produced which decayed into $\sim$ 100 hadrons and no π^0s. The average $p_t \sim 1.5 - 2$ GeV/c in this interpretation.

Further search by the Brazil-Japan Collaboration group have revealed four more events of Centauro type and one from Pamir experiment but none from Mt. Fuji and Mt. Kanbala experiments. These events, while rich in hadrons, do contain larger number of gamma rays than Centauro-I but less than the remaining events. The rate of family events in Cosmic rays is about 1 m^{-2} yr^{-1} and the observed rate of Centauro events is 10^{-2} m^{-2} yr^{-1}. Another type of hadron rich events of total energy $\sim$ few 100 TeV, called Mini-Centauro events, in which the number of particles (5 – 20) is smaller than that in the Centauros, are also found in these experiments. Fig. 4.17 shows a plot depicting the number of hadrons versus the fraction of energy carried by the hadrons. The expectations on the basis of the QGS model is also shown in the figure.

Searches for both Centauro and Mini-Centauro events at accelerators were carried at the CERN SPS collider at center of mass energies of 540 and 900 GeV, corresponding to laboratory energies of 146 and 400 TeV respectively. These searches gave totally negative results since no case of a Centauro type event was found. While

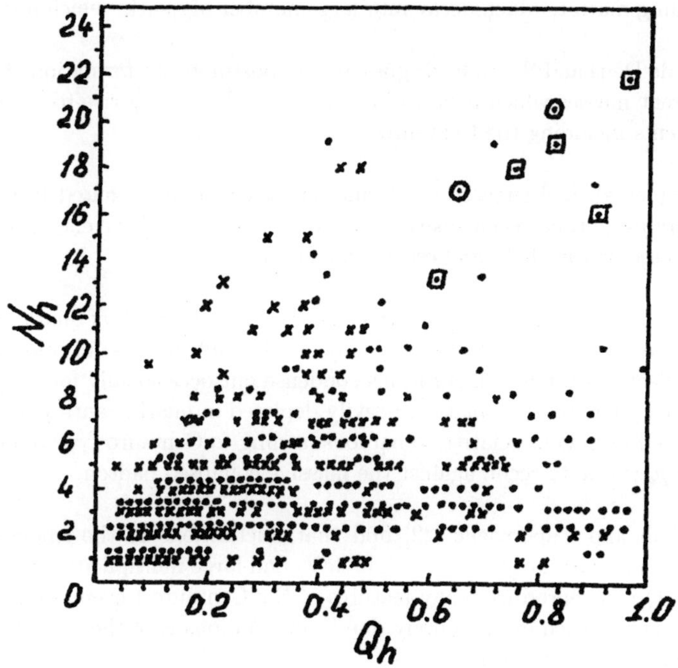

Figure 4.17 Correlation plot of number of hadrons, N_h, versus fraction of hadron energy, $Q_h = \sum E_h^\gamma / \sum E_{tot}$. The events are selected with $\sum E_{tot} = \sum E_\gamma + \sum E_h^\gamma > 100$ TeV, $N_h \geq 0, N_\gamma \geq 0, \sum E_\gamma \leq 400$ TeV.

the accelerator energies are adequate to find Mini-Centauro events, they are less than the energies of the Centauro events found in cosmic rays.

Besides the Brazil-Japan Collaboration group, efforts were made by many others to understand the Centauro-I event (see Ref. [24] and references therein). Some of the theoretical models proposed to understand the Centauro events are:

1. Bjorken and McLerran [16] postulated the existence of a new component, a dense primordial glob of nuclear matter or quark matter, in the primary cosmic radiation, which penetrates deep in the atmosphere and gives rise to Centauro-I type of event.

2. Kinnunen and Rubbia [17] consider the possibility of production of Centauro events due to hard collisions between hadronic constituents and rule out the same based upon flux considerations.

3. Sutherland [18] proposed the formation of quark hot spots in a small region of colliding matter, which decay into large number of quarks which form baryons.

4. Dias de Deus and W. A. Rodrigues Jr. [19] postulate the formation of clusters of different masses which have various decay modes, giving rise to different types of events including the Centauros.

5. Fowler et al. [20] suggest the formation of coherently excited fireballs which undergo superfluorescent-like decay. The π^0s produced have energies lower than the detector threshold and escape detection.

6. Mann and Primakoff [21] opine that a new phase of matter of quarks and gluons is formed from which clusters of many quarks (anti-quarks) in color singlet and zero-electric charge configurations condense out occasionally forming collapsed hadrons. The residual matter, heavily depleted in quarks (anti-quarks), will be depleted in mesons relative to baryons leading to Centauro type of events. The large p_t is due to recoil against the massive collapsed hadron.

7. Attallah and Capdevielle [22] find that fluctuations in the single diffractive component with large mass ($\sim$ 120 GeV) and large multiplicity ($N_{ch} = 50 - 60$, $n_\gamma \leq 1$) can reproduce many features of the Centauro-I event. The calculated frequency of such events purely due to fluctuations is of the same order as the observed Centauro events.

The possibility that fluctuated air shower cores in proton-initiated showers mimic Centauro type of events has been studied in great detail. No one has succeeded in reproducing the Centauro-I event though other events of Centauro type were reproduced. Acharya et al. [23] show, by detailed Monte Carlo simulations using scaling model with interaction cross sections rising with energy, that the Mini-Centauro events could be explained as moderate fluctuations in the cores of air showers generated by proton primaries. They also [24] have shown through Monte Carlo simulations that those Centauro events in which gamma ray content is comparable to that of hadrons, could be understood as extreme fluctuations in air shower cores using conventional physics and concluded that these events do not belong to the same class of events as Centauro-I. The possibility that a heavy nucleus interacting rather low in the atmosphere and giving rise to Centauro-I type of event was considered by Acharya and Rao [25]. They point out that the production height of the Centauro-I event was estimated on the basis of triangulation of only a few of the many jets observed. The spatial resolution ($\sim 1\mu m$) of the detector did not permit triangulation of the remaining jets. They show that such a situation arises in some of the cores of iron-initiated showers. They have shown that it is possible to reproduce all features of the event, including the fictitious production height, even though the simulated

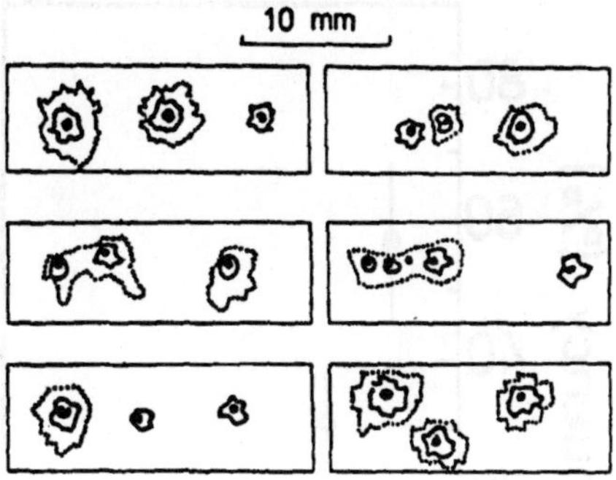

Figure 4.18 Examples of events with multicore halo, $N_c \geq 3$, observed in the Pamir experiment.

cascade actually first interacted at about a km above the detector. The total number of such events expected in all the experiments that looked for them, however, is 2×10^{-5} as against one observed event.

4.5.1.4 Halo Events

The scanning of X-ray emulsions in the Emulsion Chambers has focussed attention on some special types cascades of energy ~ 400 TeV in which the dark spots have an area of ~ 4 mm^2. In 20 to 30% of such super family events, multiple halos corresponding to multiple cores have been seen (Fig. 4.18). The number distribution of such core events is shown in Fig. 4.19.

An interesting aspect of these multi-halo events is that there is a tendency among them to align themselves along a straight line in the target plane. This would naturally indicate co-planar emission in the collision in which such high energy secondaries are produced.

Co-planar emission studies have been extended to include events of very energetic multiple cores of hadrons and gamma quanta and also those having narrow particle groups. An asymmetry parameter, λ_N, has been defined as

$$\lambda_N = \frac{\sum\limits_{i \neq j \neq k} \cos 2\Phi_{ij}^k}{N\,(N-1)\,(N-2)} \tag{4.1}$$

where Φ_{ij}^k is the angle between two straight lines connecting the kth centre with i^{th} and j^{th} centres, N is the number of centres. The parameter $\lambda_N = 1$ when all the centres are exactly aligned along with same straight line and $= 1/(1-N)$ in the isotropic cases.

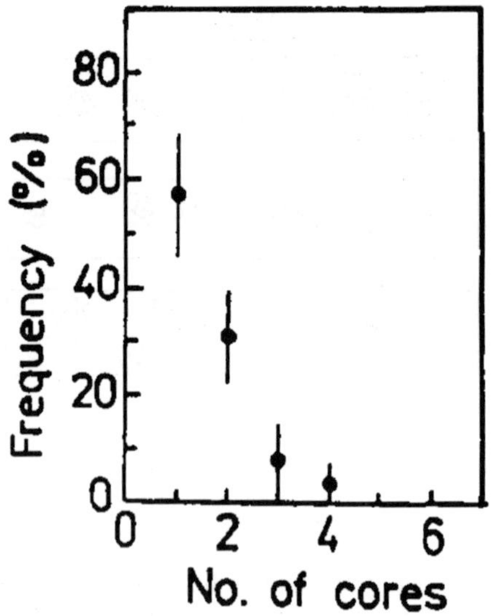

Figure 4.19 The distribution of the number of cores in halo events in the Pamir experiment.

Fig. 4.20 shows the percentage dependence of aligned events with respect to the total as a function of energy. The percentage increases from 30% at 400 TeV to 60% at 1000 TeV. This percentage at detection would mean that it may be even higher at production. What is intriguing is that alignment is practically absent below E = 100 TeV. This phenomenon of alignment does not find any explanation in the conventional models of high energy interactions.

Mukhamedshin [26] after analyzing the features of the events under a variety of models of interaction and effects of magnetic and electric fields in the atmosphere created by lightning, has come to the conclusion that the phenomenon can partly be explained by the presence of a new component at about 1% level in the primary cosmic radiation which has very special interaction characteristics (Royzch 1992 [27]) and partly by the presence of strong atmospheric electric fields. In both these cases, the alignment is expected to be more pronounced in the case of hadrons than gamma quanta, a feature that needs to be established by further experiments. Whether this phenomenon has discernible effects in extensive air shower features needs to be also investigated.

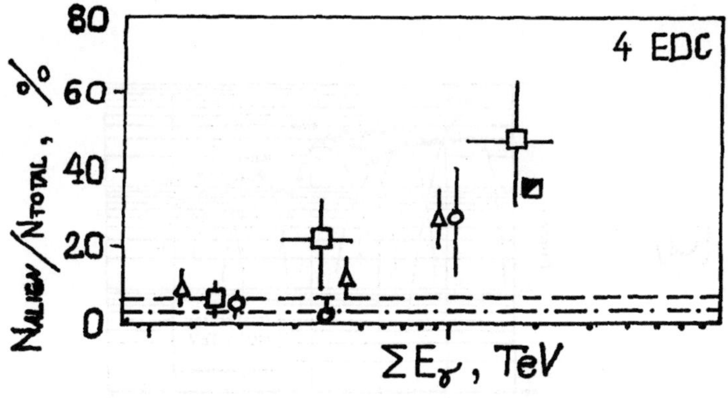

Figure 4.20 Dependence of the fraction of the aligned events on energy.

4.5.1.5 Penetrating Cascades

In the deep lead emulsion chambers several cases of very anomalous transition behavior of cascades have been registered. Three typical cases [28] are given in the Fig. 4.21.

The shower PB69-B8-S152 has not died down even after 100 cascades units of lead and shows anomalous structure. The total energy of the shower is about 400 TeV. The curves stand for trial fits to reproduce the observed profile by superposition of Pb-jet shower of different energy starting at different depths.

The PB68-S98 starts developing at a depth of 25 cm of lead of the top chamber and persists up to the end of the chamber which has 60 cm of lead.

The event PB68-B18-F8 is composed of a pair of closely separated showers, the distance between which is as small as 1.3 mm. The showers begin to be seen only at 32 cascade units in the top chamber. The visible energies are 80 TeV and 41 TeV. If the two hadrons responsible for the event are assumed to be produced in a collision at a height of about a kilometer, the $p_t \sim 40$ MeV/c. Such a low p_t process is possible only as a result of fragmentation or diffraction.

These anomalous results obtained in the Emulsion Chamber experiments on hadron and γ ray families with total energy greater than 10^{14} eV, have raised a very important question, particularly relevant to the interpretation of air shower observations, whether the characteristics of interaction do undergo a drastic change beyond the range of current accelerator energies. Hasegawa and his collaborators [29] are of the opinion that there are drastic changes. The secondaries in such high energy collisions (i.e. 10^{14} eV) are not predominantly pions, but are new type of particles christened "Chirons" which have very special characteristics. At production Chirons have a large

(a)

(b)

(c)

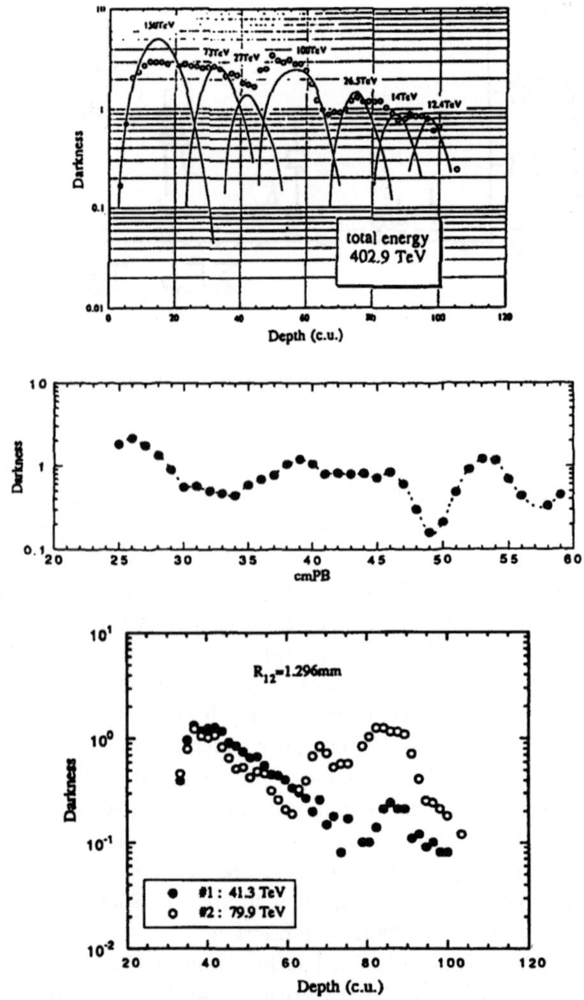

Figure 4.21 Shower transition on spot darkness, D, measured by square slit of 200 mm × 200 mm for the events (a) PB69-B8-S152; (b) PB68-S98; and (c) PB68-B18-F8.

$p_t \sim 2-3$ GeV/c, an order of magnitude higher than pions. The Chirons have shorter interaction mean free path—1/3 to 1/2 of geometric mean free path. The Chirons in their interactions give rise to extremely low p_t (~ 10 MeV/c) secondaries which are responsible for the "Mini-clusters" observed in the emulsion chamber experiments.

Evidence for possible drastic changes at these energies (10^{14} eV) is also accumulating from the hybrid experiments (Emulsion chamber + Extensive Air Shower Array) as discussed earlier.

References

1. Y. Muraki, *Proc. 21st International Cosmic Ray Conference, Adelaide* **11** 257 (1990).
2. R. Holynski et al., *Proc. 23rd International Cosmic Ray Conference, Calgary* **4** 9 (1993).
3. T. Yuda, *Proc. 22nd International Cosmic Ray Conference, Dublin* **5** 313 (1991).
4. G. B. Yodh et al., *Phys. Rev. Lett.* **28** 1005 (1972).
5. T. K. Gaisser and G. B. Yodh *Ann. Rev. Nucl. Part. Sci.* **30** 475 (1980).
6. R. J. Glauber, *High Energy Physics and Nuclear Structure* ed. S. Devons (New York: Plenum, 1970).
7. R. J. Glauber and G. Matthiae *Nucl. Phys.* **B21** 135 (1970).
8. S. C. Tonwar, *J. Phys. G: Nucl. Phys.* **5** L193 (1979).
9. M. Nagano et al., *Proc. 22nd International Cosmic Ray Conference, Dublin* **4** 189 (1991).
10. B. K. Chatterjee et al., *Proceedings of the International Conference on Cosmic Rays, London* **2** (1965) 802.
11. S. C. Tonwar and B. V. Sreekantan, *J. Phys. A: Gen. Phys.* **4** (1971) 868.
12. S. I. Nikolsky, *Proc. 23rd International Cosmic Ray Conference, Calgary* **4** 243 (1993) and references therein.
13. R. Holynski, *Proc. 21st International Cosmic Ray Conference, Adelaide* **11** (1993) 280.
14. Z. Wlodarczyk, *Proc. XXIII International Cosmic Ray Conference, Calgary, Invited, Rapporteur & Highlight Ppers* (1993) 355.
15. C. M. G. Lattes et al. (Brasil-Japan emulsion chamber collaboration), *Proc. 13th International Cosmic Ray Conference, Denver* **4** (1973) 2671.
    C. M. G. Lattes, Y. Fujimoto and S. Hasegawa, *Phys. Reports* **65** (1980) 151.
16. J. D. Bjorken and L. D. McLerran, *Phys. Rev.* **D20** (1979) 2353.
17. R. Kinnunen and C. Rubbia, *Proc. 17th International Cosmic Ray Conference, Paris* **11** (1981) 112.
18. D. G. Sutherland, *Cosmic Rays and Particle Physics* ed. T. K. Gaisser (AIP Conf. Proc. No. 49) p503.
19. J. Dias de Deus and W. A. Rodrigues Jr., *Nuovo Cimento* **A55** (1980) 34.
20. G. N. Fowler, E. M. Friedlander and R. M. Weiner, *Phys. Lett.* **B116** (1982) 203.
21. A. K. Mann and H. Primakoff, *Phys. Rev.* **D22** (1980) 1115.
22. R. Attallah and J. N. Capdevielle *J. Phys. G: Nucl. Part. Phys.* **19** (1993) 1381.
23. B. S. Acharya et al., *Phys. Rev.* **D24** (1981) 2437.
24. B. S. Acharya et al., *Il Nuovo Cimento Lett.* **31** (1981) 149.
25. B. S. Acharya and M. V. S. Rao, *J. Phys. G: Nucl. Part. Phys.* **17** (1991) 759.
26. R. A. Mukhamedshin, *Proc. 23rd International Cosmic Ray Conference, Calgary* **4**

(1993) 100.

27. I. I. Royzch, *Preprint, LPTHE 92/57, Univ. de Paris XI, France 1992* (unpiblished).
28. T. Atisara et al., *Proc. 23rd International Cosmic Ray Conference, Calgary* **4** (1993) 136.
29. S. Hasegawa, *Proc. VIII International Symposium on Very High Energy Cosmic ray Interactions, Tokyo* (1994) 115.

Chapter 5
PRIMARY ENERGY SPECTRUM

The properties of the primary particles of EAS that are of interest from the astrophysical point of view are the energy spectrum, their composition as a function of energy and the degree of isotropy of their arrival directions in space. These properties throw light on the origin, acceleration and propagation of high energy cosmic rays. While the arrival direction distribution can be directly measured, the others have to be deduced from measured EAS parameters. The primary energy spectrum is derived from measurements of shower size spectrum, Cerenkov light spectrum, muon size spectrum, density of charged particles at some large distance from the core, shower development curves obtained from fluorescent light measurements etc., using Monte Carlo simulations. The composition of the primaries is deduced from measurements of shower parameters such as the mean number as well as the fluctuations of muons, fraction of delayed hadrons, the high energy end of the hadron energy spectrum, depth of shower maximum, etc., which are sensitive to the mass number of the primary particle. As most of these parameters are sensitive to the characteristics of high energy interactions also, interpretation in terms of composition is not unique. However, any change in the behavior in these parameters as a function of shower size or primary energy can be reasonably well ascribed to change in the composition at the appropriate primary energy. In recent years, multi-component techniques are being developed to overcome these ambiguities.

Before we discuss the results from EAS studies, we shall briefly summarize the current situation on the primary energy spectrum and composition below air shower energies, but above 1 TeV, obtained from direct measurements using detectors flown in balloons and satellites. This will lay the foundation for interpreting the air shower data.

5.1 Direct Measurements

The first direct measurement of primary energy spectrum was carried out by Grigorov et al., [1] with a deep ionization calorimeter on board the Proton series of satellites. The calorimeter has 7 interaction mean free paths or 84 radiation lengths of matter in the form of lead and iron and therefore is a total absorption calorimeter. They measured the spectra of protons and α particles individually as well as all particles. The integral all particle spectrum shown in Fig. 5.1, taken from Sreekantan [2], extends from $10^{11} - 5 \times 10^{15}$ eV. They find that the spectrum can be represented

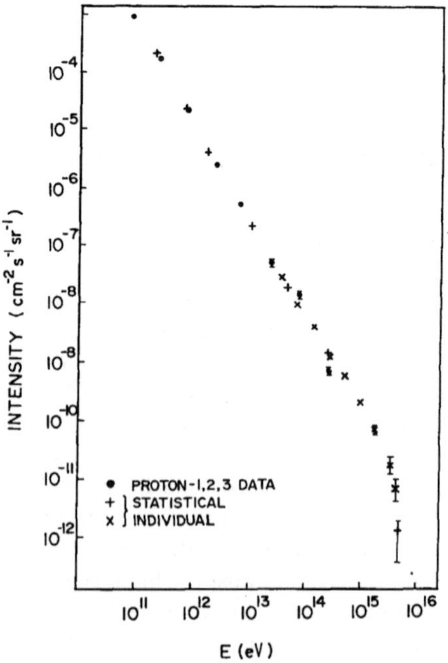

Figure 5.1 The all paticle integral energy spectrum obtained by Grigorov et al (1971) using total absorption calorimeters on board Proton satellites.

by a power law with exponent -1.6±0.01 in the energy range $5 \times 10^{12} - 10^{15}$ eV and steepens at energies higher than $\sim 2 \times 10^{15}$ eV. This steepening appears to be real since the instrumental effects make it steeper only at energies higher than 10^{16} eV. To date, this is the only direct measurement that extends beyond 10^{15} eV. The proton spectrum reported by them showed a steepening at around 2×10^{12} eV. Ellsworth et al. [3] point out that this is due to the increasing effect, with energy, of albedo in their calorimeter which could have resulted in wrongly identifying an increasing fraction of protons as heavier nuclei.

The spectra of protons and heavier nuclei are being investigated using thin emulsion chamber detectors by Ivanenko et al. [4] on a satellite, and in balloons by the JACEE group [5], Ichimura et al. [6] and Zatsepin et al. [7]. The current status of the primary energy spectrum and composition in the energy range below 10^{14} eV is summarized in the Rapporteur talk of Simon Swordy [8]. The spectra of protons and He nuclei in the energy range 10 GeV/n to few times 10^{14} eV/n and of CNO, Ne-S, and Fe groups of nuclei in the same energy range are given in Figs. 5.2 and 5.3.

At energies of interest to those concerned with air shower phenomena ($> 10^{12}$ eV), the two interesting trends seen from Fig. 5.2 are

1. the relative proportion of He compared to protons seems to increase with energy and
2. a possible spectral cut-off of protons around 10^{14} eV, though errors are large at the high energy end. An important question that arises is whether this steepening is connected with the knee of the cosmic ray spectrum (see Sec. 5.2.1 for "knee").

Fig. 5.3 shows that as far as the heavy components are concerned, the spectra tend to flatten at higher energies—the differential spectrum changes from -2.75 to more like -2.5 to -2.6 at the energies of 10^{13}eV/n. This flattening may be connected with the expected escape of particles from the galaxy at energies of $\sim 10^{12}$ eV/n.

The spectra obtained using the emulsion chamber detectors are based on the energy transferred to the soft component, $\sum E_\gamma$. The primary energy spectrum is derived by unfolding the $\sum E_\gamma$ spectrum using the mean value of the fraction of energy, K_γ, transferred to γ rays and its distribution from accelerator data or an interaction model. For this method to work satisfactorily, the number of events recorded up to the highest $\sum E_\gamma$ values should be large enough for the distribution to be sampled adequately. Near the tail end of the derived primary spectrum, this is usually not the case and the spectrum in this region is, therefore, likely to be in error.

5.2 Primary Energy Spectrum from Air Shower Studies

5.2.1 The Akeno and KGF Experiments

The Akeno group [9] estimated the primary energy spectrum from detailed measurements of the electron size spectrum and muon size spectrum with the 1 km^2 array at an atmospheric depth of 920 g cm^{-2}, shown in Fig. 5.4, consisting of 150 scintillators, each of area 1 m^2 and thickness 50 mm, and 9 muon detectors each of area 25 m^2. In a separate experiment (experiment I-B and I-C) to measure small size showers, they used 96 detectors, each of area 0.25 m^2, and 25 of 0.02 m^2 in an area of 30×30 m^2 [10]. Shower sizes were estimated by fitting the observed densities to the NKG function (Eq. 2.22). The lateral distribution of electrons obtained by them for three different shower sizes with scintillators of thickness 50 mm and 3 mm are shown in Fig. 5.5. Since there is a transition effect in thick scintillators near the core (see page 39) the shower size is, in general, overestimated. On the other hand, the NKG formula which is fitted to the densities near the core predicts lower values than are observed far from the core and hence the shower size is underestimated. By accident both effects compensate each other in the shower size estimation and the shower size

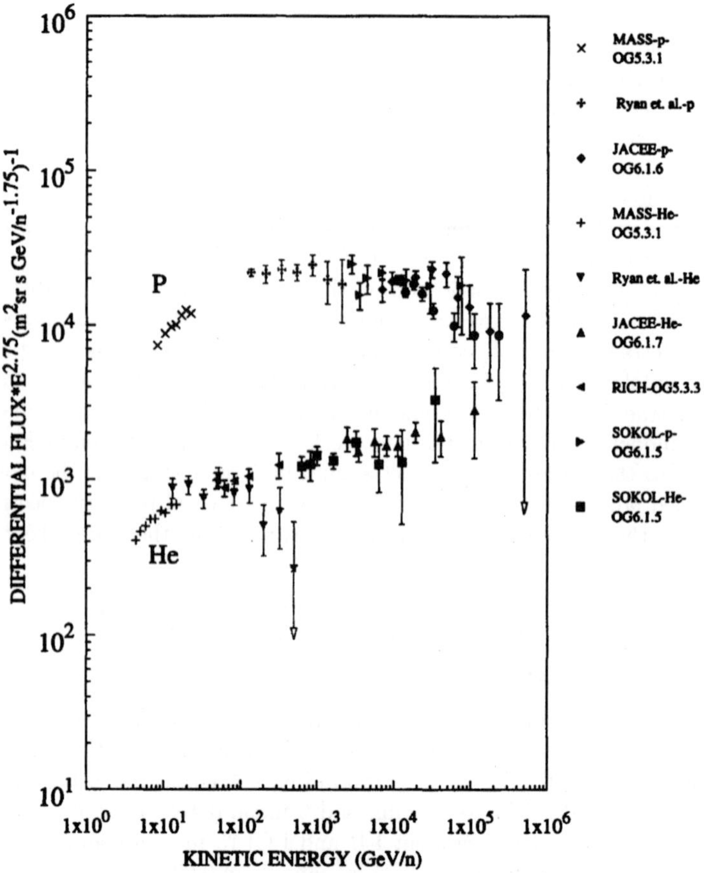

Figure 5.2 A compilation of direct measurements of the primary energy spectra of protons and He nuclei by Swordy (1993). For references see Swordy (1993).

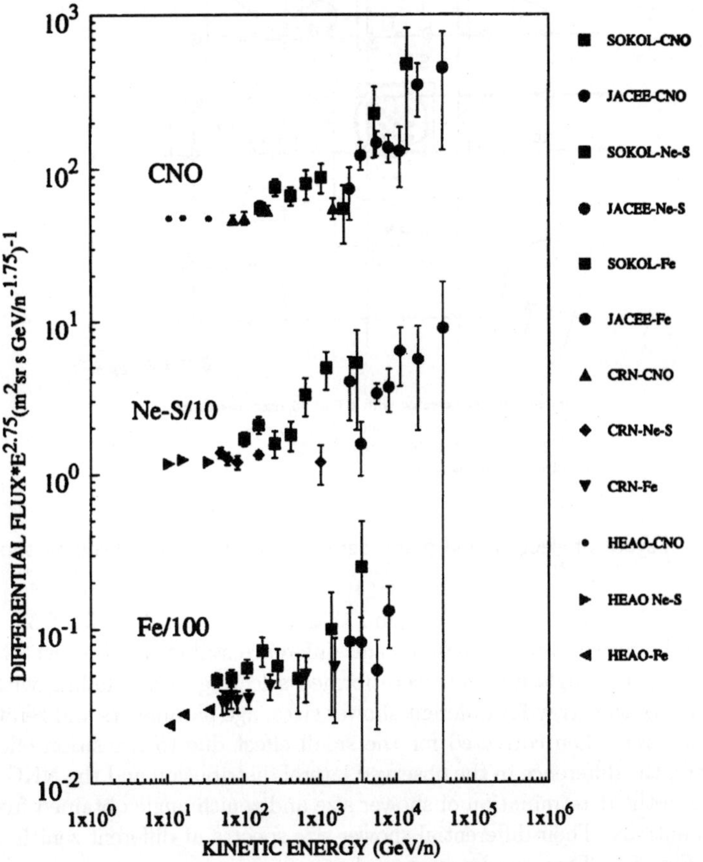

Figure 5.3 A compilation of direct measurements of the primary energy spectra of heavy nuclei compiled by Swordy (1993). Note that the fluxes of Ne and Fe group nuclei are divided by 10 and 100 respectively. For references see Swordy (1993).

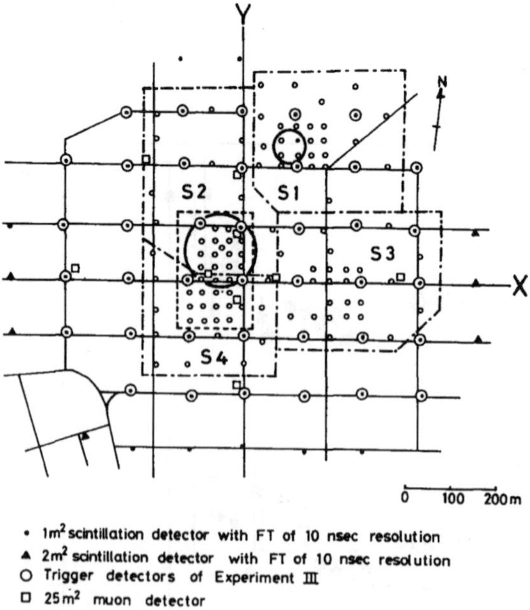

Figure 5.4 Layout of electron and muon detectors of the 1 km^2 Akeno air shower array.

fitted to the densities near the core and that fitted to densities far from the core, turn out to be nearly same though a small difference still exists. Shower size spectra at different zenith angle intervals were formed selecting events falling within 100% efficient area of the array for different shower sizes, age parameters and zenith angles. The spectra were then corrected for the small effect due to transition effect in the scintillators, the difference in the observed lateral distribution and the NKG function, and errors in the determination of shower size and zenith angle obtained from Monte Carlo simulations. Their differential shower size spectra at different zenith angles are shown in Fig. 5.6. They can be expressed as

$$J_e(N_e, \theta)\, dN_e = A_e(\theta) \left(\frac{N_e}{10^6}\right)^{-\gamma_e(\theta)} dN_e. \tag{5.1}$$

Values of the constants below and above the shower size of 10^6 for different zenith angles are shown in Table 5.1. They compared their shower size spectrum at 920 g cm^{-2} and sea level (1020 g cm^{-2}) with the spectra obtained by other groups at various atmospheric depths, which are shown in Fig. 5.7 with the fluxes multiplied by N_e^2.

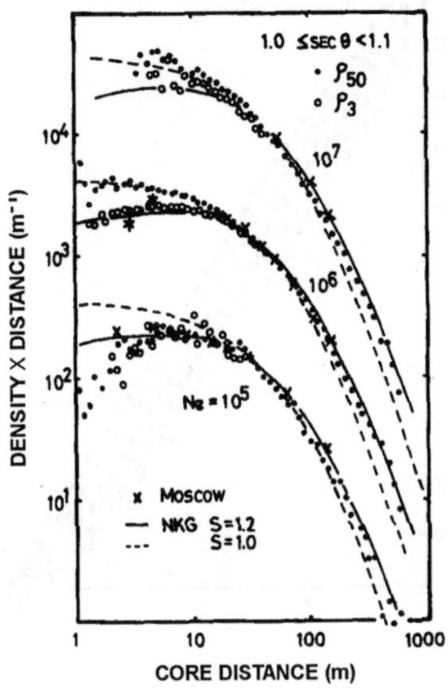

Figure 5.5 Example of lateral distribution of electrons measured with thick (50 mm) and thin (3 mm) scintillators by Nagano et al. (1984). Solid and dashed lines are the NKG function for s = 1.2 and 1.0 respectively.

Table 5.1 Values of $\gamma_e(\theta)$ for two ranges of shower size and $A_e(\theta)$ of the shower size spectra obtained by the Akeno group (Nagano et al, 1984).

$\sec\theta$	$\gamma_e(\theta)$		$A_e(\theta)$ ($\mathbf{m^{-2}\ s^{-1}\ sr^{-1}}$)
	$\mathbf{10^{5.0}–10^{6.0}}$	$\mathbf{10^{6.0}–10^{8.0}}$	
1.0	2.48 ± 0.09	2.83 ± 0.04	$(2.80 \pm 0.08) \times 10^{-13}$
1.1	2.49 ± 0.12	2.80 ± 0.12	$(1.2 \pm 0.2) \times 10^{-13}$
1.2	2.59 ± 0.19	2.74 ± 0.08	$(3.8 \pm 0.7) \times 10^{-14}$
1.3	2.63 ± 0.26	2.57 ± 0.22	$(1.2 \pm 0.3) \times 10^{-14}$
1.4	2.73 ± 0.32	2.62 ± 0.20	$(5.0 \pm 0.8) \times 10^{-15}$

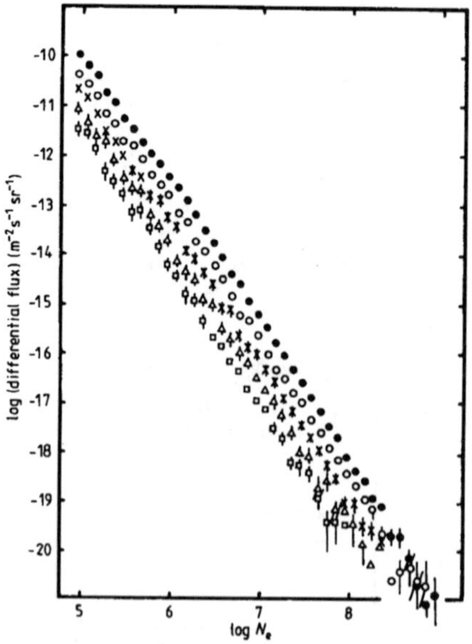

Figure 5.6 The differential shower size spectra at five zenith angles obtained by the Akeno group. $\sec\theta = 1.0$ (•), 1.1 (∘), 1.2 (×), 1.3 (△), 1.4 (□).

The only other shower size spectrum available at 920 g cm^{-2} is that by the KGF group [13, 18]. The KGF group obtained the shower size spectrum with the air shower array at KGF (see Fig. 3.20), which has a densely packed large area (2.2 m^2) detectors at its centre to measure showers of small size accurately [19], but the zenith angles could not be measured for all showers. The fluxes were obtained from the omnidirectional fluxes assuming an angular distribution of the form $\cos^7\theta$, which is consistent with the limited angular distribution data available from their experiment. The KGF fluxes are about a factor of 1.5 smaller than Akeno fluxes, though the spectral slopes are in agreement within errors. The vertical differential shower size spectrum of the KGF group is given by

$$J_e(N_e)\,dN_e = A_e\left(\frac{N_e}{N_0}\right)^{-\gamma_e} dN_e \;\; (\mathrm{m}^2\ \mathrm{s\ sr})^{-1} \tag{5.2}$$

where the constants are given by

$$\begin{array}{cccc} A_e & \gamma_e & N_0 & \text{Shower size range} \\ (1.62\pm0.06)\times10^{-8} & 2.55\pm0.01 & 10^4 & 10^4 \le N_e < 2\times10^6 \\ (2.26\pm0.20)\times10^{-14} & 3.02\pm0.12 & 2\times10^6 & 2\times10^6 \le N_e < 10^7. \end{array} \tag{5.3}$$

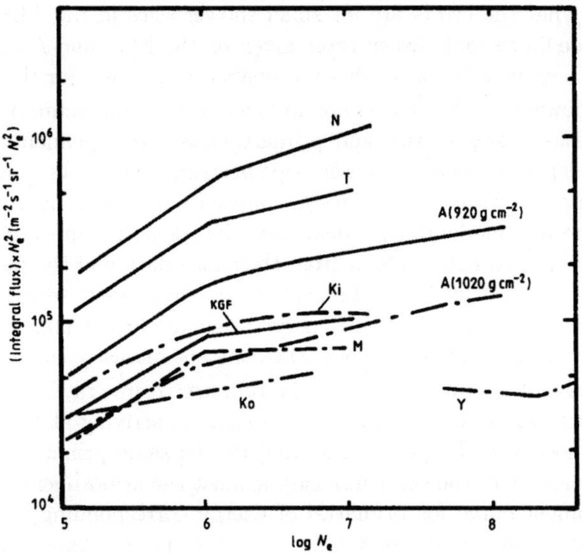

Figure 5.7 Comparison of the Akeno size spectra at $\sec\theta = 1.0$ and 1.1 with those measured by various groups at different altitudes. A, Nagano et al (1984); N, Norikura (740 g cm^{-2}; Kawakami et al 1983); T, Tien Shan (690 g cm^{-2}; Kirov et al 1981); KGF (920 g cm^{-2}; Acharya et al 1981); Ko, Kobe University (sea level; Asakimori et al 1982); M, Moscow State University (sea level; Vernov et al 1968); Ki, Kiel University (sea level; Bagge et al 1977); Y, Yakutsk (sea level; Diminstein et al 1977).

The Akeno group explain the lower KGF flux as due to the small value of the exponent of the angular distribution used instead of a value of 10 they find. The higher value of the exponent would increase the fluxes by a factor of 1.33. Another factor which influences the shower size spectrum is the area of individual detectors and their mutual separation in the array. The larger the separation the larger the core distances at which densities of particles are sampled. The KGF array has 43 detectors, each of area 2.25 m^2, in a circle of radius 20 m in the central region making up a total sampling area of $\sim$ 100 m^2, whereas the Akeno array has $\sim$ 25 m^2 of sampling area in a similar region. Detailed analysis of artificially generated showers using Monte Carlo simulations by the KGF group [19] show that systematic underestimate or overestimate of shower sizes is small even for showers of small size. In general, smaller sampling area for densities leads to an overestimate of the age parameter and hence the shower size. The Akeno group's values of age parameter are larger by about 0.15–0.2 than those of the KGF group. The published results of the Akeno group [9] show error analysis only for showers of size larger than 10^7, where the KGF data end. It is not clear what the errors are for small shower sizes in the Akeno experiment. Perhaps the true fluxes lie between those given by the KGF and Akeno groups. The sea level spectra agree with each other reasonably well except for the Kobe group.

Since the showers at Akeno level are all beyond their maximum development and the relation between shower size and primary energy is dependent on the nuclear interaction model, the observed shower size spectrum can not be converted directly to primary energy spectrum unless some assumptions are made. In order to derive the energy spectrum in a model independent way, the Akeno group estimate the shower sizes at their maximum development from their own data and the data obtained by the Chacaltaya group at an atmospheric depth of 540 g cm^{-2}, which is high enough an altitude to observe showers in this energy range at their maximum development. They do this by using what is known as the constant intensity cut method.

In the absence of fluctuations and if the primary cosmic rays are all of the same type, e.g. all protons or all iron nuclei, then showers arriving at the same intensity at different atmospheric depths are initiated by the same primary energy. Thus a plot of size of showers of the same intensity against the atmospheric depth gives the shower development curve for primaries of energy corresponding to that intensity. This method of obtaining the shower development curve is known as the constant intensity cut method.

In this method, the integral shower size spectrum at various zenith angles, corresponding to various slant depths are plotted multiplying, if necessary, by some power of the shower size to bring out the detailed features as shown in Fig. 5.8. Lines of fixed integral intensities, multiplied by the same power of shower size, are drawn and the shower sizes at the various depths at the intersection of the shower size spectrum and the fixed intensity lines are read off. The shower development curves obtained by the Akeno group in this manner are shown in Fig. 5.9.

They multiply the shower sizes at Akeno by the ratio of the shower size N_m at maximum development to that at 920 g cm^{-2} calculated from the Chacaltaya data for different intensities, to obtain the shower sizes at maximum. To convert N_m to primary energy E_0 they use the conversion factor w obtained from the Monte Carlo calculations of Hillas [22] using energy dependent hadron-air-nucleus cross sections. This conversion factor is dependent on N_m and has the values w =1.4 GeV for $3\times10^6 \leq N_m < 10^{10}$, $w = 1.45$ GeV at $N_m = 10^6$ and $w = 1.59$ GeV at $N_m = 10^5$. Extrapolating to lower shower sizes, they get the following relation between the primary energy and the shower size at Akeno level:

$$E_0 = 3.9 \times 10^{15} \left(\frac{N_e}{10^6}\right)^{0.9} \quad \text{eV}. \tag{5.4}$$

The differential primary energy spectrum, thus obtained by them, is shown in Fig. 5.10 along with the compilation of other results by Hillas. Their differential spectrum

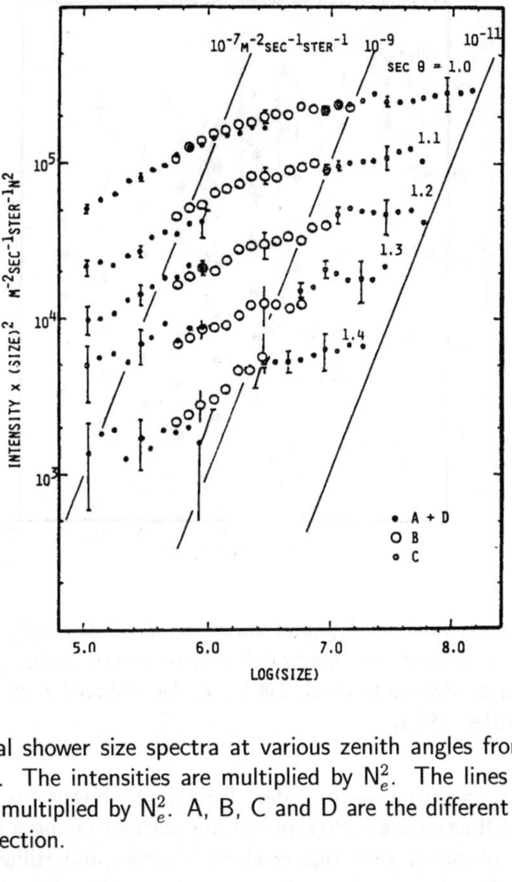

Figure 5.8 Integral shower size spectra at various zenith angles from the Akeno air shower experiment. The intensities are multiplied by N_e^2. The lines are for different integral intensities multiplied by N_e^2. A, B, C and D are the different triggering criteria used for shower selection.

is given by

$$J(E_0)\,dE_0 = (4.0 - 5.0) \times 10^{-23} \left(\frac{E_0}{10^{15.67}}\right)^{-\gamma} dE_0 \;\; \mathrm{m^{-2}\,s^{-1}\,sr^{-1}}, \qquad (5.5)$$

with E_0 in eV and $\gamma = 2.62 \pm 0.12$ in the energy range $10^{14.5} - 10^{15.67}$ eV and 3.02 ± 0.05 in the range $10^{15.67} - 10^{18}$ eV. This steepening of the spectrum in this energy region is generally known as the knee of the cosmic ray spectrum.

The lower end of the Akeno energy spectrum overlaps with the all particle spectrum measured directly by Grigorov et al. and they are in good agreement. In the higher energy region, it agrees with the Yakutsk spectrum but not with the Haverah Park spectrum.

A few remarks regarding the procedure adopted by the Akeno group in deriving

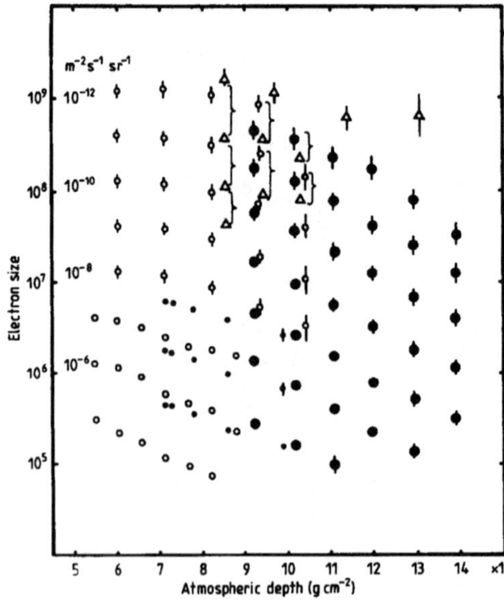

Figure 5.9 The longitudinal development curves obtained by the Akeno group from the equi-intensity cuts of integral size spectra at various zenith angles. •, Nagano et al (1984); ○, Chacaltaya (Kakimoto et al, 1981); •, Tien Shan (Kirov et al., 1981); △, Volcano Ranch (Linsley, 1973).

the primary energy spectrum are in order. Firstly, the method is strictly applicable only if there are no fluctuations. Presence of fluctuations changes the meaning of the shower size, $N_e^{\rm cut}$, obtained with this method. Gaisser and Hillas [23] have shown, using a simple model, that, under rather general considerations including that of a mixed composition with the fluctuations and the composition being independent of energy, $N_e^{\rm cut}$ is not equal to the average shower size for a fixed primary energy, but is given by

$$N_e^{\rm cut} = \left\langle N_e\left(t, E_0\right)^{\gamma-1}\right\rangle^{\frac{1}{\gamma-1}} = N_e\left(t, E_0\right)_{\rm r.m.s} \quad \text{if } \gamma = 3. \tag{5.6}$$

where $N_e\left(t, E_0\right)$ is the shower size at depth t, measured in units of $\lambda = 70$ g cm^{-2}, in an individual shower of primary energy E_0 and γ is the slope of the differential primary energy spectrum. Thus $N_e^{\rm cut}$ is always greater than the average shower size for a fixed primary energy, even at shower maximum. As a result, the primary energies and therefore primary fluxes are overestimated if the conversion factor from average shower size at maximum to primary energy, obtained from Monte Carlo simulations, is used.

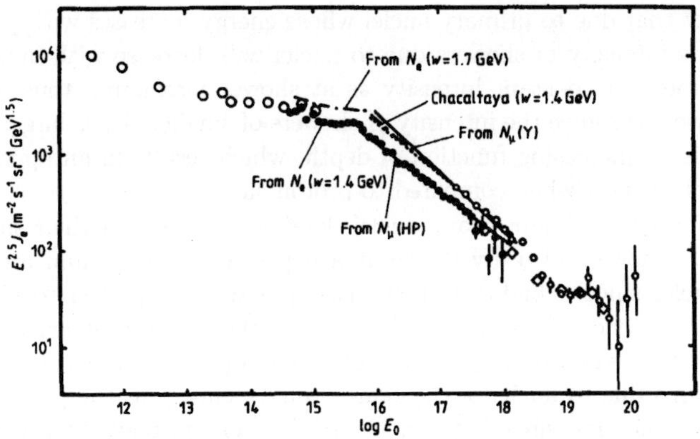

Figure 5.10 The differential primary energy spectrum obtained by the Akeno group from electron and muon size spectra (•). The other results are : ◯, Grigorov et al's satellite measurements; ○, Haverah Park (HP); ◇, Yakutsk (Y).

Also, they obtain the relation between the shower size at Akeno and primary energy from the Chacaltaya data and the Monte Carlo simulations of Hillas only for shower sizes larger than 4.6×10^6 at Akeno level and extrapolate the relation to smaller shower sizes. The shower size spectrum shows a change of slope at a shower size of 10^6. If this is due to a change in the primary energy spectrum slope, N_e^{cut} would be equal to a different moment of $N_e(t, E_0)$ and the assigned energies would be even more in error. Further more, there are indications from several other experiments that the primary composition or the characteristics of nuclear interactions may be changing in the knee region. If the composition is changing, then while the situation at shower maximum is not effected, the shower sizes and the fluxes at Akeno level are altered in a complicated way. In particular, the factor used to convert $N_e(920)$ to $N_e(\text{max})$ would be different. Thus, while the spectrum beyond the knee may not be much in error, except to the extent that Eq. 5.6 is not used, the spectrum below the knee may not be correct.

Furthermore, the constant intensity cut method may be useful in obtaining a measure of the shower size at maximum, but it does not give the correct development curve if the composition is mixed, even though independent of energy. Showers of a given size beyond the maximum development are dominated by protons and light nuclei. This is because heavy nuclei-initiated showers reach maximum higher than those initiated by light nuclei (see Eq. 2.38) and are therefore absorbed more giving rise to a smaller shower size at the at lower altitudes, even though the shower size at maximum is nearly same for all nuclei in the superposition model. Thus the intensity of showers of a given shower size comprises, not of the same primary energy for all nuclei, but of that due to primary nuclei whose energy increases with primary mass. Therefore the intensity of showers due to nuclei will decrease with increasing mass. The requirement of constant intensity as at shower maximum, thus *decreases* the average shower size since the intensity of showers of smaller size is larger. The extent of decrease is an increasing function of depth, which results in an apparently faster absorption of showers when compared to proton showers.

Above shower maximum heavy nuclei dominate because of their smaller energy per nucleon. Showers of a given size at a depth above maximum are mostly due to heavy nuclei since a higher primary is required for a proton to give the same shower size and hence of lower flux. Alternately, the integral intensity of showers of a given size expected according to the proton development curve, comprises not only of contributions due to nuclei of the same primary energy as the proton but also of lower primary energies, since heavy nuclei of the same primary energy give a higher shower size. The requirement of constant intensity as at shower maximum *increases* the shower size at that depth since showers of larger size are of smaller intensity. This results in a shift in the position of the shower maximum towards the top of the atmosphere and an apparently fast developing proton cascade curve. Therefore the cascade curve does not represent the true development curve either for protons or for nuclei and cannot be directly compared with Monte Carlo simulations.

The KGF group [24, 25] followed the direct method of comparing the shower size spectrum with that expected from models of primary spectra and composition using their Monte Carlo simulations for scaling model of high energy interactions with energy dependent interaction cross sections [26], similar to that of Hillas and giving similar results. For example, the relation between the average shower size at 920 g cm^{-2} and the primary proton energy in the KGF simulations is given by

$$E_0 = 3.3 \times 10^{15} \left(\frac{N_e}{10^6}\right)^{0.87} \text{ eV}, \tag{5.7}$$

which is nearly same as Eq. 5.4. The composition model used by them is chosen to fit their data on the variation of muons of energy > 220 GeV with shower size (see Sec. 6.3.1.2). In this model, a modified version of a model first suggested by Hillas [27], primary nuclei are lost at an energy per nucleon of 7.5×10^4 GeV due to photodisintegration in the point sources themselves and proton spectrum steepens near the knee due to leakage from some confinement region. The differential spectra are of the form

$$F(E_0)\, dE_0 = K E_0^{-\gamma} dE_0 \ \text{ m}^{-2}\text{ s}^{-1}\text{ sr}^{-1} \tag{5.8}$$

where E_0 is the total energy in GeV and $\gamma = 2.68$ for all nuclei with $E_0 < E_c$ and E_c is the total energy at which nuclei are lost and proton spectrum steepens. Beyond E_c, there are no nuclei and the proton spectrum has a slope of $\gamma = 3.18$. The constants for the various components are shown in Table 6.5. The all particle spectrum in this model agrees with that of Grigorov et al. in the overlapping region. The KGF shower size spectrum is shown in Fig. 5.11.

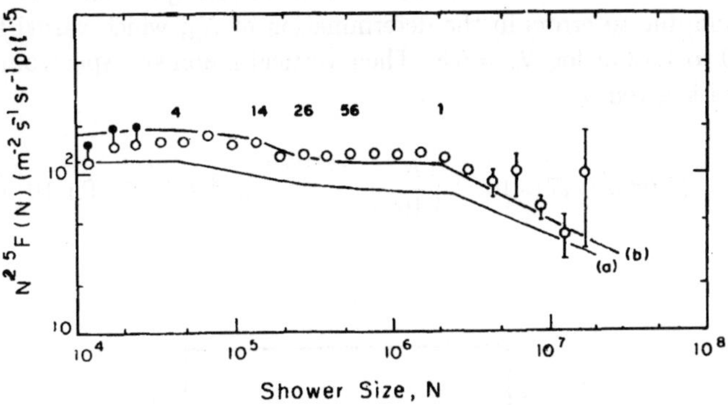

Figure 5.11 The differential shower size spedctrum measured by the KGF group at 920 g cm^{-2} compared with the expected spectra two different composition models and scaling model of nuclear interactions with energy dependent cross-sections.

The filled circles are the fluxes corrected for the detection inefficiency of old showers in the small shower size region. The numbers in the figure indicate where primaries of corresponding mass numbers are cut off. Curve (b) is for the present model. Curve (a) is for the same model, but the spectral index for all the components is 2.72 with the total intensity normalized with direct observations at 100 GeV/nucleon. Note that the steeper primary spectra result in lower fluxes for the shower size spectrum. The present model is in very good agreement with experimental data.

It is clear that, since both Akeno and KGF shower size spectrum agree with that expected from Grigorov spectrum at the lower end even though they differ by a factor of 1.5, one of the shower size spectrum must be in error. In view of the above remarks regarding the Akeno shower size spectrum and since the KGF group used the direct method, the KGF shower size spectrum seems to be closer to the true spectrum.

The Akeno group obtained the primary energy spectrum from their muon size spectrum also. The Akeno array contains nine muon stations of proportional counter arrays each of area 25 m^2. The threshold muon energy for vertical incidence is 1 GeV. The vertical muon sizes were obtained by fitting the observed muon densities with a modified Greisen formula for the lateral distribution of the form

$$\rho_\mu = C_\mu N_\mu r^{-0.75}\left(1+\frac{r}{R_0}\right)^{-2.5} \tag{5.9}$$

where ρ_μ is the muon density at a core distance of r m, $R_0 = 280$ m, N_μ is the muon size and C_μ is the normalization constant. It is to be noted that the showers were triggered by electrons. The muon densities were corrected for the proportional counter response (ratio of the number of muons determined from one layer of counters to the number of tracks determined from three layers of counters), which varied from 1.85 at $r = 50$ m to 1.6 at $r = 100$ m. The muon size spectrum was corrected for overestimation due to errors in the determination of N_μ, which varied from 1.42 at $\log N_\mu = 5.0$ to 1.18 at $\log N_\mu = 6.8$. Their vertical muon size spectrum is shown in Fig. 5.12 and is given by

$$J_\mu (N_\mu)\, dN_\mu = (3.08 \pm 0.17)\times 10^{-16} \left(\frac{N_\mu}{10^6}\right)^{-3.44\pm 0.09} \mathrm{m^{-2}\, s^{-1}\, sr^{-1}} \text{ for } 10^5 < N_\mu < 10^{6.7} \tag{5.10}$$

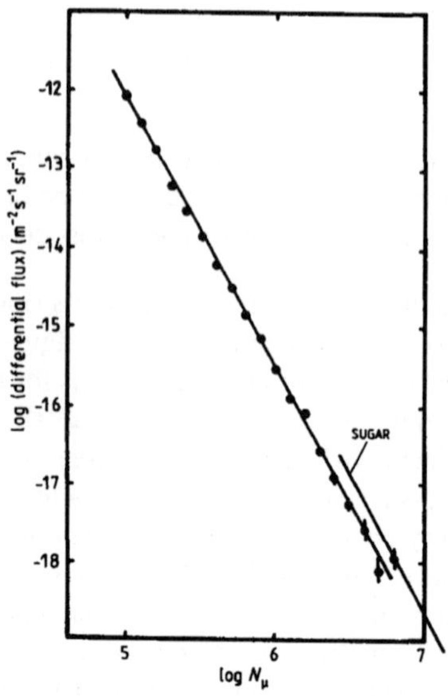

Figure 5.12 The differential muon size spectrum measured by the Akeno group (Nagano et al, 1984) at $\sec\theta = 1.0$–1.1. The results of the SUGAR experiment of Horton et al (1983) are also shown.

The spectrum from the SUGAR array [28] is about 25% higher, perhaps due to the lower muon energy threshold of 0.75 GeV in that experiment. The Akeno group use several steps to convert the muon size spectrum to primary energy spectrum. They first determine the relation between the electron density, $S\,(600)$, at a distance of 600 m form the core measured with scintillators in their own experiment at a zenith angle corresponding to an equivalent vertical depth at sea level and N_μ, which is given by

$$S\,(600) = (0.34 \pm 0.03)\left(\frac{N_\mu}{10^6}\right)^{1.20\pm0.05} . \tag{5.11}$$

They then combine this with the relation between the signal measured in water Cerenkov detectors, $\rho\,(600)$, and the scintillator density, $S\,(600)$, measured by the Haverah Park group [29] at sea level and the relation between E_0 and $\rho\,(600)$ from Monte Carlo simulations of the Haverah Park group to get the relation between N_μ and E_0. These relations are given by

$$E_0\,(\mathrm{eV}) = 3.26 \times 10^{17} S\,(600)^{1.018} \tag{5.12}$$

and

$$E_0\,(\mathrm{eV}) = 1.08 \times 10^{17}\left(\frac{N_\mu}{10^6}\right)^{1.22} . \tag{5.13}$$

In arriving at these relations, attenuation of muons between Akeno and sea levels is neglected and Eq. 5.12, which was obtained by the Haverah Park group for very high primary energies, is extrapolated to lower primary energies. The energy spectrum thus obtained is also shown in Fig. 5.10 marked as 'From N_μ (HP)'. This agrees very well with that from the electron size spectrum. They also derive the conversion factor from the Yakutsk group's [30] measurements of Cerenkov flux and $S\,(600)$ and the resulting spectrum is also shown in Fig. 5.10 marked 'From N_μ (Y)'. This spectrum is higher than the earlier one, but agrees with the spectrum derived from shower size spectrum using the conversion factor of 1.7 GeV from shower size to primary energy suggested by Linsley based on simulations using scaling model, but with energy independent cross-sections. However, this model is not realistic.

5.2.2 *The Moscow Experiment*

The Moscow State University group have updated their sea level shower size spectrum using the modified air shower array [31] consisting of 77 detectors, containing GM counters of different areas, spread up to 550 m meters from the center of the array. The central area of 50 m × 50 m is packed with 3000 GM counters. There is no transition effect in GM counters since the material in the counter walls and the container lid is negligible. When they determine shower sizes using an NKG type of lateral distribution function of the form

$$f\,(r) \sim \left(\frac{r}{r_0}\right)^{s-2}\left(1+\frac{r}{r_0}\right)^{s-3.9} \tag{5.14}$$

with $r_0 = 80$ m, they obtain a shower size spectrum that is different from that of the Akeno group. The agreement is better when they [32] use a shower size dependent function of the form

$$f\,(r) \sim \left(\frac{r}{r_0}\right)^{s-2}\left(1+\frac{r}{r_0}\right)^{s-3.9}\left(1+C_2\frac{r}{r_0}\right) \tag{5.15}$$

with

$$s\,(N_e) = 1.65 - 0.12 \log\,(N_e) \tag{5.16}$$

and

$$C_2 = 0.16 \log\,(N_e) - 0.63. \tag{5.17}$$

The shower size spectrum with the two types of lateral distribution functions are compared in Fig. 5.13 with the Akeno spectrum in the vertical direction recalculated to sea level using an absorption mean free path of 185 g cm^{-2}.It clearly demonstrates

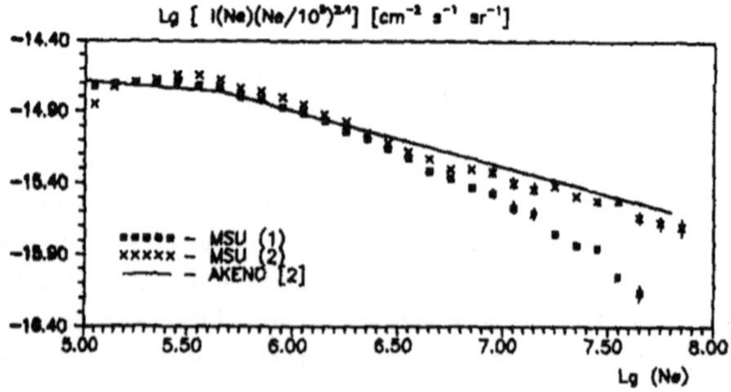

Figure 5.13 Sea level shower size spectra from Fomin et al (1991) compared with the corresponding Akeno spectrum. MSU(1) ; with coneventional NKG function; MSU(2) : with size dependent function.

the importance of the use of the correct lateral distribution in determining shower sizes.

At the Rome Cosmic Ray Conference in 1995, the Moscow group [33] presented their statistically improved shower size spectrum at sea level. They find that the knee occurs at $N_e = 4 \times 10^5$ and the differential slopes before and after the knee are 2.38 ± 0.02 for $10^5 < N_e < 4 \times 10^5$ and 2.88 ± 0.03 for $1.5 \times 10^6 < N_e < 3 \times 10^7$ respectively. The change in slope is 0.5 with a significance of 14 σ. Comparing with the spectra from the Tien Shan group [34] at 690 g cm^{-2} and the Akeno spectrum at 920 g cm^{-2} (see Fig. 5.14), they point out that the knee in the shower size spectrum occurs at different shower sizes at different altitudes. The shower size at the knee is defined as that size where the points following it deviate by more than 3 σ from a power law fit to the data below that shower size. The shower size at the knee increases with increasing altitude unlike in earlier observations. Their compilation of the dependence of the shower size at knee with atmospheric depth is shown in Fig. 5.15. The curve is the expected dependence for the rigidity cut-off model considered by them (see below). This is the first time since air shower studies began that such

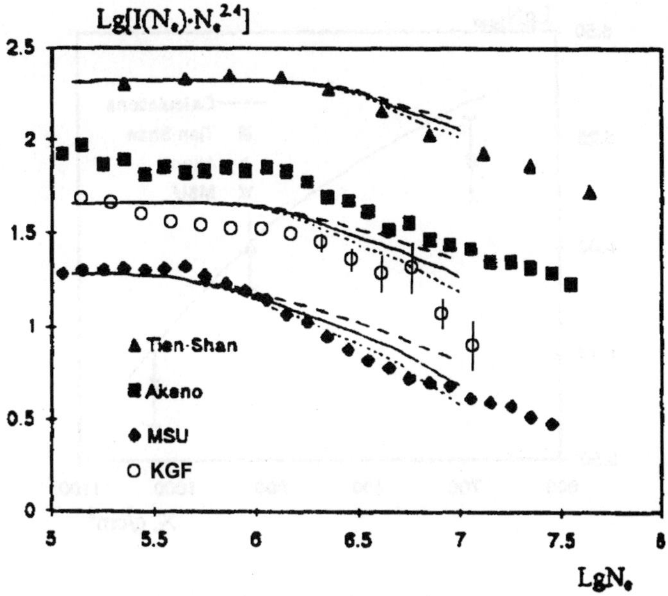

Figure 5.14 Differential size spectra at different altitudes (690, 920 and 1020 g cm^{-2}) compared with calculations using different primary composition models by Khristiansen et al (1995).

a feature, which is expected if this knee is connected with a similar knee in the primary energy spectrum, has been seen. If this is confirmed by other experiments at other altitudes, then it is a clear evidence that the knee is not due to change in the characteristics of nuclear interactions as claimed by some authors.

They also compare the shower size spectrum with the predictions of various composition models using Monte Carlo simulations with their favored quark-gluon string (QGS) model (see later). All spectra are assumed to have the same differential spectral slope of -2.60 before a critical rigidity of 3×10^{15} V. It is not clear what absolute intensities they have used. They consider three different composition models before the cut-off rigidity. In what they call the normal composition model, the relative abundances of various groups of nuclei are p : α : M : H : Fe : : 32 : 23 : 21 : 14 : 10. The other two models are those suggested by Swordy (see Table 6.1) and the JACEE group [35]. Three different values for the change in the slope, $\Delta\gamma$, of the spectra after the critical rigidity were considered. The dashed, solid and dotted lines in Fig. 5.14 correspond to $\Delta\gamma = 0.6$, 0.8 and 1.0 respectively. They find that the calculated spectra do not describe the experimental data well. For comparison, we

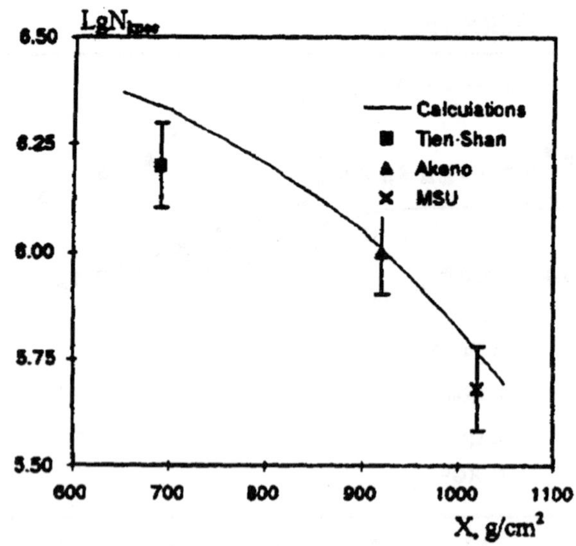

Figure 5.15 Dependence of the shower size at the knee in the shower size spectrum on atmospheric depth compiled by Khristiansen et al (1995). The curve is the expected dependence for the rigidity cut off model described in the text.

have added the KGF shower size spectrum at 920 g cm^{-2} also in the figure. This spectrum agrees with calculations better than the Akeno spectrum, though it is still not good.

5.2.3 The EAS-TOP Experiment

The EAS-TOP collaboration [36] also find evidence for the shower size at the knee to decrease with increasing atmospheric depth. Fitting the observed densities recorded with their array (see Fig. 3.21) to the standard NKG function (Eq. 2.22) they determine the shower sizes and form the shower size spectrum in various zenith angle intervals accepting events in the central area of 1.1×10^4 m^2 where the detection efficiency is $\geq$ 95%. Using the method of constant intensity cuts, they form the plots of shower size as a function of depth for different intensities and determine the absorption mean free path of shower size, which has a mean value of 221.9 ± 6.9 g cm^{-2}. Using this value they convert the shower sizes in inclined directions in a given zenith angle interval to vertical incidence and form the differential shower size spectrum at different zenith angles corresponding to different atmospheric depths.

Table 5.2 Constants in the differential shower size spectra at different zenith angles obtained by the EASTOP group.

x(g cm^{-2})	γ_1	γ_2	J_k (**m^{-2} s^{-1} sr^{-1}** N_e^{-1})	log (N_{ek})
810–850	2.06 ± 0.01	3.1 ± 0.1	$(2.4 \pm 0.4) \times 10^{-13}$	6.050 ± 0.001
850–890	2.67 ± 0.02	3.1 ± 0.2	$(2.2 \pm 0.6) \times 10^{-13}$	5.99 ± 0.06
890–930	2.64 ± 0.02	3.0 ± 0.2	$(2.0 \pm 0.8) \times 10^{-13}$	5.9 ± 0.1

The spectra are fitted to a power law of the form

$$J(N_e) = J_k \left(\frac{N_e}{N_{ek}}\right)^{-\gamma} \tag{5.18}$$

with two indexes, $\gamma = \gamma_1$ for $N_e < N_{ek}$ and $\gamma = \gamma_2$ for $N_e > N_{ek}$, where J_k is the intensity at the shower size, N_{ek}, at the knee. Their results are shown in Table 5.2. Though errors are still large, it is seen that there is a tendency for N_{ek} to decrease with depth with nearly same intensity at the knee. Since these measurements are carried out fairly deep in the atmosphere and over a small interval of depth (120 g cm^{-2}), the effect of mixed composition on the constant intensity cut method is not serious as the showers are predominantly due to light nuclei.

The EAS-TOP group compare their shower size spectrum below the knee with that expected from extrapolation of direct measurements of the primary spectra of various nuclei [35,37] using the Monte Carlo code HEMAS [38]. Their result is shown in Fig. 5.16. Contribution to the shower size spectrum from various nuclei as well as their sum are shown in the figure. The dashed lines show the effect of including $\pm 1\sigma$ errors on the Helium spectrum. The observed shower size spectrum agrees very well, below the knee, with the expected one within errors. Notice that the contribution from heavier nuclei is negligible compared to that from protons and He nuclei.

5.2.4 The Tibetan Experiment

The Tibetan air shower group [39] also determined the primary energy spectrum. The array (see Fig. 3.36) is located at a very high altitude of 606 g cm^{-2} where showers from primaries of energy in the knee region attain maximum and hence eminently suited for studying the primary energy spectrum. Their Monte Carlo simulations using the GENAS code [40] confirm that proton as well as iron initiated showers of primary energy 10^{14}–10^{17} eV are near maximum development at that altitude. Showers with zenith angle $< 25°$ are selected over a central area of 60 m $\times$ 60 m where primaries of energy > 400 TeV are detected with full efficiency and the vertical shower size spectrum is formed. Using their Monte Carlo simulations and a heavy dominant (HD1) composition (14% protons and 42% iron nuclei at 10^{16} eV) shower sizes are converted to primary energies. The resulting differential energy spectrum is shown in Fig. 5.17. The spectrum shows a gradual steepening at energies around $2{\times}10^{15}$ eV

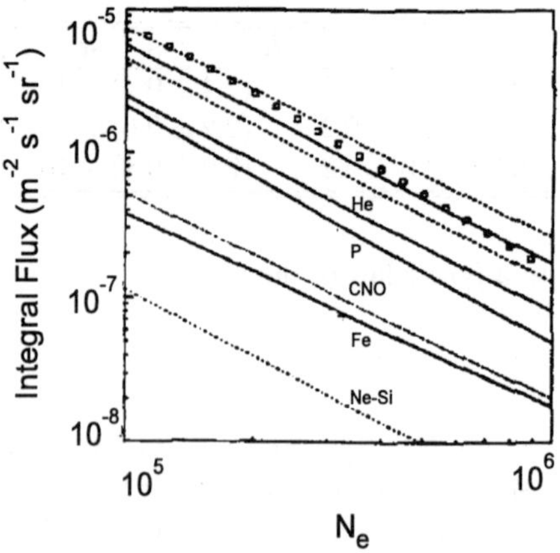

Figure 5.16 Shower size spectrum below the knee obtained by the EAS-TOP group compared with Monte Carlo simulations with that expected from primary spectra of various nuclei obtained by direct measurements.

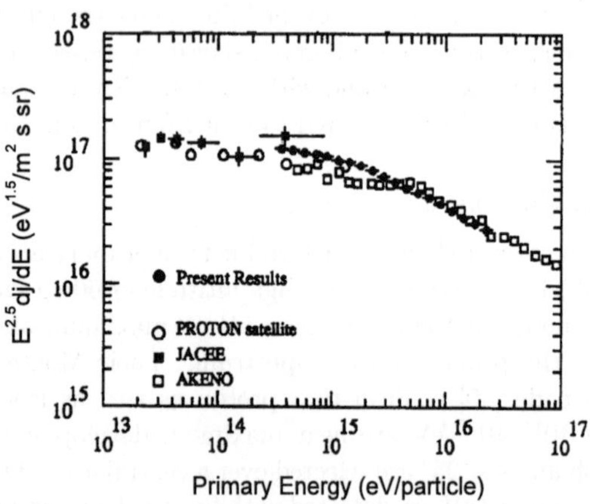

Figure 5.17 The primary energy spectrum obtained by the Tibet air shower group (Amenomori et al, 1995).

and no abrupt break as in other experiments. The exponent of the spectrum changes from 2.60 ± 0.04 below this energy to 3.00 ± 0.05 above. The spectrum connects smoothly to the direct measurements of Grigorov et al. and JACEE group in the lower energy region. The energy at which the spectrum starts steepening is lower than that in the Akeno spectrum and the fluxes at around 10^{16} eV are lower than the Akeno fluxes. A proton dominant (PD1) composition model (28% protons and 15% iron) gives about 10% lower fluxes at $\sim 10^{15}$ eV and the difference is smaller at higher energies. Thus the spectrum is almost composition model independent.

5.2.5 The Moscow Cerenkov Experiment

The Moscow group [41] obtained the primary energy spectrum from measurements of Cerenkov light, which is a better measure of primary energy than shower size since it is proportional to the track length integral of particles above the Cerenkov threshold and fluctuations are smaller. They operated an array of 21 scintillation detectors and 15 Cerenkov detectors at Samarkand (750 m above sea level) arranged as shown in Fig. 5.18. A notable feature of this experiment is that showers are

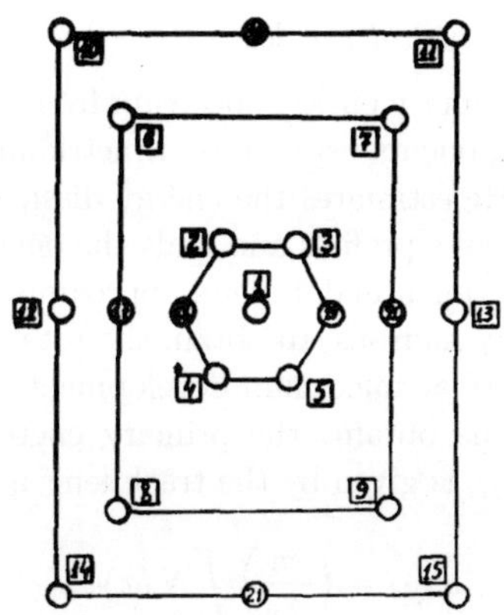

Figure 5.18 Layout of the Samarkand air ahower array. ○ - scintillators; □ - Cerenkov detectors.

triggered with a coincidence of the inner 7 Cerenkov detectors, with a threshold of 8 photons/(cm^2 eV), and therefore are relatively free from bias associated with charged particle trigger. Shower arrival directions are determined from the relative arrival times at the Cerenkov detectors. The Cerenkov photon size is determined by fitting the observed photon densities to a lateral distribution function of the form

$$q(R) = \frac{cQ}{R_0^2\left(0.075 + \frac{R}{R_0}\right)\left(1 + \frac{R}{R_0}\right)^{2.87}} \tag{5.19}$$

where $q(R)$ is the photon density at a core distance R, Q the total photon size and R_0 a characteristic radius depending on the position of shower maximum and zenith angle and c is a normalization constant. Showers are accepted over a central area of radius 60 m where the detection efficiency is > 90%. The photon size gives the energy transferred to the electron-photon component. Based on calculations corrections are made for the energy transferred to the muon and neutrino components, nuclear spallations and the energy not deposited by the electron-photon and hadron components above the observation level. The integral primary energy spectrum, thus obtained, can be represented as

$$J(>E_0) = 2.53 \times 10^{-11}\left(\frac{E_0}{3 \times 10^{15}}\right)^{-\gamma} \quad \mathrm{cm^{-2}\,s^{-1}\,sr^{-1}} \tag{5.20}$$

where $\gamma = 1.65 \pm 0.05$ for $10^{15} \leq E_0 < 3 \times 10^{15}$ eV and 2.3 ± 0.1 for $3 \times 10^{15} \leq E_0 < 10^{16}$ eV, which agrees with the Grigorov spectrum in the overlapping region.

5.2.6 The Calorimetric Method of Linsley

Linsley [42] estimated the primary energy spectrum from the world data available on the electron and muon components using a calorimetric method without recourse to Monte Carlo simulations. He estimates the energy dissipated by electrons from the experimentally observed shower profiles and adds the energy contained in the muon component, again from experimental data, and approximate estimate of energy going into neutrino and low energy hadrons, to obtain the total primary energy. From the observed intensity of showers at maximum development and the muon size spectra from various experiments, he obtains the primary energy spectrum. The energy dissipated by electrons, $E_{EM,}$ is given by the track length integral

$$E_{EM} = \left(\frac{\epsilon_0}{X_0}\right)\int N(x)\,dx \tag{5.21}$$

where $N_e(x)$ is the number of electrons at depth x g cm^{-2}, ϵ_0 is the critical energy and X_0 is the radiation length. This can also be written as

$$E_{EM} = K\left(\frac{\epsilon_0}{X_0}\right)\sigma N_m \tag{5.22}$$

where N_m is the height and σ is the width (standard deviation) of the average shower profile. The shower profile can be approximated to a Gaussian or a gamma distribution. In the case of Gaussian, $K = \sqrt{2\pi} = 2.51$, and in the case of gamma distribution, which can be fitted to observed shower profiles, it varies from 2.35 for small showers to 2.42 for large showers. Taking an average value, he obtains

$$E_{EM} = \left(\frac{\sigma}{192}\right) N_m \tag{5.23}$$

with σ in g cm^{-2} and E_{EM} in GeV. Taking σ from measured shower profiles by the Yakutsk and Fly's Eye groups (see Ref. [42] and references therein) at high energies ($\sim 10^9$ GeV) and extrapolating it to lower energies using the expected dependence of σ on primary energy, E_0, given by

$$\sigma^2 = A + BD_{10} \log E_0 \tag{5.24}$$

where D_{10} is the

per decade ($\sim$ 65 g cm^{-2}) and B is a characteristic length of order 60–70 g cm^{-2}, he obtains

$$\sigma^2 = 1.1 \times 10^4 + 4.2 \times 10^3 \log E_0 \tag{5.25}$$

and substituting in Eq. 5.23

$$E_{EM} = 0.71 N_m^{1.025}. \tag{5.26}$$

Linsley argues that the muon energy spectrum measured is nearly independent of primary energy in the energy range $3 \times 10^5 - 10^8$ GeV and hence the total energy of the observed muons, $E_{\mu,\,\text{obs}}$, is proportional to $N_\mu(> 1 \text{ GeV})$, the number of muons (at sea level) with energy > 1 GeV, where the proportionality constant is 10.0 ± 0.5 GeV. By propagating the observed muons backwards, he obtains the muon production spectrum and from it $E_\nu = 0.4 E_{\mu,\,\text{obs}}$, the energy given to neutrinos including ν_μ and ν_e. Finally he estimates the average energy deposited by low energy hadrons, $E_h = 0.4 E_{\mu,\,\text{obs}}$, from existing experimental data. Adding, he obtains the total non-electronic contribution as

$$E_{\mu\nu\text{h}} = \left(18^{+3.5}_{-1.5} \text{ GeV}\right) N_\mu \left(> 1 \text{ GeV}\right)_{\text{sea level}} . \tag{5.27}$$

Since both E_{EM} and $E_{\mu,\,\text{obs}}$ are obtained from experimental data, they are independent of primary composition. He obtains the all particle energy spectrum from the world data on N_m and N_μ spectra by adding E_{EM} and $E_{\mu\nu\text{h}}$ for the corresponding N_m and N_μ with the same integral intensity in the range 10^{-6}–10^{-12} m^{-2} s^{-1} sr^{-1} where reliable data are available. The resultant differential energy spectrum is shown in Fig. 5.19. It is in good agreement with the spectrum obtained by the Yakutsk and Haverah Park groups (see Ref. [42]).

5.3 The Knee

The all particle energy spectrum obtained by various groups is shown in Fig. 5.20.

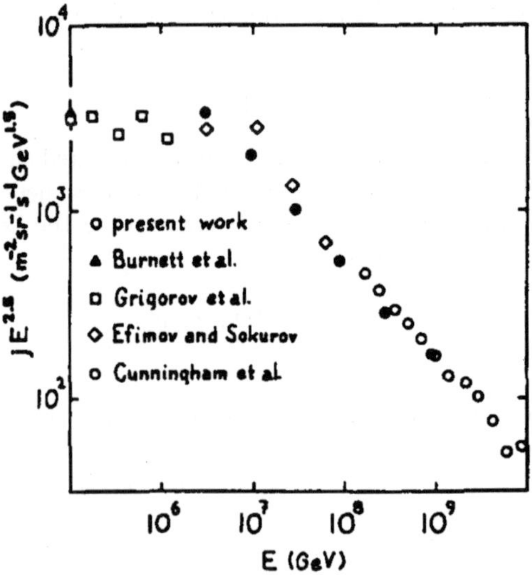

Figure 5.19 All particle primary energy spectrum obtained by Linsley (1985) using calorimetric method.

It can be seen that the intensities obtained by various groups, including the direct measurement of Grigorov et al. do not differ by more than a factor of 2. In the region beyond the knee, where the slope of the spectrum is ~ 3, this translates to about 25% difference in energy estimates, which is understandable considering the different techniques used by different groups in estimating the energies. The knee is clearly seen where there is data on either side of it. The position of the knee, though, is different in different experiments. It is clear that the knee occurs somewhere in the region $2 \times 10^{15} - 5 \times 10^{15}$ eV.

In fact, historically, investigations of shower size spectrum revealed the knee as early as the late forties. Linsley [43] mentions in his Rapporteur paper at the Bangalore Cosmic Ray Conference in 1983 that the knee has been seen by R. W. Williams [44] in the shower size spectrum derived from density spectrum measurements. It appears that the first group to have seen the knee using the conventional method of determining the shower sizes of individual showers by fitting a lateral distribution function is the Kobe University group [45] from Japan, who operated an array of 5 detectors at an altitude of 2770 m above sea level. They found that the exponent of the shower size spectrum changes from 1.55 ± 0.1 to 2.04 ± 0.2 at a shower size of 4×10^5.

Properties of some of the air shower components also were found to change their

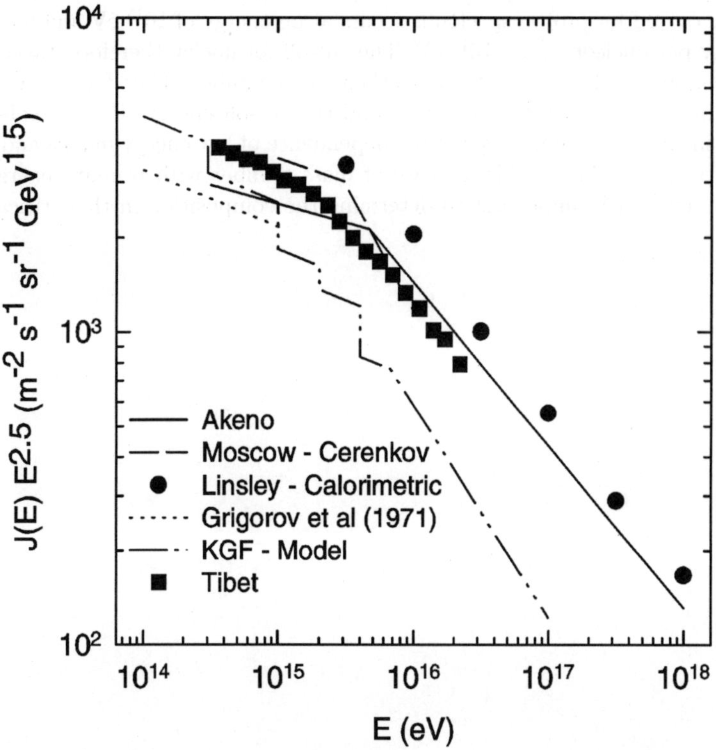

Figure 5.20 Summary of the all particle differential energy spectrum obtained by various groups(see text for references).

behavior at a certain critical shower size. Vavilo et al. [46] found that the logarithm slope of the dependence of the number of muons of energy > 1 GeV changes from 0.62 ± 0.12 to 0.95 ± 0.05 at a shower size of 4×10^5. Nikolsky et al. [47] observed that the logarithmic slope of the dependence of low energy hadrons (> 1 GeV) also changes from 0.2 to 1.0 around the same shower size. To explain these features Peters [48] proposed what is now known as the rigidity cut-off model. In this model, the composition of primary cosmic rays up to a critical rigidity of 10^{15} V is same as that observed at lower energies. Cosmic rays undergo a cut-off at this critical rigidity. Thus the cut-off for proton spectrum occurs at an energy of 10^{15} eV and for nuclei at an energy per nucleon of 5×10^{14} eV. The cut-off for nuclei, therefore, takes place at a total energy which is proportional to their mass number. Therefore the mean mass number increases with total energy beyond the cut-off energy. Peters explained the increase in the logarithmic slope of the dependence of low energy muons and hadrons on shower size as due to this increase of mass number with primary energy. It is, therefore, extremely important to determine the composition in this critical energy region.

References

1. N. L. Grigorov et al., *Proc. 12th International Cosmic Ray Conference, Hobart* **5** (1971) 1746.
2. B. V. Sreekantan, *Space Science Reviews* **14** (1972) 103.
3. R. W. Ellsworth et al., *Astrophys. & Sp. Sciences* **52** (1977) 415.
4. I. P. Ivanenko et al., *Proc. 23rd International Cosmic Ray Conference, Calgary* **2** (1993) 17.
5. K. Asakimori et al., *Proc. 23rd International Cosmic Ray Conference, Calgary* **2** (1993) 21.
6. M. Ichimura et al., *Proc. 23rd International Cosmic Ray Conference, Calgary* **2** (1993) 5.
7. V. I. Zatsepin et al., *Proc. 23rd International Cosmic Ray Conference, Calgary* **2** (1993) 13.
8. Simon Swordy, *Proc. XXIII International Cosmic Ray Conference, Calgary, Invited, Rapporteur & Highlight Papers* (1993) 243.
9. M. Nagano et al., *J. Phys. G: Nucl. Phys.* **10** (1984) 1295.
10. M. Nagano et al., *J. Phys. Soc. Japan* **53** (1984) 1667.
11. S. Kawakami et al., see Ref. [9].
12. N. Kirov et al., *Proc. 17th International Cosmic Ray Conference, Paris* **2** (1981) 109.
13. B. S. Acharya et al., *Proc. 17th International Cosmic Ray Conference, Paris* **9** (1981) 162.
14. K. Asakimori et al., *J. Phys. Soc. Japan* **51** (1982) 2059.
15. S. Vernov et al., *Can. J. Phys.* **46** (1968) S197.
16. E. R. Bagge et al., *Proc. 15th International Cosmic Ray Conference, Plovdiv* **12** (1977) 24.
17. O. S. Diminstein et al., *Proc. 15th International Cosmic Ray Conference, Plovdiv* **12** (1977) 154.
18. B. S. Acharya, *Ph. D. Thesis* University of Bombay (Unpublished) 1983.
19. B. S. Acharya et al., *Nucl. Instr. Meth. in Phys. Res.* **A270** (1988) 556.
20. F. Kakimoto et al., *Proc. 17th International Cosmic Ray Conference, Paris* **11** (1981) 254.
21. J. Linsley, *Proc. 13th International Cosmic Ray Conference, Denver* **5** (1973) 3207.
22. A. M. Hillas, *Proc. Cosmic Ray Workshop, University of Utah* (1983) 16.
23. T. K. Gaisser and A. M. Hillas, *Proc. 15th International Cosmic Ray Conference, Plovdiv* **8** (1977) 353.
24. B. S. Acharya et al., *Proc. 19th International Cosmic Ray Conference, La Jolla* **2** (1985) 226.

25. M. V. S. Rao, in *Cosmic Pathways: Contemporary Perspectives in Physics and Astrophysics*, ed. Ramnath Cowsik (Tata McGraw-Hill Publishing Company Limited, 1986) p175.
26. B. S. Acharya et al., *Proc. 16th International Cosmic Ray Conference, Kyoto* **9** (1979) 109.
27. A. M. Hillas, *Proc. 16th International Cosmic Ray Conference, Kyoto* **8** (1979) 7.
28. L. Horton et al., *Proc. 18th International Cosmic Ray Conference, Bangalore* **6** (1983) 124.
29. A. J. Bower etal., *J. Phys. G: Nucl. Phys.* **9** (1983) L53.
30. A. V. Glushkov et al., *Proc. 16th International Cosmic Ray Conference, Kyoto* **8** (1979) 158.
31. S. N. Vernov et al., *Proc. 16th International Cosmic Ray Conference, Kyoto* **8** (1979) 129.
32. Yu. A. Fomin et al., *Proc. 22nd International Cosmic Ray Conference, Dublin* **2** (1991) 85.
33. G. B. Khristiansen et al., *Proc. 24th International Cosmic Ray Conference, Roma* **2** (1995) 772.
34. T. N. Danilova et al., *Izv. RAN Ser. Fiz.* **58** (1994) 67.
35. K. Asakimori et al., *Proc. 23rd International Cosmic Ray Conference, Calgary* **2** (1993) 25.
36. EAS-TOP Collaboration, *Proc. 24th International Cosmic Ray Conference, Roma* **2** (1995) 732.
37. D. Muller et al., *Ap. J.* **374** (1991) 356.
38. C. Forti et al., *Phys. Rev.* **D42** (1990) 3668.
39. M. Amenomori et al., *Proc. 24th International Cosmic Ray Conference, Roma* **2** (1995) 736.
40. K. Kasahara and S. Torii, *Comput. Phys. Commun.* **64** (1991) 109.
41. G. B. Khristiansen and Yu. A. Fomin, *Proc. 18th International Cosmic Ray Conference, Bangalore* **9** (1983) 195.
42. J. Linsley, *Proc. 19th International Cosmic Ray Conference, La Jolla* **2** (1985) 154.
43. J. Linsley, *Proc. 18th International Cosmic Ray Conference, Bangalore* **12** (1983) 135.
44. R. W. Williams, *Phys. Rev.* **74** (1948) 1689.
45. T. Kameda et al., *Proc. of the Moscow Cosmic Ray Conference* **II** (1960) 58.
46. K. N. Vavilov et al., *Zurn. Eksp. Teor. Fiz.* **32** (1957) 1319.
47. S. I. Nikolsky et al., *Dokl. Akad. Nauk. S.S.S.R.* **111** (1956) 71.
48. B. Peters, *Il Nuovo Cimento* **22** (1961) 800.

Chapter 6
PRIMARY COMPOSITION

The primary composition at energies $> 10^{14}$ eV can, at present, be estimated only indirectly from measurements of air shower parameters and hadron and gamma ray families in emulsion chambers deep in the atmosphere. All these estimates are based on the assumption of the superposition model for nucleus-nucleus interactions. Tomaszewski and Wdowczyk [1] have shown that superposition model is valid when one considers only average properties of cascades. The widths of distributions of various air shower parameters are, however, broader than given by the superposition model. The composition derived from these measurements depend, in general, on the model of hadronic interactions assumed in the calculations. Since the sensitivity of the parameters to composition is not very large, only gross features of the composition can be inferred from these measurements. In particular, change in composition, if any, in the energy region being investigated can be reasonably well identified.

The various air shower parameters used for inferring composition are the depth of shower maximum (X_m), its variation with primary energy, the dispersion $\sigma(X_m)$ of X_m, variation of number of muons with shower size, low energy delayed hadron content, high energy end of the hadron energy spectrum, hadron and gamma ray families in emulsion chambers etc. Some of these parameters like $\sigma(X_m)$ are useful for very large showers. Here we shall confine ourselves to composition in the energy range 10^{14}–10^{17} eV, the knee region.

6.1 Direct Measurements

Before we discuss the results on composition from air shower studies, we shall summarize briefly the composition obtained from direct experiments, as in the case of primary spectrum in the previous chapter. The energy spectra of various nuclei shown in Figs. 5.2 and 5.3 are used to obtain the composition at energies $< 10^{14}$ eV, the minimum energy at which air shower can be studied. Swordy [2] has made an estimate of the composition of the primary radiation at the crucial energy of 10^{14} eV/particle, which is very important for linking up the emulsion experiments with air shower observations. His estimates are given in the Table 6.1.

The procedure that has been followed by Simon Swordy in arriving at the calculated fluxes at the same energy/particle (total energy) from the directly measured energy/nucleon spectra, is as follows. He assigns a mass to each estimated group as

Table 6.1 An estimate of the composition of primary nuclei at 10^{14} eV/particle by Swordy (1993).

Group of nuclei	p	He	CNO	Ne-S	Fe-group
A	1	4	14	24	56
E (GeV/n)	10^5	2.5×10^4	7.14×10^3	4.17×10^3	1.79×10^3
Flux estimate (GeV/n)	1.3×10^4	2×10^3	1.2×10^2	3×10^1	7
Flux calculated (GeV/particle)	1.3×10^4	2.26×10^4	1.22×10^4	7.81×10^3	8.02×10^3
Relative abundance (%)	20	36	19	12	13

in Table 6.1 and estimates the fluxes from the figures (Figs.5.2 and 5.3), at the appropriate energy/nucleon. These are then converted to energy/particle fluxes using the mass for each group. He remarks that the total estimated flux of 6.36×10^4 (particles per (m^2 s sr) $GeV^{1.75}$) is in good agreement with the proton satellite measurements of Grigorov et al. [3].

6.2 Depth of Shower Maximum

The depth of shower maximum, X_m, for showers of a given primary energy depends both on the behavior of high energy interactions and the primary mass number. X_m is a function of primary energy E and mass number A and can be expressed as

$$X_m = a \ln (E/A) + b \tag{6.1}$$

and the rate of change of X_m with the logarithm of primary energy, called the elongation rate (ER), is then

$$\frac{\partial X_m}{\partial \ln E} = a \left(1 - \frac{\partial \ln A}{\partial \ln E}\right) \tag{6.2}$$

The constants a and b depend on high energy interaction characteristics. The elongation rate, is expected to be nearly constant if these characteristics and primary composition do not change with energy. Any change in the elongation rate, if opposite to the expectation from conventional ideas of particle physics, can be attributed to change in the primary composition. The depths of shower maxima at various primary energies were measured by several groups using the Cerenkov technique and the constant intensity cut method.

6.2.1 Cerenkov Technique

In order to obtain the depth of shower maximum from Cerenkov measurements, the relation between the two has to be obtained from theory. Several workers [4, 5] have shown that the relation between the lateral distribution as well as the pulse profile of Cerenkov light and the depth of shower maximum is practically independent of the nuclear interaction model and the mass of the primary particle. Two approaches have been taken to calculate this relation. One is to formulate the problem of cascade development analytically and obtain numerical solution. This involves assumptions regarding the energy and angular distributions of electrons as functions of radial distance and atmospheric depth. Though this method does not require much computational time, the accuracy of the final results depends on the validity of the assumptions. The other method is complete 4-dimensional Monte Carlo simulations. In this case, the assumptions involved are much less since all the electromagnetic processes can, in principle, be taken into account. This method takes a considerable amount of computational time as a large number of showers have to be generated to achieve statistically significant results. However, as Kamata [6] pointed out, the most important parameter in this context is the angular spread of electrons and since direct experimental evidence of the angular spread is very scarce, a careful study is necessary. Another important factor to be considered is the absorption and scattering of Cerenkov photons due to Rayleigh and Mie scattering processes. These processes may alter not only the lateral distribution but also the pulse profiles through changes in the path length distributions.

6.2.1.1 Theoretical Results

The first detailed calculations of the pulse profiles were carried out by Fomin and Khristiansen [7]. They used the same method as Zatsepin and Chudakov [8], who calculated the lateral distribution of Cerenkov light in a detailed way. They, however, made certain simplifications which, they say, are valid for large distances from the core. They use Monte Carlo method only for selecting the interaction point and inelasticity of leading nucleon interactions and make use of the average nuclear-electromagnetic cascade curves calculated by Dedenko [9] for the CKP model of nuclear interactions. They neglect (i) the radial spread of electrons during the shower development, (ii) the Cerenkov angle of emission, (iii) the curvature of the shower front and (iv) the finite thickness of the shower disc. They use, for electrons, an angular distribution of the form $\exp(\theta^2/\theta_0^2)$, where $\theta_0^2 = 0.245(1+32E)^{-2}$ and an energy spectrum of the form $(1+32E)^{-2}$, independent of the age of the shower. They showed that the width of the Cerenkov pulse increases with core distance and, more important, that the shape of the pulse at a core distance of 2 km depends on the shape of the cascade curve. Their results on the pulse shape are shown in Fig. 6.1a for the three different cascades shown in Fig. 6.1b.

Kalmykov et al. [10,11] extended these calculations to showers initiated by various primary nuclei incident at various zenith angles and obtained the following relations between the full width at half maximum (FWHM), τ, and the core distance, R, and between τ_{300}, the FWHM at $R = 300$ m, and H_m, the height of shower maximum:

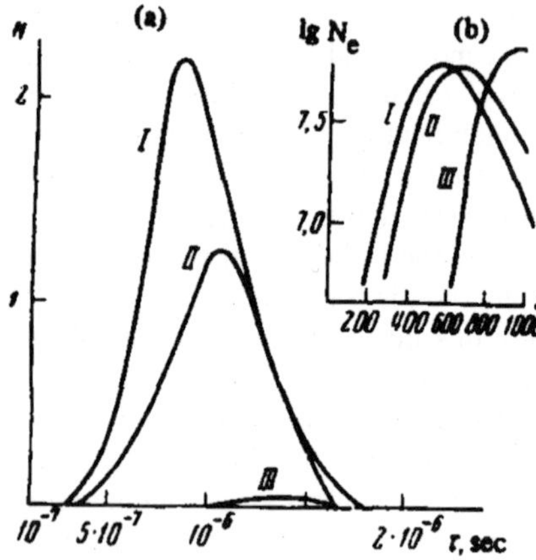

Figure 6.1 Cerenkov pulse profiles (a) for three different cascades (b) calculated by Fomin and Khristiansen (1972).

$$\tau \propto R^{n(H_m)} \text{ for } 300 \text{ m } < R < 600 \text{ m} \tag{6.3}$$

$$n = 1.5 + 0.1H_m \tag{6.4}$$

$$H_m = 17.05 - 9.17 \log \tau_{300} \text{ for } H_m < 6 \text{ km} \tag{6.5}$$

with R in meters, H_m in kilometers and τ_{300} in nanoseconds. Scattering of photons was not considered in these calculations.

Hara et al [12] also calculated the lateral distribution function (LDF) and pulse profiles in a similar way. They use, for the energy spectrum of electrons as a function of atmospheric depth, a modified version of the approximate formula given by Greisen [13]. Angular distribution is assumed to be same as by Fomin and Khristiansen. The authors claim that both these distributions are in agreement with Monte Carlo simulations by Messel and Crawford [14]. They, however, neglect the radial spread of electrons, shower front curvature and disc thickness. Absorption of photons is taken into account but not scattering. They find that the Cerenkov light intensity and pulse shape are strongly dependent upon the longitudinal development of showers. They also find that the peak of Cerenkov pulse lies always above the shower maximum. They have subsequently [15] carried out a complete Monte Carlo simulation of cascades of primary energy $E_0 = 10^{15}$ eV and found support for the angular distribution assumed in the earlier work. In their later calculations, they find that the maximum corresponding to the Cerenkov pulse peak lies above the depth of shower maximum, X_m, by an amount which varies from 150 g cm^{-2} to 210 g cm^{-2} as the core distance at which the measurements are made varies from 1 km to 2 km (Fig. 2 of Ref. [15]) confirming the earlier work. It is important to study this behavior at distances less than 1 km, a region where synchronized time measurements at different distances have been used to reconstruct the longitudinal shower profile. The authors have included only Rayleigh scattering in their calculations. Their preliminary estimate of the effect of Mie scattering is to shift X_m by about 40 g cm^{-2}.

Ivanenko and his collaborators from the Moscow State University have investigated extensively the relation between the various Cerenkov light parameters and the longitudinal shower profile, using a similar method. They use two extreme models of nuclear interaction to study the model dependence of this relation. While they [16] find that the LDF and FWHM can yield X_m independent of the model, their sensitivity increasing with core distance R, only the rising portion of the Cerenkov pulse maps the longitudinal profile independent of the model. According to them [17] the FWHM starts becoming sensitive to X_m only for $R \geq$ 200 m. They also investigate the effect of various approximations made in the calculations on the pulse shape and find that neglecting the lateral spread of electrons leads to a narrower pulse at small core distances. Galkin et al. [18] have shown that the pulse shape is narrower and amplitude smaller if the Cerenkov emission angle, θ_c, is neglected. According to them, the pulse shape becomes narrow if the curvature and thickness of the shower front are not considered in the calculations. These considerations show clearly the importance of detailed calculations to obtain the pulse width to shower maximum relation, particularly at small core distances ($\sim$ 200 m) where measurements in small showers are necessarily made, because of flux limitations of Cerenkov photons.

Protheroe and Turver [5] have carried out detailed Monte Carlo simulations in order to identify Cerenkov light parameters sensitive to the longitudinal shower development. They followed the complete hadronic cascade and electron-photon cascades of energy above 75 GeV under approximation A in one dimension. For energies below 75 GeV, they draw upon a bank of previously generated cascades in 4 dimensions, injected at various atmospheric depths. This method is free from assumptions regarding the lateral, energy and angular distributions of electrons and shower front curvature and thickness etc. They however, consider the Cerenkov light to be originating from the midpoint of the electron track instead of all along the track. Ozone, aerosol molecular absorption are taken into account, but not scattering. They find that the lateral distribution of photons, integrated over all angles, can be fitted by a power law of the form $(R + 50)^{-\delta}$ and δ is linearly related to X_m independent of the interaction model, primary energy and mass number, as shown in Fig. 6.2. Their results on pulse profiles include the temporal response of their detectors used in the Haverah Park experiment [19]. The dependence on core distance of rise time (10–90)%, FWHM and fall time (90–50)% is shown in Fig. 6.3. The variation with X_m, of rise time, FWHM, width of the top of the pulse (90–90)% and the fall time at different core distances is shown in Fig. 6.4. The FWHM, which is the most used parameter to derive X_m, has a strong and well behaved dependence in both R

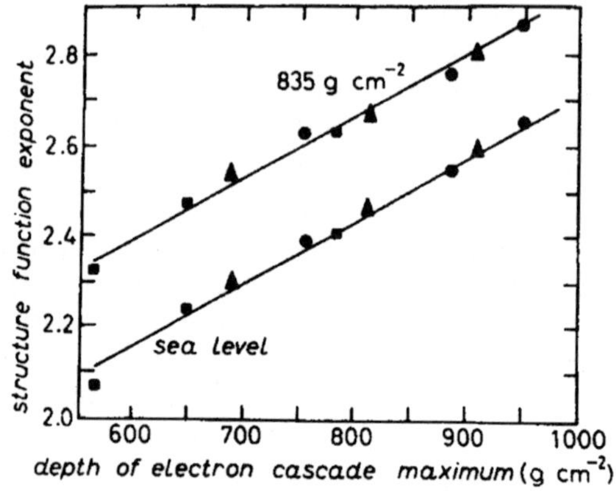

Figure 6.2 Relation between the exponent of the lateral distribution function of Cerenkov radiation and the depth of shower maximum at two altitudes from Monte Carlo simulations of Protheroe and Turver (1979); • protons, ▲ α-particles, ■ iron nuclei.

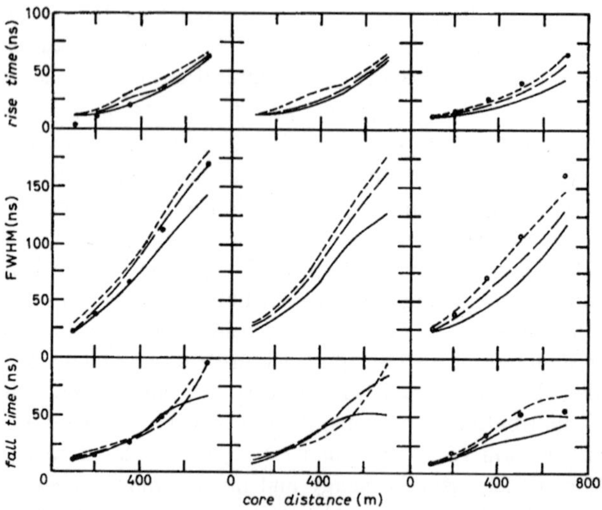

Figure 6.3 Cerenkov pulse parameters, calculated by Protheroe and Turver (1979), as a function of core distance, R, in showers of primary energy $E_0 = 10^{16}$ eV (full line), 10^{17} eV (long dashes), and 10^{18} eV (short dashes) initiated by protons (l.h.s.), α-particles (middle) and iron nuclei (r.h.s.) initiated showers.

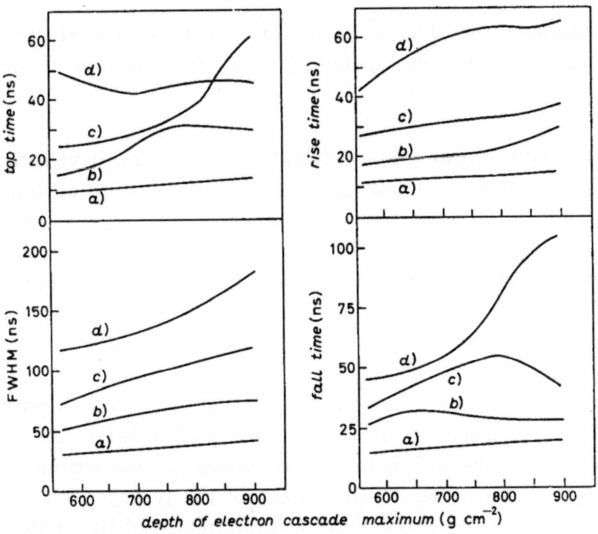

Figure 6.4 Cerenkov pulse parameters as a function of X_m at $R = 200$ m (a), 350 m (b) and 700 m (c) from Monte Carlo simulations of Protheroe and Turver (1979).

and X_m, as shown by other calculations also. It is important to bear in mind, while normalizing measurements to a fixed distance, that the relation between FWHM and R is dependent on X_m.

The sensitivity of LDF and FWHM to X_m is, thus well established. The quantitative relationships obtained by different workers, however, differ from each other. Grigorev et al. [20] find that the values of X_m obtained by using the relations given by Kalmykov et al. [10] and Ivanenko et al. [21] differ by 75 g cm^{-2}. Inoue et al. [22] also find that the relations given by Kalmykov et al. [11], Ivanenko et al. [21] and Hara et al. [12] yield values of X_m which differ from each other by as much as 150 g cm^{-2}. A large part of these differences is due to the different assumptions made in different calculations, particularly regarding lateral, angular and energy distributions of electrons. An important point to note is that the derived elongation rate (rate of change of X_m with the logarithm of E_0) also depends on the calculations used as pointed out by Inoue et al. [22]. Also, Eqs. 6.3 and 6.4 are not consistent with the curves shown in Ref. [20] as well as the results of Protheroe and Turver [5]. Obviously these differences have to be resolved before consistent values of X_m and ER can be obtained. Furthermore, the effect of Rayleigh and Mie scattering, which have not been considered so far in detail, should be investigated.

6.2.1.2 Experimental Results

Detailed measurements of LDF and FWHM have been carried out mainly by the Australian, U.K., U.S.S.R. and Japanese groups. We shall now describe the results obtained from these experiments.

The Adelaide Experiment The Adelaide group [23, 24] have measured the LDF up to $R = 300$ m in showers of size $(2 - 50)10^5$ at sea level using nine 170 mm EMI 9623 photomultipliers, with an acceptance angle of 45°, located at 20 m to 300 m from the centre of the Buckland Park Conventional air shower array which provided shower size, core location and zenith angle. The LDF is fitted to a function of the form $A\exp(-bR/10^4)$. They find that b is related to H_m by $bH_m =$ const. if they estimate H_m from the relation between X_m and E_o given by Linsley and Watson [25] for protons, thus suggesting that the LDF is sensitive to X_m in accordance with theory. To measure FWHM, they use a single Mullard XP2040 photomultiplier whose output is fed to a Tektronix transient recorder with a system FWHM of 5.3 ns. This detector, whose opening angle is also 45°, is moved to various distances from the centre of the array to cover distances up to 350 m. Some of the recorded pulses shows a clear bumpy structure, which perhaps reflects the bumpiness of the shower development. Since showers initiated by protons and, to some extent α particles, are subjected to larger fluctuations than heavy primary showers and a certain fraction of them can have bumpy structure, the percentage of such pulses can give an estimate of their fraction in the primary beam. Gaisser [26] estimates, from Monte Carlo simulations, that about 1% of the showers have bumpy structure in non-scaling models and about 0.2–0.3% for scaling model.

The Adelaide group [27] find that, after allowing for the system response, FWHM $\propto R^n$ where n is a function of H_m which they estimate from a simple minded theory. This result is consistent with theoretical expectations. In a later paper [28], however, they fit their data with a constant value of $n = 1.4 \pm 0.2$, independent of shower size. They use this value to estimate FWHM(300 m) from their measurements at various distances, for further analysis. If both LDF and FWHM are related to X_m, as predicted by theory, then they should be related to each other. Fig. 6.5 shows their plot [24] of FWHM against the LDF parameter, b, which are clearly correlated, the correlation coefficient being 0.72. From the FWHM(300), they [29] obtain X_m as a function of the shower size, N_e, using Kalmykov et al.'s theory (Eq. 6.5), which, they showed, is applicable in their case. They find that $dX_m/d\ln N_e$, the elongation rate in N_e is 60 g cm^{-2}, which is much larger than 32 g cm^{-2} expected from the ER theorem [30] for constant composition and interaction model in which scaling is preserved in the fragmentation region. They suggest that the primary composition is changing from predominantly heavy to mostly protons in the shower size region 10^5–10^7. Their results, as quoted by Linsley and Watson [31], are shown in Fig. 6.6, which is a composite plot of X_m vs E_0 from various experiments. Apart from the points raised by Orford and Turver [32] in their criticism, which essentially affect the absolute

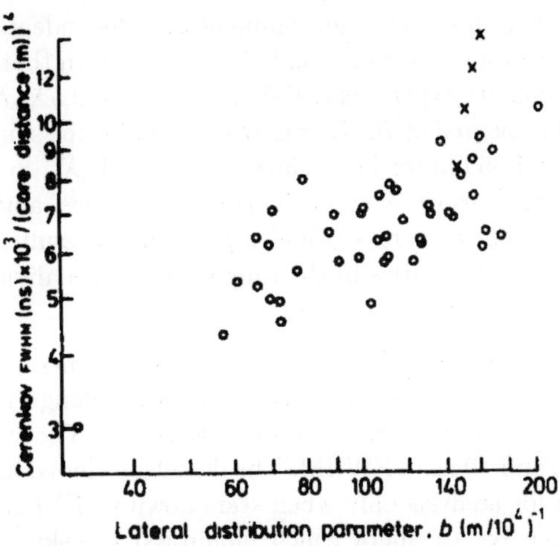

Figure 6.5 Correlation between the pulse shape parameter, FWHM, and the exponent of the lateral distribution of Cerenkov light observed by Kuhlman et al (1981).

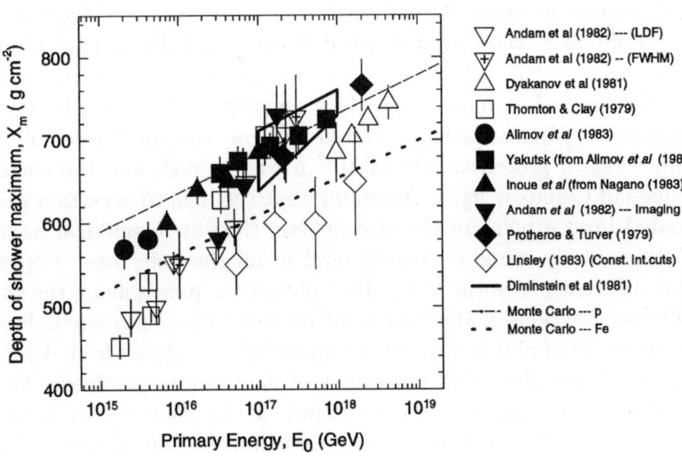

Figure 6.6 Summary of measurements of X_m as a function of E_0 from various experiments compared with Monte Carlo simulations.

values of X_m, there are two factors which might alter the elongation rate. One is that theory predicts the radial dependence of FWHM to be a function of X_m and therefore N_e, whereas the authors use a constant value of n, independent of N_e. The other is due to the non-linear propagation of errors. It can be shown that, for the parameters applicable to the Adelaide experiment, the error $\delta X_m = 0.5 X_m \delta\tau(R)/\tau(R)$, where $\tau(R)$ is the FWHM measured at R. Taking $R = 150$ m for the small showers for which $X_m = 450$ g cm^{-2} and 350 m for large showers for which $X_m = 670$ g cm^{-2}, δX_m is 43 g cm^{-2} and 9 g cm^{-2} respectively in the two cases. Thus, any systematic error in FWHM will alter the elongation rate. Finally, preliminary results from their improved experiment [33] show that n varies in the range 0.9–3.5 in individual showers.

The Durham Experiments The Durham group [19] operated eight 5″ diameter RCA 4522 photomultipliers spread over an area of 1 km^2, along with the Haverah Park air shower array. Pulses were recorded on oscilloscopes. The system response time was 9 ns in rise time and 19 ns in FWHM. The detectors viewed almost the entire sky. Data were accepted for analysis only when stars down to 7th magnitude were visible on photographs taken by a camera which monitored the sky. The detectors were calibrated using a standard radioactive light pulser. From the oscilloscope records they determined the rise time (10% – 90%), top time (90% – 90%), FWHM, fall time (90% – 50%), pulse area which gives the photon density and the pulse arrival time at the 10% (D_{10}, which is found to correlate with X_m) level relative to a time marker, in the distance range 100 m–600 m for showers of primary energy 10^{17}–10^{18} eV. An interesting and powerful technique they [34] developed is to reconstruct the shower development curve from the relative arrival times of light at the same percentile level of pulses recorded in detectors located at different places in the array, Their simulations showed that the earliest observed light originates high in the atmosphere and the pulse shape is clearly a measure of the cascade development. Fig. 6.7 shows the Monte Carlo generated electron cascade curves of eight sub-showers, into which a proton shower of 10^{17} eV is divided, and the corresponding eight sub-pulses of Cerenkov light. Assuming that the light at a certain point in the pulse originated from a point in the atmosphere, they fit a spherical wave front to the sets of times at the same percentile level in all the detectors. Repeating this procedure for different percentile levels, they obtain the points along the shower axis at which different fractions of the peak light intensity have originated. Examples of two showers constructed in this way, are shown in Fig. 6.8. Apart from demonstrating the method, the figure illustrates the extent of fluctuations involved. They obtain the depth of shower maximum from the depth of Cerenkov light maximum using computer simulations. Not only the shower development, but also the core position and arrival direction can be determined using this imaging technique. As each point on the shower axis can be determined to an accuracy of about 100 m, the accuracy achieved in the arrival direction is $\sim 0.25°$, an order of magnitude better than the conventional method. With the help of computer simulations they [5] also estimate

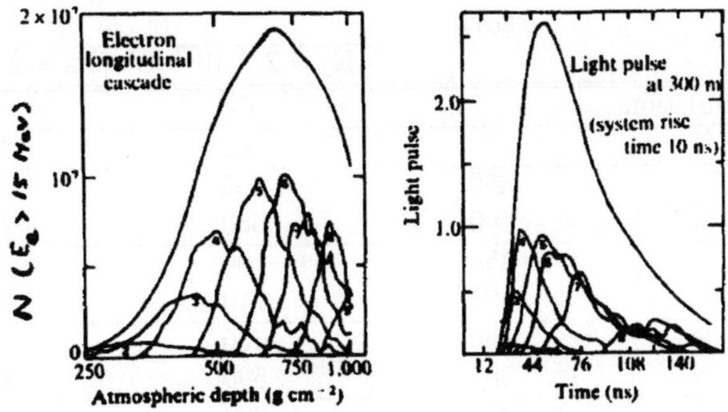

Figure 6.7 Shower development curves due to 8 sub-showers and the corresponding Cerenkov sub-pulses from Monte Carlo simulations of Orford and Turver (1976).

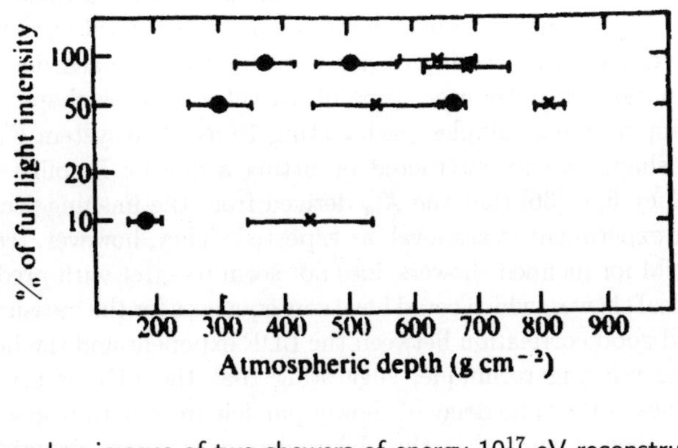

Figure 6.8 Cerenkov images of two showers of energy 10^{17} eV reconstructed by Orford and Turver (1976).

Table 6.2 Depth of shower maximum for two primary energies obtained by Protheroe and Turver (1979) from several Cerenkov light parameters.

Parameter	Core distance (m)	X_m (g cm^{-2}) $E_0 = 2 \times 10^{17}$ eV	X_m (g cm^{-2}) $E_0 = 2 \times 10^{18}$ eV
Structure function	—	650	800
D_{10} (g cm^{-2})	—	730	910
Rise time	350	800	850
	500	680	800
Top time	350	700	680
	500	710	670
Fall time	350	600	>650
	500	590	670
FWHM	350	700	900
	500	650	730

X_m from LDF, FWHM and pulse rise time etc. measured earlier by them [19]. These values, shown in Table 6.2, have a spread of $\sim$ 200 g cm^{-2}. The average values of X_m are 681 ± 20 g cm^{-2} and 766 ± 31 g cm^{-2} for showers of primary energy 2×10^{17} eV and 2×10^{18} eV respectively. These values are also plotted in Fig. 6.6.

The Durham group [35] continued their investigations at Dugway (atmospheric depth: 862 g cm^{-2}) and extended them to smaller showers in the energy range 2×10^{15}–5×10^{17} eV. The array was operated with three different spans at different periods to cover the wide energy range. Though the photomultipliers used were the same as in the earlier experiment, the electronics has been improved. Each detector provided digitized information on pulse area, time of arrival and pulse shape via five or six partially overlapping pulse samples, each lasting 10 ns. The system FWHM was 6.7 ns. The pulse shape was reconstructed by fitting a quartic B-spline to the digital information. They find [36] that the X_m derived from the imaging technique agrees with the earlier experiment at sea level, as expected. They, however, remark that the measured FWHM for inclined showers does not seem to agree with predictions for an equivalent vertical shower, which could have consequences for the measured elongation rates. They find good correlation between the LDF exponent and the height of origin obtained by the imaging technique, suggesting that the LDF is sensitive to X_m. From their studies of the time delay of shower particle front with respect to Cerenkov light front [37], they suggest that this delay also gives a measure of X_m. This, if established, is very useful to study small showers, since the delay is expected to increase with decreasing shower size. With the help of computer simulations, they estimate X_m from measurements of LDF as a function of zenith angle [38] over the entire energy range and from FWHM [39] and imaging at higher energies. They have revised [40] their values since the Paris Cosmic Ray Conference and these values are shown in Fig. 6.6 along with measurements of other workers. The authors, finding that the elongation rate is greater than the expectation for constant composition, suggest a possible interpretation that the mean primary mass number is close to 56 at 10^{16} eV, decreasing to ≤ 10 at 2×10^{17} eV.

The Durham group [41] also studied the fluctuations, $\sigma(X_m)$, in X_m to be able to put a stronger constraint on the composition, since X_m at a given energy depends on both composition and interaction model. They estimate $\sigma(X_m)$ from values of FWHM (250 m) obtained by the interpretation of 72 selected individual showers after correcting for systematic variations with zenith angle and primary energy and measurement errors. The value thus obtained is $\sigma(X_m) = 80 \pm 11$ g cm^{-2} for showers of $E_0 = 2 \times 10^{17}$ eV for which the mean value of X_m is estimated to be 675 ± 17 g cm^{-2}. They compare these values with simulations using nine different interaction models which preserve scaling in the fragmentation region and differ in the energy dependence of cross section and multiplicity in the central region. Their results are shown in Fig. 6.9 which is a plot of $\sigma(X_m)$ vs X_m. The multiplicity dependence

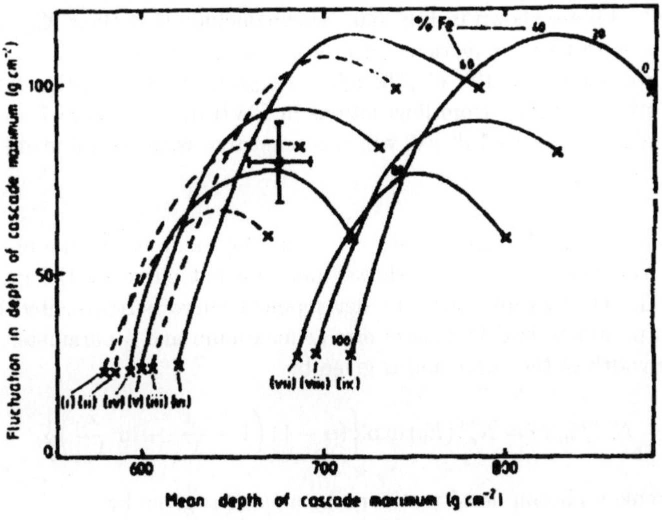

Figure 6.9 Experimentally measured σ_{X_m} and $\langle X_m \rangle$ compared with Monte Carlo simulations from Chantler et al. (1982).

is $E^{0.33}$ for curves (i), (ii) and (iii), $E^{0.25}$ for (iv), (v) and (vi), and scaling law for (vii), (viii) and (ix). The cross section dependence is $\ln^2 S$ for (i), (iv) and (vii), $\ln S$ for (ii), (v) and (viii) and constant for (iii),(vi) and (ix) . Each curve traces the percentage of iron in the two-component composition model of protons and iron nuclei. They conclude that an interaction model with multiplicity $\propto E^{0.25}$ and cross section $\propto \ln^2 S$ and a composition with 30–40% Fe is favored, though a model with $E^{0.33}$ multiplicity and lighter composition is not ruled out.

The Moscow Experiment The U.S.S.R. group [42] operated five Cerenkov detectors along with the Yakutsk air shower array at sea level. The central detector is a FEU-65 photomultiplier with a photocathode area of 180 cm^2, the three located at 250 m from the centre are each a group of 7 FEU-110 photomultipliers of total area 200 cm^2 whose outputs are added and the fifth one has 19 FEU-110's of total area 800 cm^2 located at 500 m. The system FWHM of the central detector is 14 ns and 23 ns for the others. Signals from these detectors are displayed on oscilloscopes and photographed. Synchronized time measurements are also available. Core position and arrival direction are determined both from the particle array and the Cerenkov array. LDF, FWHM and pulse profile measurements have been carried out for showers of primary energy $> 10^{16}$ eV up to a core distance of 1 km. Kalmykov et al. [11] estimate X_m from the FWHM measurements using their own simulations for showers in the energy range 3×10^{16} eV–6×10^{17} eV. Their values as quoted by Alimov et al. [43] are shown in Fig. 6.6. They conclude that their X_m values are in disagreement with their predictions for protons using scaling model. The agreement is good for iron showers at 10^{17} eV, though the experimental elongation rate is slower. They [44] estimate $\sigma(X_m)$ from fluctuations in FWHM, to be 66 ± 7 g cm^{-2} in the energy range $2 \times 10^{16} - 2 \times 10^{18}$ eV, which is consistent with the value of the Durham group.

The Yakutsk Experiment Dyakonov et al. [45] made use of the observed LDF and N_e at Yakutsk to derive the shower development curve by the inverse transform method. They assume that the development curve is represented by gamma-distribution parameterized by shower size at maximum and a parameter, α, characterizing the width of the curve and is given by

$$N_e\left(E_0, x\right) = N_e^m\left(E_0\right) \exp\left\{(\alpha - 1)\left(1 - \frac{x}{X_m} + \ln\frac{x}{X_m}\right)\right\}. \tag{6.6}$$

The Cerenkov photon density at core distance R is given by

$$Q\left(E_0, R\right) \simeq \int_0^{x_0} G\left(R, x, \cdots\right) N_e\left(E_0, x\right) dx \tag{6.7}$$

where x_0 is the depth of the observation level. They fit the observed lateral distribution function to that expected from Eq. 6.7 by minimizing the function

$$\sum_i \left[Q\left(R_i, E_o, X_m, N_e^m, \alpha\right) - Q\left(E_0, R_i\right)\right]^2 \tag{6.8}$$

and solve for X_m, N_e^m and α. They estimate X_m for four primary energy values, in the range 10^{18} eV–5×10^{18} eV, which are determined from photon densities at 400 m from the core. Their values are also shown in Fig. 6.6. They [46] also estimate $\sigma(X_m)$, from fluctuations in the ratio of photon densities at 200 m and 550 m, to be 79 ± 4 and 63 ± 5 g cm^{-2} for $E_0 = 7.5 \times 10^{17}$ eV and 4×10^{18} eV respectively. The value at the lower energy is consistent with other measurements. The authors conclude that there is a tendency for σ (X_m) to decrease rapidly with E_0, which may be due to change in composition.

Diminstein et al. [47] estimate X_m from the ratio Q/N_e, where Q is the total Cerenkov flux , obtained by integrating the LDF, and N_e determined from the particle array at Yakutsk. They find that Q/N_e decreases with E_0, as expected. They use the relation between X_m and Q/N_e given by the calculations of Ivanenko et al. [48] and obtain $Xm = 660$ g cm^{-2} at $E_0 = 10^{17}$ eV where $Q/N_e = 1.04\times10^5$. Fig. 6.10 shows the plot of X_m vs Q/N_e calculated by Ivanenko et al. for different interaction models for $E_0 = 10^{17}$ eV (closed symbols) and 10^{18} eV (open symbols). Combining the results

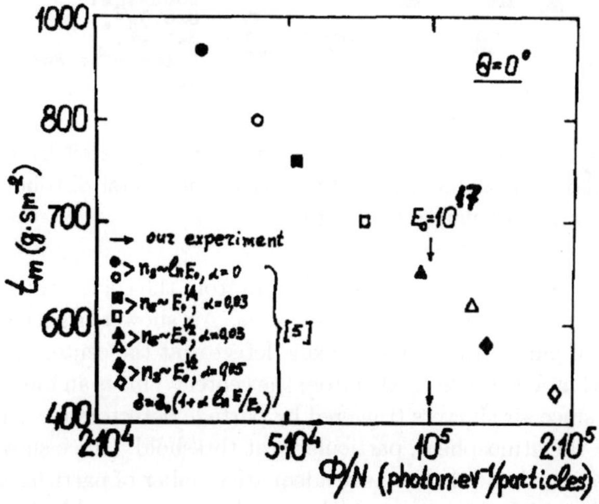

Figure 6.10 Relation between depth of shower maximum and Q/N from the calculations of Ivanenko et al as quoted by Diminstein et al (1981).

obtained from the Yakutsk experiment using various methods (LDF, FWHM etc.), they obtain the following relation between X_m and E_0:

$$X_m = (730 \pm 30) + (55 \pm 20)\,(\log E_0 - 18) \tag{6.9}$$

This is shown in Fig. 6.6 as a box with thick solid line.

The Samarkand Experiment The Samarkand group [49, 50] carried out LDF and FWHM measurements at Samarkand (940 g cm^{-2}) with an array consisting of both particle and Cerenkov detectors as shown in Fig. 6.11. The array consists of 13

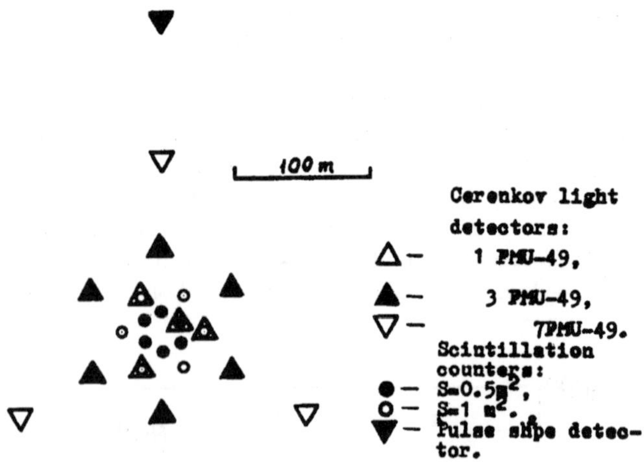

Figure 6.11 The Cerenkov and particle detector array operated by the Moscow-Samarkand collaboration at Samarkand to measure the lateral distribution and pulse shape of Cerenkov light (Kalmykov et al, 1983).

FEU -49B photomultipliers located at 30–120 m from the centre of a conventional air shower array. The system FWHM is about 10 ns. Air shower trigger was provided by the coincidence of pulses from the Cerenkov detector at the center of the array and 6 detectors located at a distance of 60 m from the center. This is an important feature of the experiment since air showers triggered by particle detectors record showers which develope late in the atmosphere, particularly at threshold, (since showers that develope early are absorbed and do not have adequate number of particles to be triggered) and therefore are enriched in protons, whereas showers triggered with Cerenkov light do not have this bias since Cerenkov light is approximately proportional to the total primary energy. Of particular interest is the Cerenkov pulse shape detector, which consists of 13 photomultipliers (XP-2041) with a 11 cm photocathode and $\sim$ 5 ns proper resolution. The pulses from the 13 photomultipliers were summed up in a 7-input electronic summator. 12 photomultipliers in pairs had a common anode load. Photomultipliers in each pair operated with the same delay accurate to within ± 1 ns. The delays from different pairs of photomultipliers were equalized precisely by selecting the lengths of the connecting cables to an accuracy of ± 1 ns. This detector was located at 220 m from the center of the array. An additional single photomultiplier was also used at the same location to increase the measurement range. Pulses from these two detectors were transmitted to the center of the array, amplified and displayed on an oscilloscope. The resultant resolution time was 8 ns for the multi-photomultiplier detector and 7 ns for the single photomultiplier detector. Showers with a detection efficiency of more than 90% only were used for further analysis.

Note that there are two main differences between this experiment and the earlier ones: one is the Cerenkov light trigger, which has negligible bias for heavy primary generated showers compared to proton showers, and the other is the large area, nearly 5,000 cm^2, of the pulse shape detector, which enables the pulse shapes to be measured accurately.

Using the calculations of Kalmykov et al. [11], they calculate FWHM (200 m) from measurements at any distance and obtain X_m from this. This experiment is designed to study showers of energy $10^{15} - 10^{16}$ eV. Their result [43] based on 35 events is shown in Fig. 6.6. It is clearly much larger than any other result in this energy region.

The Akeno Experiments Inoue et al. [22] operated along with the Akeno air shower array (935 g cm^{-2}) two stations of Cerenkov detectors, separated by 450 m, each consisting of four 5″ Hamamatsu R1250 photomultipliers whose outputs are combined. Signals are displayed on 400 MHz storage oscilloscopes. The system rise time and FWHM are 3.6 ns and 6.6 ns respectively. The trigger and the conventional air shower parameters are provided by the Akeno array. They measure rise time and FWHM in the distance range 200–500 m in showers of primary energy $5 \times 10^{15} - 10^{17}$ eV. They determined values of X_m from FWHM (300 m) at four different primary energies, using relations given by Kalmykov et al. [11], Ivanenko et al. [48] and Hara et al. [12]. The different calculations yield different values of X_m. The values obtained using the relation of Kalmykov et al. are plotted in Fig. 6.6 along with other measurements.

Hara et al. [15] have installed three Cerenkov telescopes at a distance of 1.8 km from the centre of the Akeno air shower array. Each telescope, shown in Fig. 6.12, consists of a concave mirror of 2 m in diameter, 19 photomultipliers (HTV R877) which are set at the focal plane of the mirror and the recording system. Each photomultiplier has a hexagonal field of view of $4.5° \times 5.2°$ as shown in the figure. The fields of view of the mirrors overlap each other. Signals from the photomultipliers are logarithmically amplified (integration time 100 ns), digitized and recorded. Trigger is provided by the Akeno array if the shower size exceeds 3×10^6. As each photomultiplier tube views a different part of the sky, they can estimate the amount of light received from different atmospheric depths by simple geometry. They determine X_m from the depth of maximum Cerenkov light using their simulations. Based on 8 events they estimate $X_m = 600 \pm 100$ g cm^{-2} for showers of size 10^7–10^8 observed at 930 g cm^{-2}. This technique is similar to Fly's eye [51] with the difference that the viewing angle is small. Even so, contribution from fluorescence light may not be negligible at lower

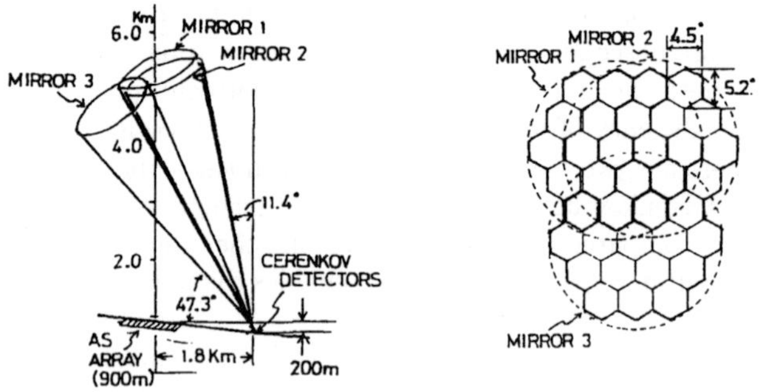

Figure 6.12 Arrangement of the three Cerenkov telescopes of Hara et al (1981).

altitudes and has to be accounted for.

The HEGRA Experiment The HEGRA collaboration [52] studied the fluctuations in X_m at a fixed primary energy in small showers. The fluctuations in proton induced showers are expected to be a factor of ~ 3 larger than those in iron induced showers. The height of shower maximum is directly related to the parameter λ of the Cerenkov lateral distribution given by

$$\varrho(r) = A \exp(\lambda r) .$$

Using Monte Carlo showers based on the Corsica program [53] λ is found to be linearly related to the parameter $P = N_e/E_0$. λ is determined from the lateral distribution measured with the AIROBICC array (see Sec. 3.2.3). N_e is determined from the particle array and E_0 estimated from this relation. Uncertainty in E_0 is about 40% and nearly independent of primary mass. Detector response is included in the simulations. It is found that the distribution of λ is approximately Gaussian for a given primary energy and mass number. The observed of λ is fitted to that expected for a mixed composition of p, O and Fe with an assumed differential primary energy spectrum of $E_0^{-2.7}$ with their relative abundances as free parameters. Their result is shown in Fig. 6.13. It was not possible to discriminate oxygen and iron. Table 6.3 shows their estimate of relative abundances of protons and heavy nuclei for different primary energy intervals. Presumably protons include He nuclei and heavy refers to all nuclei above He. The relative abundances agree with the compilation of Swordy. The authors conclude that there is no change in composition up to 1 PeV. Note that the energies involved are below the knee of the primary spectrum.

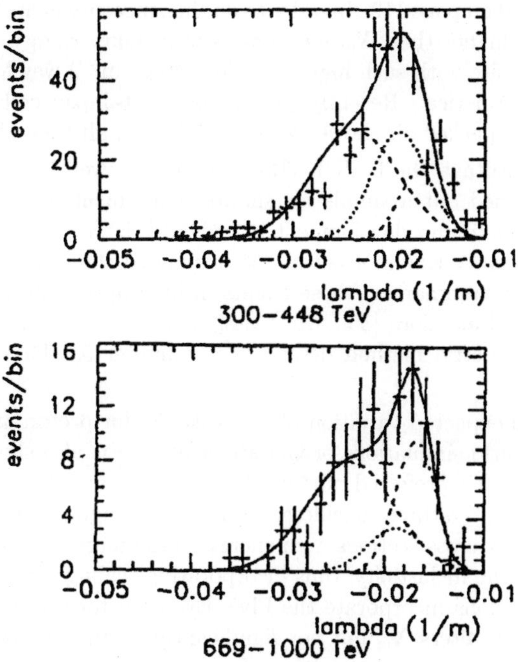

Figure 6.13 Distribution of λ, observed by the HEGRA group, compared with Monte Carlo simulations for the three primary components P (— — —), O (···) and Fe (— · —) for two primary energy intervals.

Table 6.3 Relative abundances of protons and heavy nuclei estimated by the HEGRA group for different primary energy intervals.

Primary Energy (TeV)	300–669	448–669	669–1000
protons	56 (±4, ±9, ±3, −8)	48 (±5, ±12, ±2, −14)	59 (±4, ±7, ±2, −6)
heavy	44 (±6)	52 (±9)	41 (±6)

6.2.2 Constant Intensity Cut Method

The position of shower maximum can be obtained from the constant intensity cut method as described on page 112. In order to do this, measurements have to be carried out at an altitude higher than X_m for showers of primary energy of interest. For a long time the only site with such high altitude (540 g cm^{-2}) was Mount Chacaltaya in Bolivia, South America. Recently the China-Tibet-Japan collaboration started operating a densely packed air shower array in Tibet an altitude of 600 g cm^{-2}.

At Mount Chacaltaya the BASJE (Bolivian Air Shower Joint Experiment) collaboration [54] carried out a simple air shower experiment to measure the shower size spectrum at maximum development in order to determine the primary energy spectrum. Subsequently Kakimoto et al. [55] expanded the array and continued the experiment. Linsley summarized these results in his Rapporteur talk at the Cosmic Ray Conference at Bangalore [56]. He reassigned primary energies for the events based on his calorimetric method described in Chapter 5. These results are also shown in Fig. 6.6.

In Fig. 6.6 the expected variation of X_m with E_0 for proton and iron primaries based on Monte Carlo simulations of Gaisser et al. [57] is also shown. The interaction model used is their preferred model called the Kopelievich-Nikolaev-Potashnika (KNP) model among the three considered. It uses the language of QCD Pomeron. It relates the increasing cross sections with energy to the increase in multiplicity in the central region and the inelasticity, thus incorporating scaling violation in the central region. The simulations incorporate the Fly's Eye detector response for which they are carried out. This alters X_m but not significantly, so the simulations can be used to see the trend in change of composition, if any.

It is seen from the figure that the experimental points scatter widely. Part of the reason for this is probably due to different assumptions made in the calculations for relating the observed Cerenkov parameter to X_m. Data obtained using the constant intensity cut method are the lowest in the corresponding energy region. While individual groups came to different conclusions regarding the composition, no firm conclusions can be drawn when all the data are put together. It is interesting to note that results obtained using the Cerenkov imaging technique and from experiments which use large area detectors, thus having the least bias, show a trend of the composition changing from a mixed one as seen at lower energies (the filled points) to almost pure protons at around 10^{16} eV. More work is needed to resolve this.

6.3 Muons

The average total number of muons, N_μ, in proton showers varies with shower size, N_e, as

$$N_\mu = K N_e^\alpha. \tag{6.10}$$

The number of muons in showers generated by primaries of mass A is given, under the superposition model, by

$$\begin{aligned} N_\mu(A) &= KA\left(\frac{N_e}{A}\right)^\alpha \\ &= KN_e^\alpha A^{(1-\alpha)} \end{aligned} \tag{6.11}$$

For low energy muons (up to a few tens of GeV) $\alpha \geq 0.8$, while for high energy muons ($>$ 100 GeV) it is ~ 0.7. Thus high energy muons are more sensitive to composition than lower energy muons. Moreover, low energy muons spread over large distances (characteristic distance, $r_0 \sim$ 300 m) and it is difficult to measure the total number of muons over a wide primary energy range without having to extrapolate the lateral distribution to either small or large distances. Perhaps the muon density at a fixed core distance is a better parameter. High energy muons, however, are contained within 100 m from the core and therefore their total number can be estimated better. Thus high energy muons are better suited for composition studies from the point of view of both experimental determination and sensitivity. Like other components, muons are also sensitive to interaction model, but to a lesser extent.

The composition at lower energies mentioned in Sec. 6.1 leads to a mean mass number of 15 at a fixed total energy. This translates to a mean mass number of 6 for a fixed shower size in the lower third of the atmosphere since the primary energy required for a heavy mass primary to yield a given shower size at these altitudes is larger than for a proton. This might seem to indicate that if shower size is chosen as the parameter to study muons rather than primary energy, the sensitivity to composition is lost. However, this is not so since the variation of N_μ with N_e is slower than variation with E_0. If

$$N_\mu \propto E_0^\beta \ ; \ \ E_0 \propto N_e^\delta \ \text{ and } \ N_\mu \propto N_e^\alpha$$

with $\alpha, \beta < 1$, then

$$N_\mu \propto \left(N_e^\delta\right)^\beta \propto N_e^{\delta\beta} \ ; \ \ \alpha = \delta\beta < \beta$$

At maximum development, E_0 is nearly proportional to N_e and $\delta = 1$. Beyond shower maximum showers are increasingly nearer to shower maximum as E_0 increases, so N_e increases faster with E_0 than at shower maximum and $\delta < 1$. For high energy muons, $\beta = 0.8$, $\delta = 0.87$ and $\alpha = 0.7$. If the composition changes from a mixed one to pure protons then the factor by which N_μ is altered compared to no change in composition is

$$\begin{aligned} R_{E_0} &= \left(\frac{N_\mu(A=1)}{N_\mu(A=15)}\right)_E = \left(\frac{1}{15}\right)^{0.2} = 0.58 \ \text{ for fixed } E_0 \text{ and} \\ R_{N_e} &= \left(\frac{N_\mu(A=1)}{N_\mu(A=6)}\right)_{N_e} = \left(\frac{1}{6}\right)^{0.3} = 0.58 \quad \text{for fixed } N_e, \end{aligned} \tag{6.12}$$

and if it changes to pure iron the corresponding factors are

$$
\begin{aligned}
R_{E_0} &= \left(\frac{N_\mu (A=56)}{N_\mu (A=15)}\right)_E = \left(\frac{56}{15}\right)^{0.2} = 1.30 \text{ and} \\
R_{N_e} &= \left(\frac{N_\mu (A=56)}{N_\mu (A=6)}\right)_{N_e} = \left(\frac{56}{6}\right)^{0.3} = 1.95.
\end{aligned}
\tag{6.13}
$$

Thus shower size seems to be a better parameter to classify showers to study change of composition.

Fluctuations in the muon component are also used to estimate primary composition. For example, the width of the muon number distribution in showers generated by a heavy nucleus of mass A is smaller approximately by $\sqrt{A}$ compared to that in proton showers. The observed width is compared with that expected from Monte Carlo simulations for primaries of different mass numbers to derive the composition. However, the width depends also on the high energy interaction model used for the simulations. Moreover, while the superposition model is adequate for comparing the average properties of showers with experimental data, it is not suitable for fluctuation studies as it underestimates the width of the distributions. The fragmentation model, which is an improvement of the superposition model, makes use of the details of the nucleus-nucleus interactions observed at low energies extrapolated to air shower energies. While this model is closer to reality, the validity of this extrapolation is yet to be established. Because of these reasons the composition obtained from fluctuation studies is not very reliable.

6.3.1 High Energy Muons

Among all the air shower parameters used for composition measurements the most sensitive is the high energy muon component as discussed above. Even in this case one has to take recourse to comparison of experimental data with Monte Carlo simulations to derive composition. Under certain conditions, however, study of correlations of various air shower parameters with high energy muon component leads to estimate of composition in a model independent way.

6.3.1.1 Model Independent Method

Such a method was developed by the KGF group [58]. The method is based on the assumption that the efficiency of production of high energy muons depends essentially on the ratio of the energy per nucleon of the primary particle to the energy of the muon, whereas the shower size at the observation level depends on its total energy (both the superposition and the fragmentation models support this assumption). Consider, for example, muons of energy, E_μ, larger than 200 GeV in showers of size 2×10^4 at an atmospheric depth of 920 g cm^{-2}. The primary energy, E_0, required is about 10^{14} eV if the primary is a proton. The ratio of the energy per nucleon of the primary to the energy of the muon, $(E_0/A)/E_\mu$ is 500. If the average inelasticity is 0.5, the primary particle can produce, up to eight generations, pions energetic enough to yield muons of the required energy. If they decay and if the interaction mean free path is $\sim$60 g cm^{-2} these muons could be produced up to an atmospheric depth of about 500 g cm^{-2}. If the same shower of size 2×10^4 were to be produced by an iron nucleus, the primary energy required is about 3×10^{14} eV; higher primary energy, compared to proton, would be required because of faster development. In this case, the ratio $(E_0/A)E_\mu$ is only about 25 and therefore muons can be produced only up to 4 generations of primary interactions, i.e. up to a depth of 250 g cm^{-2}. Thus, it can be seen that the average production height of muons in iron showers is much higher than in proton showers. Since these arguments are based only on energy considerations, this feature is independent of the interaction model. Calculations based on scaling model [59, 60] give values of 11 km and 20 km for the average production height of muons of energy > 220 GeV in proton and iron showers of size 2×10^4 respectively. This difference in production height results in a flatter lateral distribution of muons in iron showers compared to proton showers. Because of this the ratio between muon densities in iron and proton showers increases with core distance. Calculations mentioned above, show that this ratio increases from $\sim$4 at 10 m to more than 20 at 60 m. Thus, showers in which high energy muons are recorded at large core distances are enriched in heavy nuclei compared to those in which the muon is recorded at small distances if the primary composition is a mixture of protons and nuclei. These heavy nuclei enriched showers should then show larger number of low energy muons and flatter electron lateral structure compared to the others. Fig. 6.14 shows the correlation of density of muons of energy > 1 GeV at 27 m from the core with the core distance of high energy muons obtained from Monte Carlo simulations for three types of compositions: all protons, all iron and mixed composition. The correlation in the case of mixed composition is obvious.

6.3.1.2 The KGF Experiment

The KGF group carried out an experiment at Kolar Gold Fields. India, to obtain the composition using the above model-independent method as well as to study the detailed properties of high energy muons. The experimental arrangement, shown in Fig. 3.20 consisted of 70 scintillators spread over a circle of 100 m radius, out of which 37 were of area 2.25 m^2 each and located within 15 m from the centre of the array in a highly symmetric densely packed hexagonal pattern with a detector separation of 5 m to enable small showers of size $\sim10^4$ to be measured accurately. There were five fast timing detectors to measure the arrival directions and a muon detector of area 9 m^2 to measure low energy muons (> 1 GeV). Vertically below the array, at a depth of 270 m underground, 4 scintillators, each of area 1 m^2, were operated to measure muons of threshold energy 220 GeV. Two of the detectors were surrounded by a neon flash tube telescope, shown in Fig. 3.20, to locate the muon track. The

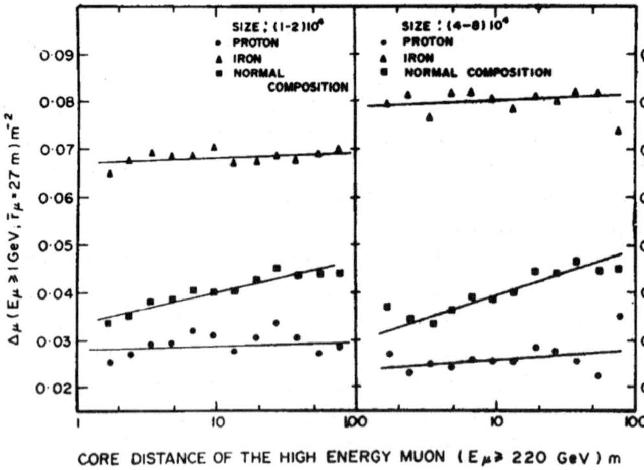

Figure 6.14 Correlation, expected from Monte Carlo simulations of the KGF group (Acharya et al., 1979), of low energy muon density at a core distance of 27 m with the core distance of high energy muons at 920 g cm^{-2}.

spatial angle of the muon could be determined to an accuracy of better than 15$'$. The overall accuracy of core distance of high energy muons was 2.5 m and the worst case accuracy of shower size estimation was ~20%.

Table 6.4 shows the correlations obtained in this experiment for two shower size groups. In columns 2 and 3 the density $\Delta_\mu (r_\mu = 25$ m) of low energy muons ($>$ 1 GeV) is shown for two shower size groups and in columns 4 and 5 the ratio α of the number of showers $N(s > 1.2)$ with age parameter $s > 1.2$ (flat showers) to the number $N(s < 1.2)$ with $s < 1.2$ (steep showers) are shown for two shower size groups $(1-2) \times 10^4$ and $(2-4) \times 10^4$. In rows 3 and 4 these parameters are shown for showers in which the high energy muon ($E_\mu > 220$ GeV) was recorded at core distances $R_\mu < 20$ m and $R_\mu \geq 20$ m respectively. Row 5 shows the difference in the values of the parameters, row 6 the difference in terms of number of standard deviations and finally the last row shows the probability that this difference is purely due to chance. If the composition was pure, the expected values of D and α are 0. It can be seen that for both shower size groups Δ_μ and α density of low energy muons and the relative fraction of flat showers are both larger in showers in which the high energy muon ($>$ 220 GeV) was recorded at $>$ 20 m, than in those with the high energy muon at $<$ 20 m. The probability that each of the results individually is due to chance is a few per cent. However, the probability that all of them are due to chance is 4.4×10^{-5} since they are all statistically independent measurements. This

Table 6.4 Correlation of the density, Δ_μ, of muons of energy $>$ 1GeV at a core distance of 25 m and the ratio, $\alpha = \frac{N(s>1.2)}{N(s<1.2)}$ of the number of flat showers to the number of steep showers, with the distance, R_μ, of muons of energy $\geq$ 220 GeV.

Parameter	Δ_μ (r_μ= 25m)		$\alpha = \frac{N(s>1.2)}{N(s<1.2)}$	
Shower Size	$(1-2)\times10^4$	$(2-4)\times10^4$	$(1-2)\times10^4$	$(2-4)\times10^4$
$R_\mu < 20$ m	0.02 ± 0.013	0.053 ± 0.02	2.2 ± 0.74	1.64 ± 0.52
$R_\mu \geqslant 20$ m	0.09 ± 0.03	0.115 ± 0.04	6.36 ± 2.7	2.80 ± 0.38
D = Row 4-Row 3	0.07 ± 0.033	0.062 ± 0.045	4.16 ± 2.8	1.16 ± 0.64
D/σ	2.14	1.39	1.49	1.8
Probability	0.016	0.068	0.082	0.036

shows, from the arguments given earlier, that the primaries of showers of size up to a few times 10^4 are a mixture of protons and nuclei. This is the first time that the composition is shown to be mixed from air shower studies.

Lateral Distribution The KGF group [61] studied the lateral distribution of muons of energy $>$ 220 GeV in small showers where it is sensitive to composition as shown above. The densities of muons were obtained from the neon flash tube data. The accuracy of the core distance of the muons was 2.5 m. Their results for four groups of shower size between 10^4 and 10^5 are shown in Fig.6.15 where they are compared with their Monte Carlo simulations using the scaling model for high energy interactions with cross sections rising with energy for three different composition models *viz* all protons, all iron and mixed composition in which the relative abundances of nuclei of mass numbers 1, 4, 14, 32 and 56 are 0.49, 0.27, 0.12, 0.05 and 0.07 respectively. Their interaction model is based on accelerator data from CERN ISR at energies equivalent to a laboratory energy of 2×10^3 GeV, nearly same as the energy per nucleon for an iron initiated shower of the lowest shower size studied. The predictions of the simulations for iron primaries are, therefore, reliable. For protons, the extrapolation is only by a factor of $\sim$ 50. It is seen from the figure that the lateral distribution does not agree with any model. Clearly pure iron composition can be ruled out. The predictions are the closest for mixed composition. The content of heavy nuclei in their model is too small compared to Swordy's compilation of direct measurements. Also, there is evidence that the cross section for jet production increases with energy in this energy region. These jets acquire a large transverse momentum and the resulting muons will be spread out to large distances. Agreement with mixed composition will be better if these two aspects are taken into account.

Variation of Muon Number with Shower Size In the same experiment the Tata group [62, 63] determined the variation of the number of muons, N_μ, of energy $>$ 220 GeV with shower size. They used two different methods to obtain N_μ. In the

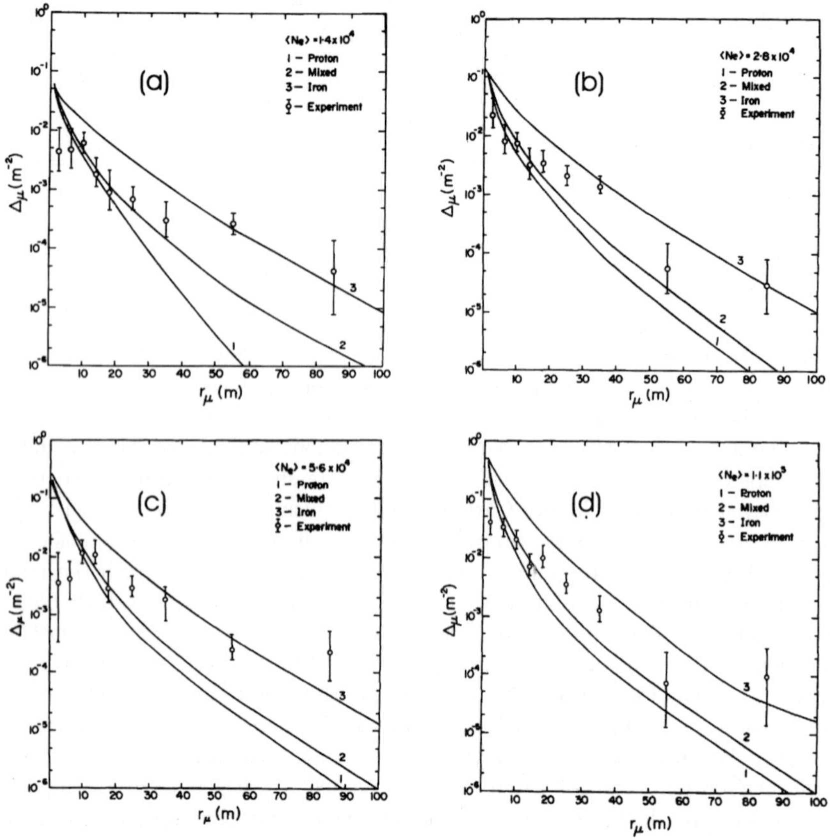

Figure 6.15 Lateral distribution of muons of energy > 220 GeV for various size groups measured by the KGF group (Acharya et al, 1981).

first method, the experimentally determined lateral distribution was integrated and N_μ obtained. This method, however, could not be used for showers of size larger than 10^5 because of limited statistics. In the second method, they used the experimentally determined probability of recording a muon in their underground detector of area 4 m^2 (association probability) in showers selected over an area of the surface air shower array where the efficiency of recording showers is nearly 100%. The association probability, P_μ, is given by

$$P_\mu = \frac{S_U(N_e)\,dN_e}{S(N_e)\,dN_e}, \tag{6.14}$$

with

$$S(N_e)\,dN_e = F(N_e)\,dN_e \int_{A_1}\int_{\Omega} \cos^{n+1}\theta dA_1 dA_2,$$

$$S_U(N_e)\,dN_e = F(N_e)\,dN_e \int_{A_1}\int_{A_2} \frac{\cos^{n+2}\theta}{l^2}\left(1 - e^{-\Delta_\mu(r_\mu)a\cos\theta}\right) dA_1 dA_2$$

and

$$\Delta_\mu = \frac{N_\mu}{2\pi r_0^2} e^{-r_\mu/r_0}$$

where $S(N_e)$ and $S_U(N_e)$ are, respectively, the rates of all showers of size N_e and those in which at least one muon is recorded, $F(N_e)$ is the shower size spectrum, $n = 7$ is the exponent of the angular distribution which is determined from a sample of showers for which fast timing information was available, A_1 is the area on the surface over which showers are accepted, A_2 the infinite horizontal plane through the muon underground muon detector, l the distance between the two area elements A_1 and A_2, Ω the solid angle, a the area of the underground muon detector, $\Delta_\mu(r_\mu)$ is the muon density at a core distance of r_μ and r_0 is the characteristic length of the lateral distribution determined to be 10 m from the neon flash tube data and assumed to be the same for higher sizes also. The probability, P_μ depends only on N_μ, r_0 and the surface-underground detector geometry. It is evaluated numerically as a function of N_μ and r_0 and found to be independent of r_0 in the range 6–16 m for $N_\mu < 300$. N_μ is then obtained as a function of N_e from the measured values of P_μ. Their result is shown in Fig. 6.16 where it is compared with the predictions of their Monte Carlo simulations for the three different models of primary composition. In the mixed composition used here, the relative abundances of nuclei with mass numbers 1, 4, 14, 32 and 56 are 0.44, 0.20, 0.11, 0.10 and 0.15 respectively—somewhat richer in heavy nuclei than that used in the case of lateral distribution—and is obtained using the proton and Helium spectra of the JACEE experiment [64] and the spectra of Simon et al. [65] for other nuclei.The interaction model used here is the same scaling model with increasing cross sections mentioned earlier. Obviously the experimental data cannot be fitted with a single power law over the entire shower size range. N_μ increases with N_e up to $N_e = 4 \times 10^5$, remains practically independent of the N_e between $4{\times}10^5$ and $2{\times}10^6$ and increases again beyond $N_e = 2 \times 10^6$. The best fit to data is obtained

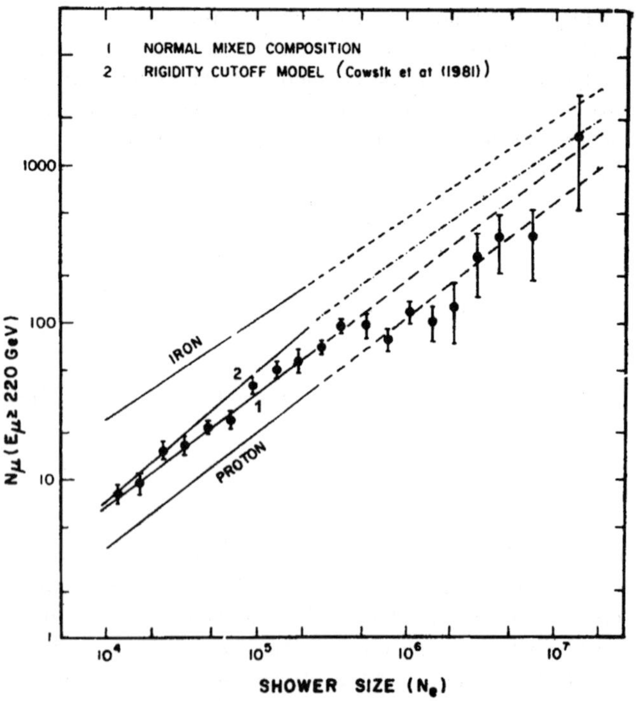

Figure 6.16 Variation of the number of muons of energy > 220 GeV with shower size at 920 g cm^{-2}, compared with Monte Carlo calculations for various models of primary composition using an interaction model with inelastic cross sections increasing with energy (Acharya et al., 1983).

with two lines of the form

$$N_\mu = K\left(N_e/10^4\right)^\alpha, \tag{6.15}$$

with

$$\begin{aligned} K &= 7.4 \pm 0.6, \quad \alpha = 0.70 \pm 0.03 \quad \text{for} \quad 10^4 \leq N_e \leq 5 \times 10^5 \\ K &= 2.7 \pm 1.9, \quad \alpha = 0.79 \pm 0.11 \quad \text{for} \quad N_e \geq 10^6 \end{aligned}$$

Curve 1 in Fig. 6.16 is for the mixed composition independent of energy. Curve 2 is for the composition model of Cowsik et al. [66] in which all components steepen at a critical rigidity of 10^5 GV and a second component of pure protons comes in at 10^9 GeV. Obviously both these models do not agree with the data. They draw following conclusions from the figure:

1. The composition up to about 10^6 GeV is neither predominantly protons nor predominantly iron. This conclusion is rather strong since the interaction characteristics are fairly well known up to 10^5 eV from accelerators and the range over which extrapolation is needed is not very large.

2. In the shower size region $10^4 - 4 \times 10^5$, the mean mass number is nearly constant and equal to that at lower energies, since the best fit line through the data points is nearly same as that due to normal mixed composition in this region.

3. The mean mass number seems to decrease to almost 1 in the shower size region $4 \times 10^5 - 2 \times 10^6$.

4. Above a shower size of 10^6, the data are consistent with pure proton composition.

Using Eq. 6.11 and the $N_\mu - N_e$ relation obtained from Monte Carlo simulations for proton showers, the mean mass number, $< A >$, is estimated as a function of N_e, which is shown in Fig. 6.17.

Fig. 6.18 shows the shower size spectrum obtained in this experiment (see Sec. 5.2.1) along with the $N_\mu - N_e$ dependence. The shower size spectrum steepens at a shower size of $\sim 2 \times 10^6$ as seen in several other experiments. Notice that the flattening of the $N_\mu - N_e$ curve occurs at a shower size much smaller ($\sim 4 \times 10^5$) than that at which the shower size spectrum steepens. This suggests that the two are not related.

The results discussed above suggest that protons continue to be present right up to the knee in the primary spectrum and beyond, since the mean mass number is nearly constant up to a few times 10^6 GeV and equal to that obtained at lower energies (a few hundred GeV/nucleon) and reduces to nearly 1 beyond. The fact that the flattening of the $N_\mu - N_e$ dependence occurs below the knee in the shower size spectrum, suggests that all nuclei are removed before the knee in the primary energy spectrum. These features are already present in the model proposed by Chatterjee [67, 68]. In this model while the galactic component, consisting of protons and nuclei, is removed due to rigidity cut-off at an energy of a few times 10^5 GeV, an extragalactic proton component with a steeper spectrum starts contributing well before the energy at which iron is removed. In this, as well as other models in which protons are retained beyond the iron cut-off, the mean mass number remains nearly constant even though the detailed composition is continuously changing with energy. However, it might be difficult to understand the observed energy dependence of the anisotropy of arrival directions (see Fig. 7.21) since the second component is extragalactic and is expected to be isotropic. Also, the two proton components have to be matched to give a smooth overall spectrum.

An alternative model suggested by Hillas [69] seems to be more attractive. In this model, all nuclei are removed at nearly the same energy per nucleon due to

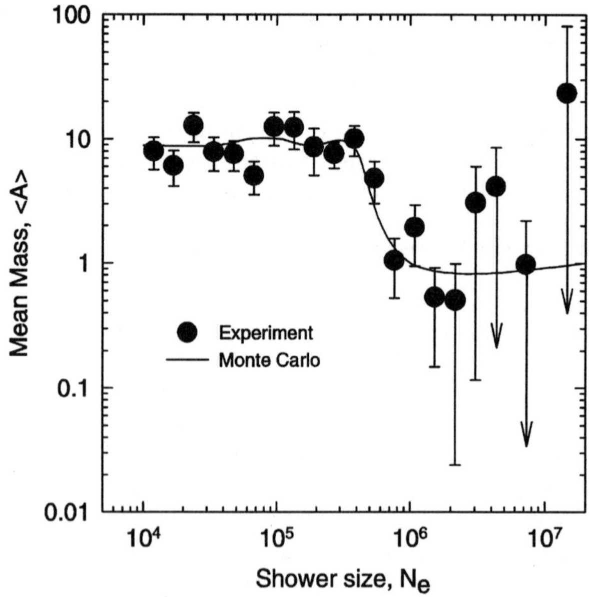

Figure 6.17 Mean mass number as a function of shower size estimated from the KGF data on the measured number of high energy muons and Monte Carlo simulations for proton showers. The curve is expected from Monte Carlo simulations for the KGF composition model.

photodisintegration at the source itself and severe energy losses set in for protons at about a factor of 10 higher in energy due to photo-pion production. This would place the bends in the spectra of protons and iron nuclei at about the same total energy. However, it may be difficult to remove iron earlier than protons, as the $N_\mu - N_e$ dependence together with the bend in the shower size spectrum seems to require. Again, the anisotropy is difficult to understand.

Alternately, one can identify the knee in the shower size spectrum as due to leakage of protons due to rigidity cut-off and the discontinuity in the $N_\mu - N_e$ dependence at $N_e = 5 \times 10^5$ as due to the removal of iron nuclei due to photodisintegration. In this model the Tata results as well as the anisotropy can be understood. The knee is due to the proton spectrum changing slope by 0.5 and the amplitude of anisotropy increases as $E^{0.5}$ as roughly expected from leakage processes [70]. Particles up to 10^9 eV are of galactic origin. The second component needed in the pure rigidity cut-off model and the necessity of matching the two components is done away with in this model. The astrophysical conditions under which this model can be vindicated have

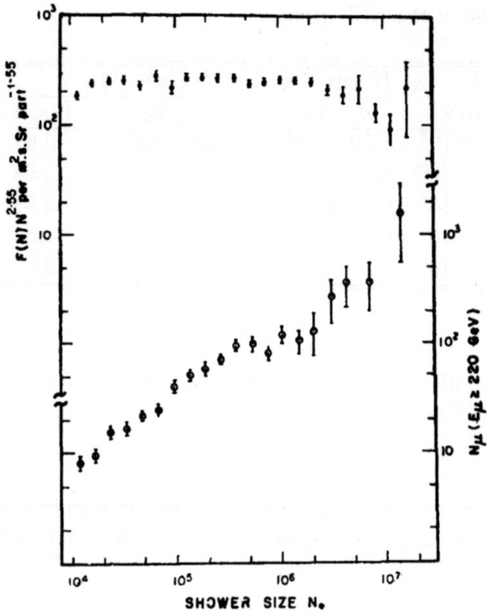

Figure 6.18 The $N_\mu - N_e$ variation of high energy muons and the differential size spectrum at 920 g cm^{-2} obtained by the KGF group (Acharya et al., 1983). Note that the discontinuity in $N_\mu - N_e$ dependence occurs at a smaller size than the knee in the size spectrum.

to be examined.

Using the scaling model with rising cross sections, they [71, 72] constructed the primary spectra for the various components. The differential spectral index of protons changes from 2.68 to 3.18 at 6.7×10^6 GeV. All nuclei have the same spectral index of 2.68 and are completely removed at an energy per nucleon of 7.5×10^4 GeV. The spectra, below a total energy of 3×10^5 GeV, of protons and various nuclei except iron are taken from the JACEE experiment [64, 73]. The JACEE fluxes of CNO and Si group, which are based on meagre statistics, are somewhat higher than extrapolation from lower energy measurements [65]. The flux of iron is estimated as the difference between the all particle flux at 10^5 GeV given by the Tien Shan group [74], which agrees with the Grigorov all particle spectrum [3], and the sum of these fluxes. Table 6.5 shows the resultant spectra of various components and their relative abundances below a total energy of 3×10^5 GeV. E_c is the critical energy at which the spectrum steepens in the case of protons, and in the case of nuclei it is the energy at which

Table 6.5 Values of the constants in the diffrential spectra of variuous components in the composition model of the KGF group.

Mass A	E_c (GeV)	K (m^{-2} s^{-1} sr^{-1} GeV^{-1}) $E < E_c$	$E > E_c$	γ $E < E_c$	$E > E_c$
1	6.7×10^6	1.29×10^4	3.36×10^7	2.68	3.18
4	3.0×10^5	0.76×10^4	–	2.68	–
14	10^6	0.90×10^4	–	2.68	–
26	2×10^6	0.35×10^4	–	2.68	–
56	4×10^6	0.57×10^4	–	2.68	–

they are removed due to photodisintegration. Fig. 6.19 shows the spectra of various components and the all particle spectrum according to this model with idealized abrupt cut-offs. The Tien Shan fluxes above 3×10^5 GeV are higher than the model

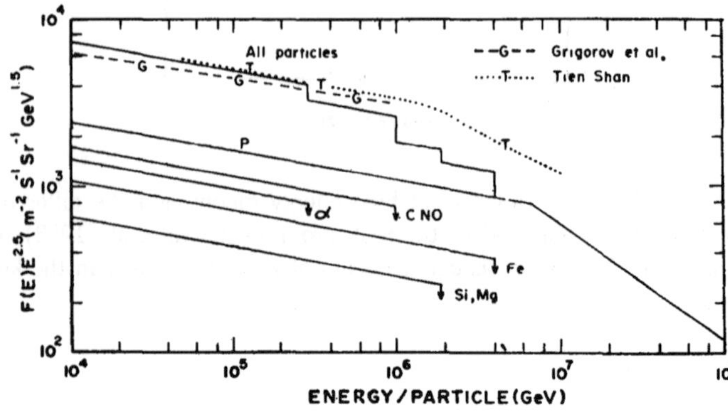

Figure 6.19 Energy spectra of protons and nuclei according to the model of the KGF group (Acharya et al., 1983) in which protons are lost due to leakage from the galaxy and nuclei due to photo-disintegration in the sources.

because they use nearly constant composition over the entire energy region. If the changing composition predicted by the model is used, their fluxes would come down since the shower size to primary energy conversion factor at a given atmospheric depth increases with mass number, and the agreement will be better. The mean mass number and percentage of iron in this model are shown in Fig. 6.20a as a function of primary energy. The mean mass number changes only from 15 to 22 in the energy region where composition is changing and becomes 1 beyond. The percentage of iron increases from 15 at 10^5 GeV to 32 at 4×10^6 GeV and becomes

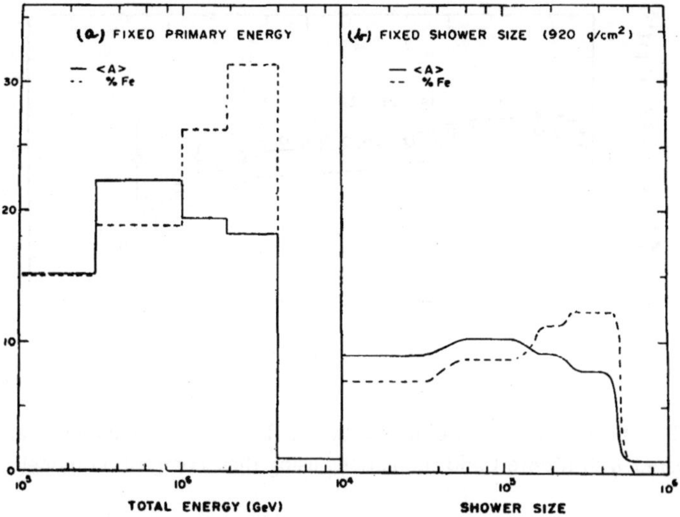

Figure 6.20 Mean mass number $\langle A\rangle$ and iron abundance as a function of (a) total primary energy and (b) shower size at 920 g cm^{-2} according to the photo-disintegration model of the KGF group (Rao 1986).

zero beyond. The same quantities are plotted in Fig. 6.20b as a function of shower size at KGF (920 g cm^{-2}). The mean mass number is nearly independent of shower size up to 5×10^5 where it decreases from about 9 to 1 rather rapidly. The rapid decrease is due to the small fluctuations in showers due to iron primaries which are abruptly removed. The percentage of iron steadily increases with shower size and becomes zero beyond 5×10^5. The expected mean mass as a function of shower size is also shown in Fig. 6.17. It can be seen that it agrees very well with data.

As mentioned above, the variation of the number of muons of energy > 220 GeV with shower size at 920 g cm^{-2} agrees with the mixed composition for sizes less than 4×10^5. In this case the mean mass number was 6.7 for fixed shower size. The present model, in which the abundances of nuclei are somewhat higher, gives a mean mass of 9 and the muon numbers are about 10% larger, but still in agreement with data. Above a shower size of 10^6 also, the model agrees with the data since only protons are present in this region. Fig. 6.21 shows the shower size spectrum calculated form this model compared with the KGF data. The left most three data points have been corrected for detection efficiency of flat showers and shown as filled circles. Curve (a) is for the composition used in calculating the $N_\mu - N_e$ dependence in Fig. 6.16 mentioned above, but with cut-off energies the same as in the present model. The

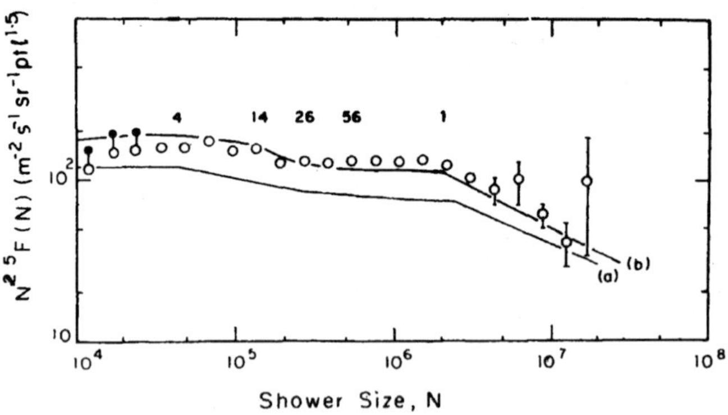

Figure 6.21 The size spectrum obtained by the Tata group (Acharya et al, 1985) at KGF compared with Monte Carlo simulations using their compsotion model.

spectra are somewhat steep in this case. Curve (b) is for the present model. The excellent agreement between the present model and the data is obvious. Thus the model is in very good agreement with the KGF data.

6.3.1.3 The Moscow Experiment

The Moscow State University group [75, 76] determined the various features of high energy muons in EAS with an underground magnetic spectrograph, shown in Fig. 3.10, operated along with an air shower array at sea level. Their result on the variation of muons of energy > 200 GeV is shown in Fig. 6.22, where the KGF result is also plotted. The shower sizes at KGF level were converted to sea level sizes by the Moscow group using an absorption mean free path of 180 g cm^{-2}. While the first two and the last points of the Moscow group are in good agreement with the KGF data, their number in the shower size region where the KGF data shows no dependence on shower size ($2 \times 10^5 - 10^6$) is larger. Notice that the Moscow data is classified into much broader shower size intervals than the KGF data presumably because of limited statistics. Nevertheless, the Moscow point differs only by about 2 standard deviations from the average value of the KGF data in that region. Thus the two results are not inconsistent with each other.

6.3.1.4 The EASTOP-MACRO Experiment

The EASTOP-MACRO collaboration also studied high energy muons to determine the composition with the EAS-TOP air shower array operated with the underground MACRO detector in Gran Sasso, described in Sec. 3.2.1.3. They determined the

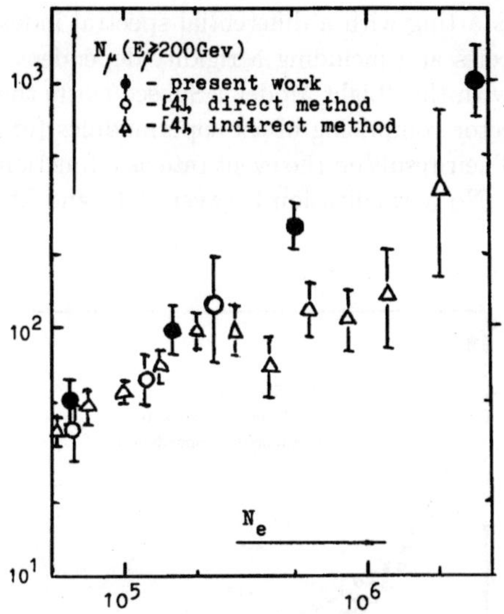

Figure 6.22 Variation of number of muons of energy > 200 GeV with shower size at sea level from Bazhutov et al (1985), compared with KGF results. Points marked 'present work' are from Bazhutov et al and 'direct method' and 'indirect method' are two different methods used by the KGF group to derive the number of muons.

composition using the underground data alone as well as data from the surface and underground detectors operated together *a la* the KGF experiment. The underground data are not as sensitive to composition as the surface-underground data, particularly when one wants to study composition as a function of primary energy, since the former integrates information over a much wider range of primary energies than the latter.

Ahlen et al. [77] used the underground data alone, collected with one (1SM) and two (2SM) supermodules over 1013 and 1195 hrs of live time respectively, to estimate the composition. Their multiplicity distribution of the true number of muons is obtained from the observed multiplicities in the two projected views and compared with expectation from Monte Carlo simulations. The parameterization of the features of underground muons given by Forti et al. [78] from their Monte Carlo simulations, using an interaction model which incorporates results from the CERN proton-antiproton collider (up to $\sqrt{s} = 0.9$ TeV) and nucleus-nucleus interactions from fixed target experiments, is used to calculate the expected rates. They consider two different composition models. The "Light Model" is a low energy composition (LEC) model starting with a differential spectral index of 2.71 for all the species, with a flux adjusted addition of a pure proton component above 20 TeV and a fixed energy cutoff at 3,000 TeV. The other is the "Heavy Model" which is an extrapolation of the Maryland model [79] starting with a differential spectral index of 2.36 for iron and 2.71 for the other species and including a rigidity-dependent cut-off at $100Z$ TeV. They find agreement with the "Light Model". Subsequently they collected data with the full MACRO detector comprising of six supermodules (6SM) and carried out a similar analysis [80]. Their result on the event rate as a function of muon multiplicity is shown in Fig. 6.23. No normalization between data and Monte Carlo simulation

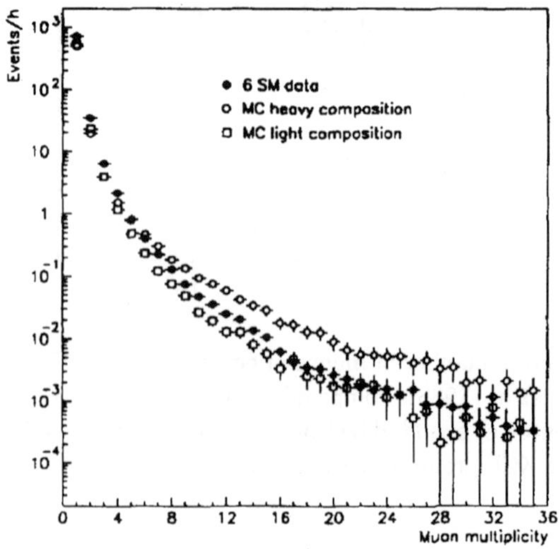

Figure 6.23 Comparison of event rate as a function of muon multiplicity with Monte Carlo simulations from the MACRO experiment (The MACRO Collaboration, 1993).

has been applied. Notice that multiplicities extend up to 35 whereas the highest multiplicities in the 1SM and 2SM samples are about 10 and 20 respectively. Even up to these high multiplicities the light composition model is in very good agreement whereas the Maryland composition predicts consistently large rates at multiplicities greater than 10, where bulk of the contribution comes from heavy nuclei. The primary energy range over which the experiment is sensitive is obtained from Monte Carlo simulations and is shown in Fig. 6.24. The central bold line in the figure gives the mean primary energy whereas the shaded region corresponds to detection of 90% of events. Notice that at low multiplicities the experiment is sensitive over the primary energy range of direct measurements. At lower multiplicities up to about 15, the observed rates are somewhat higher than predictions of the light composition model. This is probably due to the steep spectra with an assumed slope of 2.71.

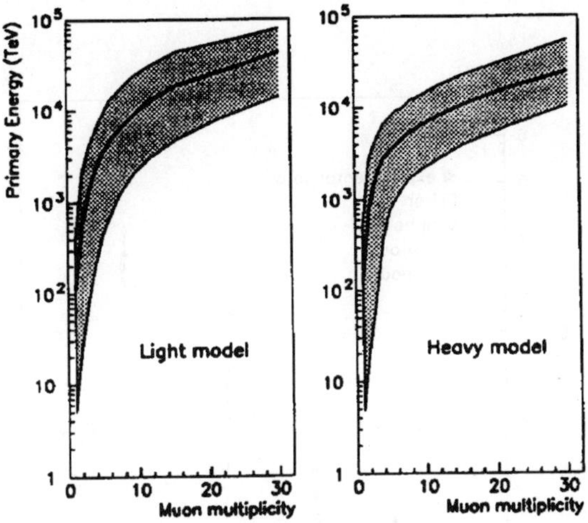

Figure 6.24 Calculated range of primary energy, over which the MACRO experiment is sensitive, as a function of muon multiplicity (The MACRO Collaboration, 1993).

Agreement will be better if the slope is taken to be 2.66–2.68, which is well within the range of errors of direct measurements, as in the model used by the KGF group.

The EAS-TOP Collaboration, Aglietta et al., [81] use the data obtained from the underground MACRO detector in coincidence with the EAS-TOP air shower array on the surface (see Sec. 3.2.1.3) to determine the composition in the energy range 10^{14}–10^{16} eV. Shower size, core location and arrival direction are obtained from the EAS-TOP array and the muon information comes from the MACRO detector. The two experiments are triggered separately and event coincidence, within $\pm$ 2 μs of the coincidence peak, is established off-line using the absolute time given by atomic clocks with an accuracy better than 1 μs. The average number of muons and the muon number distribution are obtained form the data. The dependence of the average number of muons on shower size is shown in Fig. 6.25. Monte Carlo predictions are also plotted for different primary nuclei and a mixed composition model, which they call Σ model, consisting of extrapolated proton and He spectra from JACEE data and heavier nuclei spectra from the CRN data. The integral fluxes and the exponents of the spectra of various nuclei of this model are shown in Table 6.6. The nuclear interaction model is the same as the one used for interpreting the underground MACRO data alone. It can be seen form the figure that the data in the first two shower size bins is not much sensitive to primary mass, presumably because showers from heavier nuclei are not efficient enough to produce TeV energy muons as their primary energy/nucleon is not high enough. In the shower size range $10^{5.25} - 10^{5.75}$,

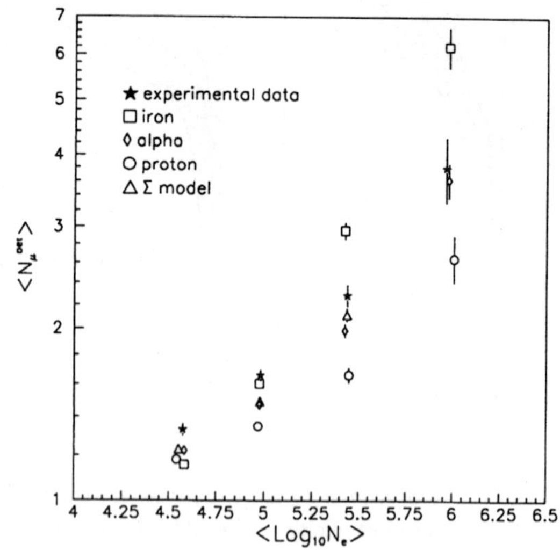

Figure 6.25 Variation of average number of muons, $\left\langle N_\mu^{\mathrm{Det}}\right\rangle$ with logarithm of shower size, $\langle\log_{10} N_e\rangle$, compared with Monte Carlo simulations from the EASTOP-MACRO collaboration (Aglietta et al., 1994).

Table 6.6 Integral flux at 100 TeV (m^{-2} s^{-1} sr^{-1}) and spectral indices of various nuclei in the model of the EASTOP-MACRO Collaboration.

Mass group	Integral Flux	Exponent
p	$(1.33 \pm 0.31) \times 10^{-5}$	1.86 ± 0.07
He	$(2.28 \pm 0.44) \times 10^{-5}$	1.72 ± 0.09
CNO	$(7.09 \pm 0.48) \times 10^{-6}$	1.67 ± 0.06
Mg-Si	$(2.48 \pm 0.15) \times 10^{-6}$	1.82 ± 0.11
Fe	$(6.94 \pm 0.63) \times 10^{-6}$	1.60 ± 0.11

which is just below the knee, the experimental point is inconsistent with either pure proton or pure iron composition but consistent with the predictions of the Σ model as well as pure Helium composition. However, the muon multiplicity distribution for this shower size group, shown in Fig. 6.26, does not agree with that expected for pure He composition. The Σ model predicts a lower tail in the high multiplicities than the

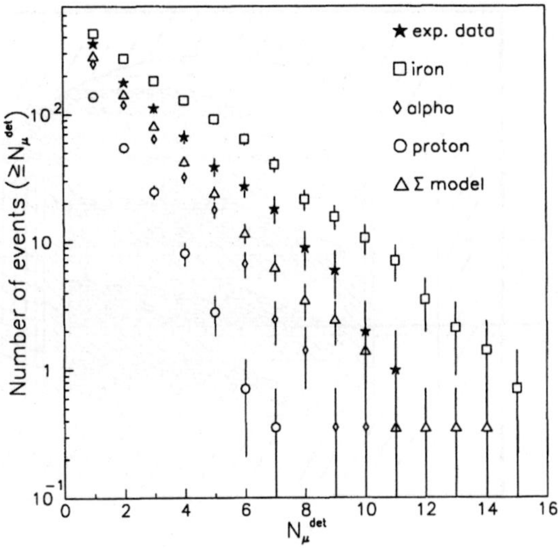

Figure 6.26 Measured and expected integral distribution of detected muon multiplicities, for $10^{5.25} \leq N_e \leq 10^{5.75}$, from the EASTOP-MACRO Collaboration (Aglietta et al., 1994).

experimental data. The authors opine that the need to increase the amount of heavy primaries is not compelling if the systematic uncertainty in the extrapolated fluxes is allowed to be as large as 30%. They note that the same uncertainty can also explain the event rates.

At $N_e > 10^{5.75}$ also, *i.e.* just above the knee, where only pure compositions are considered, pure proton or pure iron compositions are incompatible with data. An intermediate mass composition is again indicated.

Note that the shower size intervals taken in this analysis are very large (two per decade) presumably because of limited statistics. In order to see detailed features of the variation of the composition with primary energy, as in the case of the KGF experiment, the shower size intervals have to be finer. This is perhaps the reason why the detailed features of $N_\mu - N_e$ dependence seen in the KGF experiment are not yet seen in this experiment.

6.3.1.5 The Baksan Experiment

The Baksan group [82] studied high energy muons with a liquid scintillator telescope in a cave under a mountain slope. Though there is no air shower array in this experiment, the primary energies involved in their study is in the air shower region. The telescope is shown in Fig. 6.27. It consists of more than 3,132 scintillators, each

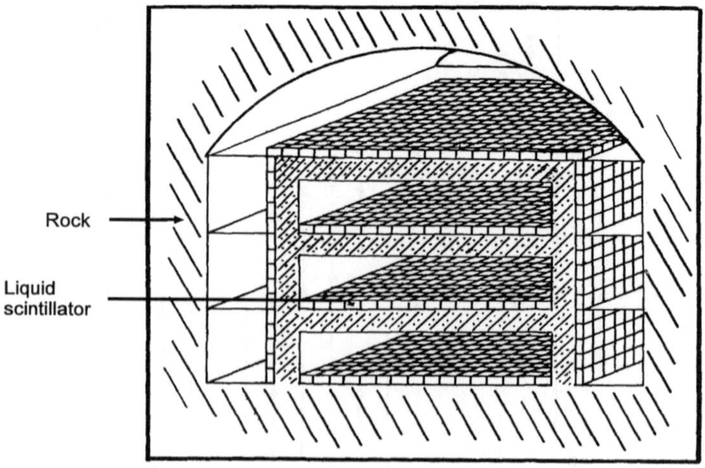

Figure 6.27 The liquid scintillator telescope under a mountain slope operated by the Baksan group (Chudakov et al, 1979).

of dimensions $70 \times 70 \times 30$ cm^3, arranged in four vertical and four horizontal layers in a nearly cubic shape of 16×16 m^2 area and 11 m height. The minimum overburden of rock corresponding to 30° zenith angle is 850 hg cm^{-2}. The experiment records the number of muons and the energy deposited by each of them in the scintillators. According to the superposition model the energy spectrum of muons in air showers depends essentially on the energy per nucleon of the primary. For a given energy per nucleon, the muon multiplicity is directly proportional to the primary mass number. Thus, by measuring the energy spectrum and multiplicity of muons one can determine the mean primary mass as a function of primary energy per nucleon.

The energy per nucleon is estimated in several steps [83]. The muon energy distribution responsible for an energy deposit of E_c in the detector is

$$P_1(\epsilon_\mu, E_c) = P(\epsilon_\mu)\, W_\mu(\epsilon_\mu, E_c) \tag{6.16}$$

where $P(\epsilon_\mu) = (0.22 + \epsilon_\mu)^{-3.7}$ is the differential energy spectrum of muons at that depth and $W_\mu(\epsilon_\mu, E_c)$ is the probability of energy deposit E_c by a muon of energy

ϵ_μ taking into account bremsstrahlung, knock-on, photonuclear interaction and e^+e^- production processes. Taking into account muon energy losses in traversing the overburden $x = 910$ hg cm^{-2} (hg stands for hectograms and is equal to 100 g) given by the expression

$$E_\mu = \epsilon_\mu e^{bx} + \frac{ae^{bx} - 1}{b} \tag{6.17}$$

with $a = 2$ MeV g^{-1} cm^2 and $b = 3.95 \times 10^{-6}$ g^{-1} cm^2, the muon energy spectrum at the surface $P_2(E_\mu, E_c)$ is then obtained.

The energy distribution of the primary nucleons is then given by

$$P_3(E_0, E_c) = \int P_2(E, E_c) \frac{d\overline{n}(E_0, E)}{dE} E_0^{-\gamma} dE \tag{6.18}$$

where E_0 is the primary energy per nucleon and $\overline{n}(E_0, E)$ the average number of muons with energy $> E$ generated by nucleons of energy E_0 in a primary nucleus of mass number A. The function $\overline{n}(E_0, E)$ is obtained from Monte Carlo simulations [84] using an interaction model incorporating scaling violation and cross sections rising with energy. It is given by

$$\overline{n}(E_0, E, \theta) = \frac{1.87 \times 10^{-2} A x(\theta) E_0}{E^{0.9+0.1 \log E}\ E} \left(\frac{E_0}{E_0 + E}\right)^d \tag{6.19}$$

where $d = 11.3/\log(10 + 0.5E_0)$, $x(\theta) = (1 + 0.36 \ln\cos\theta)/\cos\theta$ and θ the zenith angle. To simplify the calculations, they use the average values of threshold energy E_t of 0.22 TeV and zenith angle 35°. The final result is approximately described by

$$P_3(E_0, E_c) = \left(1 - \frac{C}{E_0}\right)^{1.7} \left(\frac{E_0}{C}\right)^{-\gamma+0.77} \left[1 + \left(\frac{E_0}{3 \times 10^3}\right)^2\right]^{-0.2} \tag{6.20}$$

where $C = 2.57E_c + 0.4$, with E_0 and E in TeV. The last term takes care of the knee in the primary energy spectrum.

Effective primary energy E_0^{eff} for a given E_c is determined taking into account the increase of multiplicity with E_0. First the average multiplicity $\langle \overline{n}(E_c) \rangle$ given by

$$\langle \overline{n}(E_c) \rangle = \frac{\int \overline{n}(E_0, E_t) P_3(E_0, E_c) dE_0}{\int P_3(E_0, E_c) dE_0}$$

is calculated using Eqs. 6.19 and 6.20. Then E_0^{eff} is obtained as solution of Eq. 6.19 for E_0.

The observed muon multiplicities are corrected for loss of muons due to the finite size of the detector and for muons lost in the core of the cascade generated in the detector. Both depend on the lateral distribution, which is taken to be of the form

$$\begin{aligned} \rho_\mu(r, E_\mu, \theta) &= E_\mu^{-0.7} e^{-\left(\frac{r}{r_\mu}\right)^{0.73}}, \\ r_\mu &= \frac{0.22 r_0}{E_\mu \cos\theta} \end{aligned}$$

where $\rho_\mu(r, E_\mu, \theta)$ is the density of muons of energy E_μ at a core distance r, θ the zenith angle and $r_0 = 8.3$ m. $\cos\theta = 0.82$, which gives a good fit to the observed decoherence curve.

The observed mean muon multiplicity as a function of E_0 is shown in Fig. 6.28, where it is compared with the expected from Eq. 6.19. They define the mean mass

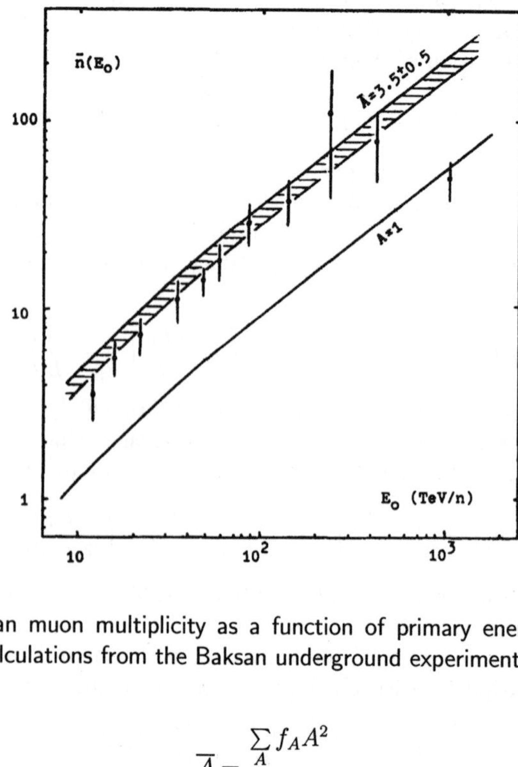

Figure 6.28 Mean muon multiplicity as a function of primary energy per nucleon compared with calculations from the Baksan underground experiment.

number $\overline{A}$ as

$$\overline{A} = \frac{\sum\limits_A f_A A^2}{\sum\limits_A f_A A}$$

with f_A =0.939, 0.055, 0.0009, 0.0035, 0.0011 and 0.0003 are the relative abundances of nuclei with mass numbers 1, 4, 9, 14, 28 and 56 respectively for a fixed energy per nucleon. This composition, which gives $\overline{A} = 3.5 \pm 0.5$, is from direct measurements at around 1 GeV/nucleon. The authors conclude that there is no visible change in the composition in the energy range 1–10^6 GeV/nucleon. The authors believe that the two events in the last energy bin are induced by protons. They estimate that the probability of this happening for the normal composition is about 0.5.

The numbers of events in the last three energy bins are 17, 6 and 2 respectively. These three bins cover the energy range > 200 TeV/nucleon, where several experiments have shown change of composition. Since more than 90% of the primaries at a fixed energy per nucleon are protons, sampling of nuclei may not be adequate with such small number of events. More statistics are needed to draw firm conclusions.

The Baksan group also compared the observed multiplicity distribution for a given E_c with calculations [85]. The calculated multiplicity distribution $M(n)$ for mixed composition is given by

$$M(n) = \sum_A f_A A M_A(n)$$

where

$$M_A(n) = \int B_A(n, \overline{n}, k) P(E_0, E_c) dE_c.$$

Their results are shown in Fig. 6.29. The curve marked NC is for the normal compo-

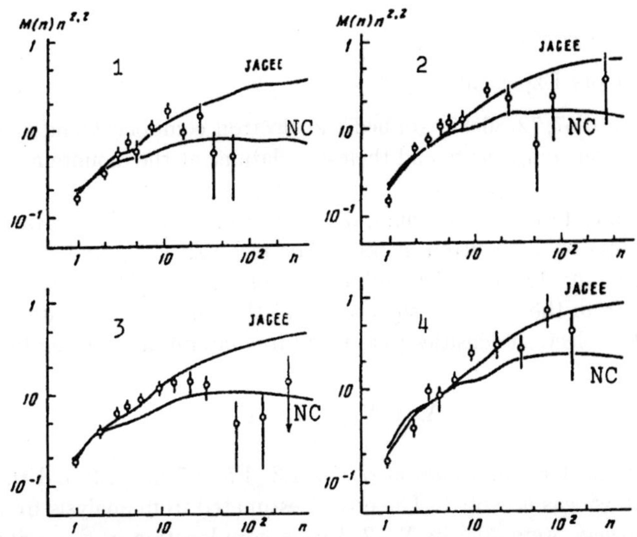

Figure 6.29 Multiplicity distributions of muons accopanying cascades of given energy E_c from the Baksan experiment: 1 - 0.95 TeV; 2 - 1.29 TeV; 3 - 2.10 TeV; 4 - 3.32 TeV.

sition discussed above. The curve marked JACEE is for the composition suggested by the JACEE experiment. In this case, the relative abundances are 0.8823, 0.1048, 0.00774, 0.00454 and 0.00058 for mass numbers 1, 4, 14, 28 and 56 respectively and the proton spectrum has a knee at $E_0 = 40$ TeV with $\gamma = 2.7$ before the knee and 3.1 after the knee. A χ^2-test was carried out for both the composition models, the results of which are shown in Table 6.7. The authors comment that multiplicities greater than 10 are likely to be in error because of the poor spatial resolution of 0.7

Table 6.7 Results of χ^2-test for the two composition models, JACEE and NC, from the Baksan experiment (Bakatanov et al, 1995). The last two columns show the probabilities in %.

ΔE_c (TeV)	E_c (TeV)	E_0 (TeV)	N	JACEE	NC
0.80–0.96	0.95	44	835	32.0	0.001
0.96–1.66	1.29	54	1119	30.7	0.11
1.66–2.39	2.10	77	456	4.7	0.26
2.39–4.95	3.32	113	257	49.0	7.8

m of the detector. They conclude that data do not rule out the heavy composition indicated by the JACEE experiment.

6.3.2 Low Energy Muons

6.3.2.1 The Ooty Experiment

The Tata group [67, 72, 86–88] studied the electron component and the low energy hadron and muon components and their correlations at the mountain station, Ootacamund (800 g cm^{-2}). The array consisted of 12 scintillation detectors spread over a circle of radius 40 m, 4 fast timing detectors to measure the arrival directions, 9 hadron detectors of the neutron monitor type (see Fig. 3.11) and 3 muon detectors each of area 0.6 m^2, two of which had a threshold energy of 0.6 GeV and located at 18 and 25 m from the center of the array and the third 1GeV located at 8 m from the center. The electron densities were fitted to a lateral distribution function of the form

$$\Delta_e \propto N_e \left(\frac{r}{r_0}\right)^{-\alpha_e} e^{-\frac{r}{r_0}}, \tag{6.21}$$

with $r_0 = 107$ m, for four values of α_e viz 1.3, 1.5, 1.7 and 1.9 and the value that gave the least χ^2 was accepted. The errors, estimated from analysis from artificially generated showers, were 40% in N_e, 2–3 m in core location in the central region of the array and 5–10 m at about 40 m from the center, and 5° in the shower arrival direction.

The following results were obtained from this experiment

1. In the shower size range 10^5 to 10^6, the lateral structure of hadrons is positively correlated with the lateral structure of electrons; i.e. showers with steep electron lateral structure (young showers) exhibit steep structure for hadrons those with flat electron structure (old showers) depict flat hadron lateral structure. Beyond 10^6, this correlation, if any, is small.

2. Fluctuations in all the three components decrease beyond a shower size of 10^6.

3. To study the variation of number of muons with shower size, the muon density Δ_μ at a fixed distance from the core was chosen rather than the total number because of the difficulties involved in its determination as discussed in Sec. 6.3. To minimize errors in muon densities due to errors in core location, showers with zenith angles $< 30°$ falling within 25 m from the center of the array only, where the core location errors were $\backsim$ 5 m, were selected. Muons in the core distance range 15–25 m were used to obtain Δ_μ at 20 m. The variation of Δ_μ with N_e for muons of the two threshold energies is shown in Fig. 6.30. The logarithmic

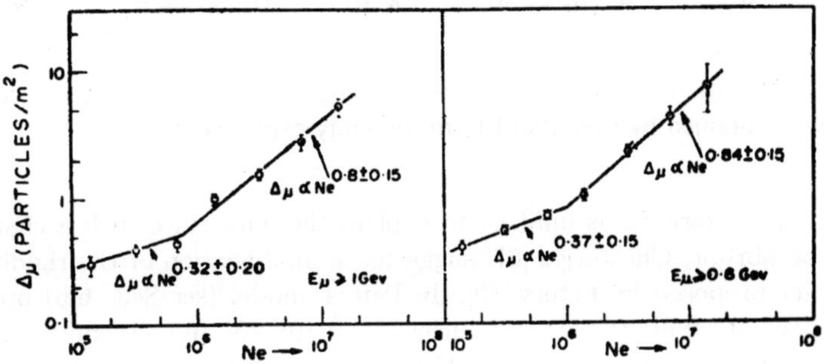

Figure 6.30 Variation of Δ_μ $(r = 20)$ with N_e for muons of the two threshold energies 0.6 GeV and 1 GeV obtained by the Ooty group (Rao, 1986).

slope of the dependence for $N_e > 10^6$ is about 0.8 as expected for a pure or constant composition (independent of primary energy). The slope at $N_e < 10^6$, however, is smaller. Densities in this region are larger than extrapolation from higher sizes. This feature is seen for muons with the two different threshold energies (0.6 GeV and 1 GeV) using different detectors.

4. The $\Delta_\mu - N_e$ dependence was also studied for flat and steep showers separately. Fig. 6.31 shows this dependence for muons of energy > 1 GeV. It is seen that the dependence for flat showers is similar to that for all showers, while in steep showers there is no change of slope. Data for steep showers can be fitted with a single slope of 0.75 through out the entire region, the slope being same, within errors, as that for flat showers for $N_e > 10^6$. Furthermore, while the muon densities in flat showers are nearly same as those in steep showers for $N_e > 10^6$, they are larger for $N_e < 10^6$. Muons of energy > 0.6 GeV also show a similar behavior.

Obviously the data cannot be understood if hadronic interactions and primary composition do not change in the relevant energy region. Arguing that change in

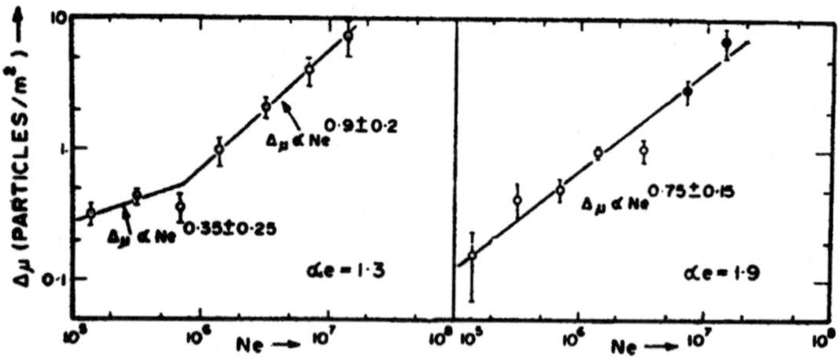

Figure 6.31 Variation of density of muons of energy > 1 GeV, Δ_μ, with N_e for flat and steep showers obtained by Rao(1966) from the Ooty experiment.

interaction characteristics is unlikely to explain the data since such a change is required to be abrupt, Chatterjee [67] suggested a modification of the rigidity cut-off model earlier proposed by Peters [89]. In Peters' model (see Sec. 6.6) protons undergo rigidity cut-off at 10^{15} eV and nuclei at 2×10^{15} eV and a second source starts contributing from an energy of 10^{16} eV with an integral slope of -1.6. Chatterjee suggested that the cut-off for protons is at 3×10^{14} eV and the second source, which is identified to be extragalactic, starts contributing from 2×10^{15} eV with an integral slope of -2.1. A schematic of the energy spectra of the different components is shown in Fig. 6.32. Notice that the extragalactic protons start at an energy less than that at which iron is cut off. The knee is due to protons. Beyond the iron cut-off energy, all are protons. According to this model, the composition is changing with N_e at $N_e < 10^6$, the mean mass remaining nearly constant, and beyond this shower size all are protons and thus the logarithmic slope of the $\Delta_\mu - N_e$ dependence, which is about 0.8, agrees with the expectation for protons. This model also explained most of the air shower data reported at that time as well as the steady state spectra of hadrons and γ rays observed at various altitudes in the atmosphere.

6.3.2.2 The Moscow Experiment

The Moscow group [90] determined the lateral distribution of muons of energy $>$ 10 GeV with an underground muon detector of area 45.4 m^2, shown in Fig. 3.10, operated along with an air shower array, at sea level, consisting of a large number of GM counter trays, each having several counters. Shower size, core location and zenith angle of each shower were determined. Lateral distribution of muons was determined, in the core distance range 15–200 m, for two broad groups of showers ($(1-10) \times 10^5$ and $(1-5) \times 10^7$) with zenith angles less than 30°. Densities in individual showers were normalized assuming $\Delta_\mu \sim N_e^{0.78}$. Using this lateral distribution, total number

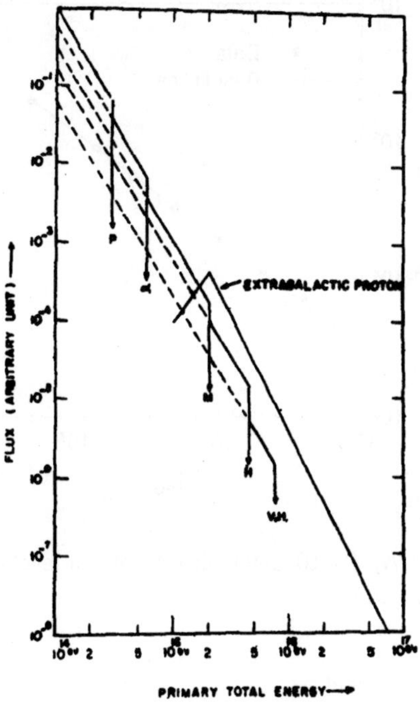

Figure 6.32 Schematic of the primary energy spectra of the various components in Chatterjee's model (Chatterjee, 1964)

of muons was obtained for each shower. The average number of muons N_μ thus obtained is shown in Fig. 6.33 as function of N_e. The data can be represented by the relation

$$N_\mu = (3.24 \pm 0.22)\, 10^3 \left(\frac{N_e}{10^5}\right)^{0.78\pm0.01}, \tag{6.22}$$

which is also shown in the figure as a straight line. There is no evidence for change of slope and the data are consistent with composition independent of primary energy.

Based on their high energy muon data the KGF group (see Sec. 6.3.1.2) suggested that the composition above $N_e = 5 \times 10^5$ at 920 g cm^2 is all protons and below this shower size the mean mass is larger than 1. This shower size corresponds to 3×10^5 at sea level, taking the shower attenuation mean free path to be 200 g cm^2. In the Moscow group's data there are only two points below this shower size. While these two points are somewhat larger than expected from the best fit line, the statistical significance is small. The Moscow data is, thus, not inconsistent with the KGF data.

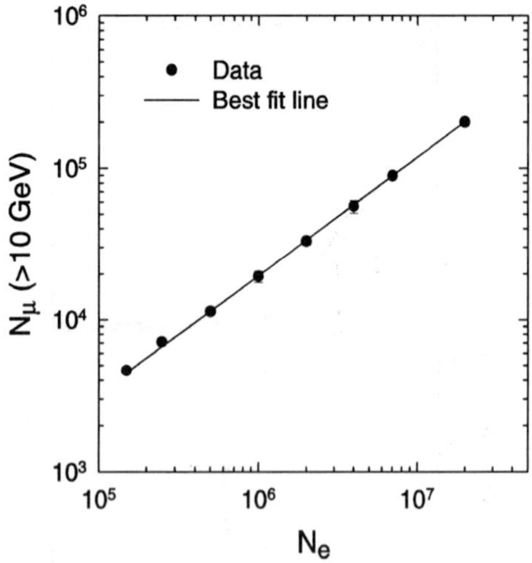

Figure 6.33 Variation of N_μ (> 10 GeV) with N_e at sea level obtained by the Moscow group (Khristiansen et al, 1971).

6.3.2.3 The Akeno Experiment

The Akeno group [91] studied muons of energy > 1 GeV with the air shower array at an atmospheric depth of 930 g cm^2 described in Sec. 5.2.1 and shown in Fig. 5.4. Shower size , core location and zenith angle are obtained from the scintillator array and muon densities from 9 muon detectors each of area 25 m^2. N_μ is obtained by fitting the muon densities to the Greisen function (Eq. 2.50) but the characteristic length, instead of being 320 m, is made variable. The dependence of N_μ on N_e for vertical showers is represented by

$$N_\mu = (2.75 \pm 0.15)\, 10^5 \left(\frac{N_e}{10^7}\right)^{0.83\pm0.01} \qquad \text{for } 5\times10^5 < N_e < 10^9 \tag{6.23}$$

There is no evidence for change of slope and therefore no change of composition in this shower size region.

Since Akeno and KGF are located at nearly same altitude, features of composition inferred from the two experiments can be directly compared in terms of shower size. KGF data showed that the change of composition occurs at sizes below 5×10^5, the smallest shower size at which N_μ is measured in the Akeno experiment. According to the KGF group the composition above this shower size is all protons and the Akeno data are consistent with this interpretation.

6.3.2.4 The Tien Shan Experiment

The Tien Shan group [92, 93] measured muons of energy > 5 GeV with their air shower array located at an altitude of 700 g cm^2. The array, shown in Fig. 6.34 consists of 64 scintillators (each of area 0.25 m^2), which are regularly spaced over an

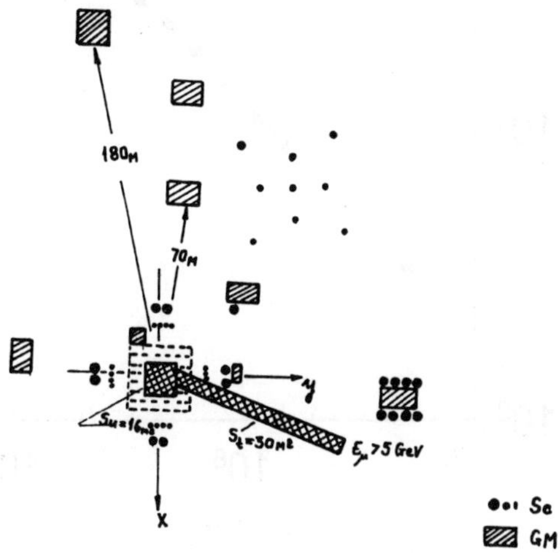

Figure 6.34 The air shower array operated by the Tien Shan group (Kirov et al, 1980) at an altitude of 700 g cm^2: Sc - Scintillators of different areas; GM - GM counter group.

area of 110 m^2 at the center of the array and groups of GM counters located at 20, 38, 73 and 180 m from the center to measure N_e and core location. Arrival directions were measured with 4 fast timing scintillators located at 10 m and 10 more at 70 m from the center. Muons were measured using groups of GM counter hodoscopes of total area 30 m^2 located at 20 m.w.e. underground.

Shower cores were located using the 64 central and a few nearby detectors. age parameter s was obtained from density measurements at 6, 20 and 73 m from the core using the NKG function with $r_0 = 120$ m. N_e was estimated using the density measurement at 70 m. N_μ was obtained using lateral distribution function of the type $r_\mu^{-0.7} \exp(r_\mu/80)$. The dependence of N_μ on N_e in the shower size range 10^5–10^7 is shown in Fig. 6.35. The data is fitted to the function

$$N_\mu = (1.26 \pm 0.02)\, 10^4 \left(\frac{N_e}{10^6}\right)^{0.80\pm0.01} \tag{6.24}$$

It can be seen that there is no evidence for a change of slope in the shower size range studied. The authors conclude that the data are in agreement with a constant composition in this shower size range and do not agree with a rigidity cut-off in the

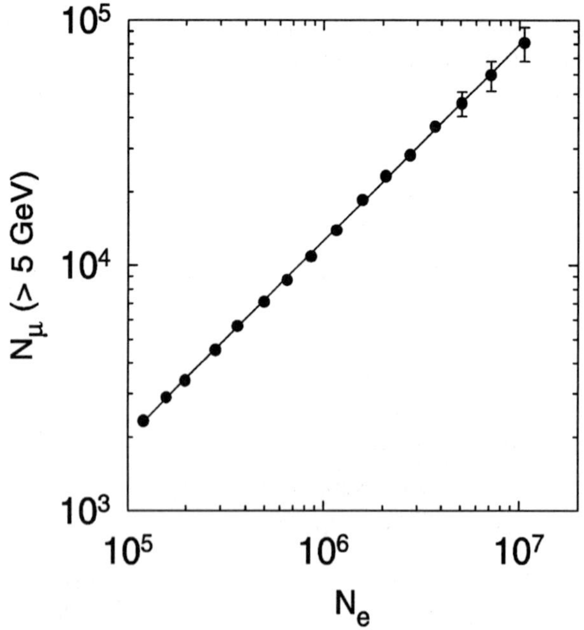

Figure 6.35 Dependence of N_μ ($>$ 5 GeV) on N_e obtained by the Tien Shan group (Kirov et al, 1980).

primary energy range 10^{15}–10^{16} eV as in Peters' model since it predicts a change in the $N_\mu - N_e$ dependence in the shower size range studied.

6.4 Hadrons

The number of hadrons in a shower is nearly proportional to the primary energy or the shower size and therefore not sensitive to composition. However, low energy hadrons delayed with respect to the shower front are predominantly nucleons and their content in the showers should be a measure of the primary mass, unless baryon-antibaryon production in hadronic interactions is very large compared to the extrapolation of accelerator data. Also, the very high energy tail of the hadron energy spectrum should be sensitive to the primary mass since the energy per nucleon of heavy primaries is smaller than that of a proton for the same shower size.

6.4.1 Delayed Hadrons

The Maryland group [79] measured the fraction of events in which a hadron is recorded in a scintillator in their hadron calorimeter with an energy deposit equivalent to 5 particles and delayed by more than 15 ns with respect to the shower front, among all events satisfying certain triggering criteria which includes a total energy deposit of 50 GeV in their calorimeter. Comparing their result with detailed Monte Carlo simulations using the scaling model, similar to the one used in the KGF work, they [66] suggest a model of composition in which the rigidity spectral index of all nuclei increased by 0.5 at 10^5 GV and the extragalactic component, consisting mainly of light nuclei, dominates at 10^8 GeV. In this model, the percentage of iron increases with energy above 10^5 GeV becoming 40% at 10^6 GeV, the percentage of protons *decreases* and the mean mass number increases with energy in the region $10^5 - 10^6$ GeV. Curve 2 in Fig. 6.16 is the expected $N_\mu - N_e$ dependence for this model which is clearly in disagreement with the data.

Their conclusions regarding composition critically depend upon the details of Monte Carlo simulation. Low energy hadrons, particularly their response in detectors, are difficult to simulate and extreme care has to be taken to consider all processes. Though the authors have taken great care in carrying out their simulations, they have not considered the following:

1. Nuclear target effects were neglected. The enhanced multiplicity observed in hadron-nucleus interactions would lead to a larger number of low energy nucleons and antinucleons, which would enhance the fraction of delayed hadrons in proton as well as nucleus initiated showers. This would reduce the abundance of iron required.

2. Hadrons of energy less than 3 GeV were discarded in the simulations. The effect would be same as in (1).

3. Mincer et al. [94] exposed an identical calorimeter to accelerator beams of various energies and found that "low energy hadrons do produce large signals (> 50 equivalent particles) in single scintillators at a surprisingly large rate" which was also pointed out earlier by Bhat et al. [95]. Of particular importance in this context is the finding that 2 GeV protons, which were neglected in the simulations, produce pulse heights greater than 5 equivalent particles with a substantial probability, apart from the fact that this probability is underestimated for higher energy hadrons in their simulations. Obviously, the abundance of iron required would be reduced if these effects are included in the simulations. In view of these findings, points 1 and 2 become even more important.

4. The triggering requirement of 50 GeV energy deposit in the calorimeter is considered to be met in the simulations by 30 GeV hadronic energy, the rest assumed to be due to the electron-photon component. Apart from fluctuations

in the relative energy contents of hadronic and electron components, they are expected to be different in proton and iron showers. Furthermore, a larger fraction of the energy deposit in inclined showers will be due to the electron component since the shielding is less in the inclined directions. This would affect the calculated total event rates and the relative contribution from proton and iron showers.

All these effects tend to decrease the abundance of iron required to explain the data. The KGF composition model, discussed in Sec. 6.3.1.2, predicts an iron abundance of 25% at 10^6 GeV and seems to be consistent with the delayed hadron data if these factors were taken into account. The main differences in the two models are that in the KGF model the fraction of protons increases and the mean mass number remains nearly constant with energy whereas the former decreases and the latter increases with energy in Cowsik et al.'s model.

6.5 Fractal Analysis

A new method of factorial moments and multifractal structures of the density fluctuations of charged particles near air shower cores has been developed by Kempa [96] to distinguish between individual proton initiated and heavy nuclei initiated showers. The fractal analysis is applied to the density fluctuations in the same core distance range but in different azimuthal regions [97]. In order to establish the method, they used proton and iron initiated showers generated by the Monte Carlo program CORSIKA.

The azimuthal range is divided into M bins of width $\delta = 360/M$. Let k_i be the number of particles in the ith bin. Since there may be bins that have no particles at all, M is defined as the number of non-empty bins which constitute a fractal set. The multifractal moments are defined as

$$G_q = \sum_{j=1}^{M} p_j^q$$

where $p_j = k_j/N$ with $N = k_1 + k_2 \cdots + k_M$, q is a real number and the summation is carried out over non-empty bins only. If the detected particles exhibit self-similar behavior, the moments show a power law relation

$$G_q \sim \delta^{\tau_q}.$$

The fractal moments G_q are calculated for q from +6 to -6. The τ_q is determined form G_q. The function $f(\alpha)$ is calculated by the Legendre transformation

$$\alpha_q = \frac{d\tau_q}{dq} \qquad \text{and} \qquad f(\alpha) = q\alpha_q - \tau_q,$$

where α_q is the Lipschitz-Hölder exponent.

It is the $f(\alpha) - \alpha$ spectra that distinguish between the two types of showers. Fig. 6.36 shows the spectra for showers initiated by gamma rays, protons and iron nuclei. Gamma ray showers were generated with the NKG function and Poisson fluctuations are imposed on the densities. The difference between showers initiated by proton

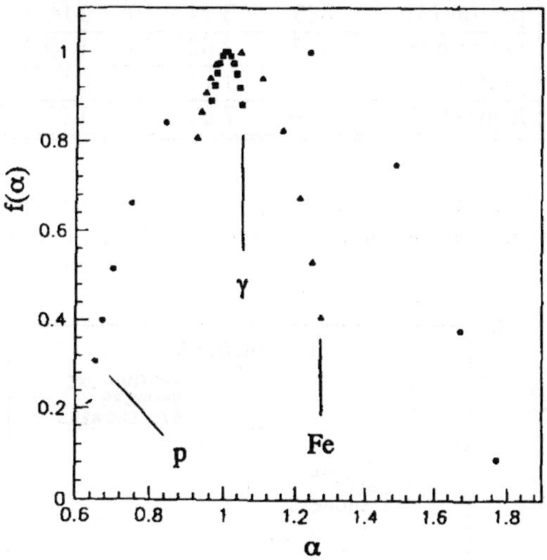

Figure 6.36 The $f(\alpha) - \alpha$ spectra for showers initiated by three different primary particles calculated by Kempa (1994): γ-ray (■); proton (•); iron nucleus (Δ).

and iron can be clearly seen. The authors claim that proton and iron showers can be identified with 90% and 70% confidence respectively.

They [98] applied this method to the neon flash tube data collected with the Kiel air shower array [99] located at sea level. The array consists of 28 scintillators to detect and determine shower size and core location, 11 fast timing detectors to determine shower arrival directions with an accuracy of 1° and a 31 m^2 neon flash tube hodoscope (see Figs. 3.1 and 3.2) to study electron core structure. The hodoscope consists of $420 \times 420 = 176,400$ neon flash tubes of 1 cm diameter each in a rectangular lattice configuration with lattice constant of 1.326 cm. Data from 1,044 showers with well located core positions are used for this analysis.

For each shower, the annular region between 0.2 m – 6 m from the core is divided into several bins in azimuth as described above. Using the $f(\alpha) - \alpha$ spectrum, individual showers are classified as proton-like or heavy-like. The results are shown in Table 6.8. The parameter s is the age parameter for the distance range 10–100 m. Proton-like showers are steeper than iron-like showers in both shower size groups, as

Table 6.8 Results of fractal analysis of Kiel data by Dudarewicz et al. (1994) to obtain primary composition.

Range of shower size N_e	Type	Number of showers	$< N_e >$	$< s >$	Un-recognised
$10^5 - 5 \times 10^5$	proton-like	385	3.14×10^5	1.05 ± 0.01	175
	heavy-like	207	2.62×10^5	1.19 ± 0.01	
$5 \times 10^5 - 5 \times 10^6$	proton-like	186	1.14×10^6	0.98 ± 0.01	55
	heavy-like	36	7.53×10^5	1.05 ± 0.03	

expected. The age distributions of the two groups of showers are shown in Fig. 6.37. Integral shower size spectra for these two groups of showers are shown in Fig. 6.38.

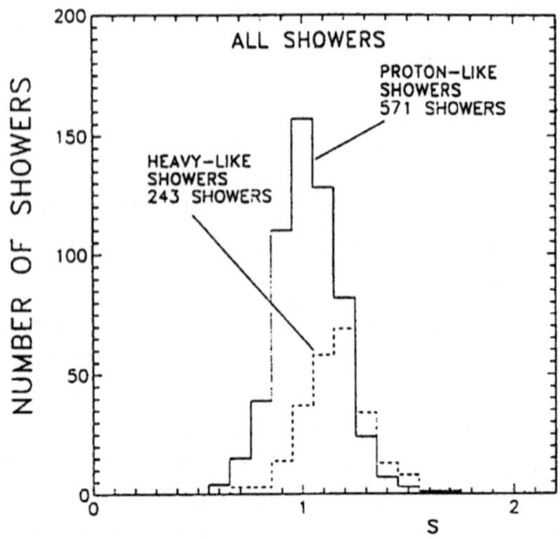

Figure 6.37 Age distribution of proton-like and heavy-like showers identified from fractal analysis by Dudarewicz et al. (1994).

The relative abundances of the two types of nuclei as a function of shower size are also estimated as shown in Fig. 6.39. They are also listed in Table 6.9 for the shower size range below and above the knee.

The authors draw the following conclusions:

- Neither before the knee nor after the knee, heavy primaries dominate in the primary spectrum (but the composition before the knee is 'heavier' than after

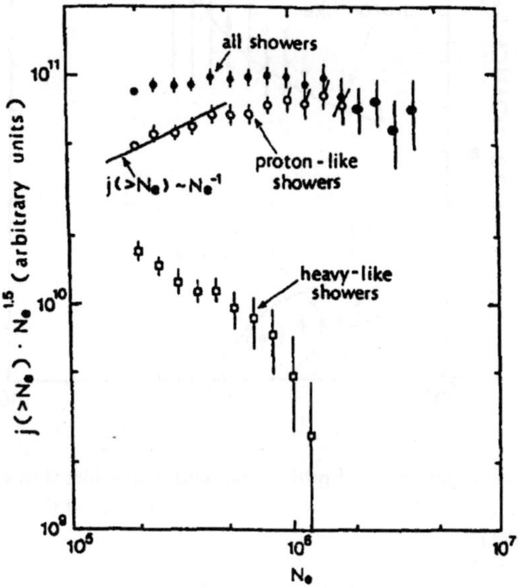

Figure 6.38 Integral shower size spectra for proton-like and heavy-like showers from fractal analysis of Kempa et al. (1995).

Table 6.9 Relative abundances (in %) of proton and heavy components before and after the knee form the fractal analysis of Kempa et al. (1995).

$\mathbf{10^5 < N_e < 5 \times 10^5}$ **before knee**		$\mathbf{5 \times 10^5 < N_e < 5 \times 10^6}$ **after knee**	
p	50 ± 8	p	67 ± 13
heavy	27 ± 5	heavy	13 ± 4

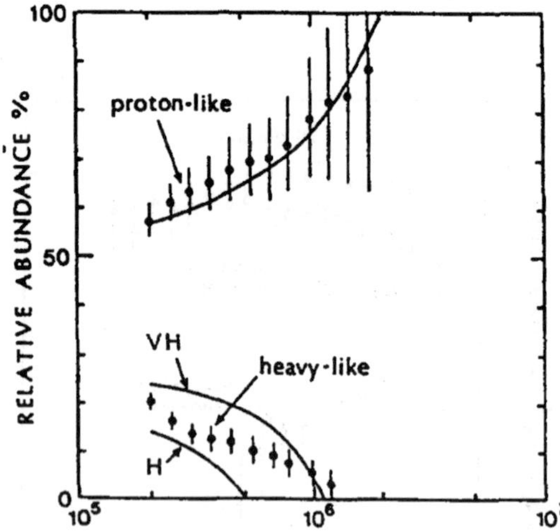

Figure 6.39 Relative abundances of proton-like and heavy-like showers with $N_e < 2 \times 10^5$ estimated by Kempa et al (1995). Solid lines are predictions of the two component model of Gawin et al. (1984).

the knee).

- After the knee the fraction of heavy particles decreases to a value of $14\% \pm 4\%$ while the fraction of protons increases to $67\% \pm 13\%$.
- Shower size spectrum for iron-like showers is much steeper than for proton-like showers.

They comment that the data are in relatively good agreement with the two-component model of Gawin et al. [100], which is similar to the models suggested by Chatterjee [67] and Fichtel and Linsley (see Sec. 6.6) as far as the variation of composition with primary energy is concerned.

We note form Fig. 6.38 that the all showers spectrum shows the knee around 1.3×10^6. Above this shower size, the flux of all showers is nearly equal to the flux of proton-like showers within errors and so the entire flux seems to be due to protons. Thus it appears that the knee is due to protons as suggested by the KGF group (see Sec. 6.3.1.2) from their data on high energy muons. Also, the heavy-like spectrum is steeper than the proton-like spectrum even near the smallest shower size and steepens further, but still before the knee. This aspect of the spectra is also in agreement with the conclusions of the KGF group.

6.6 Composition Models

Historically the first model for primary composition was put forward in the early sixties by Peters [101, 102] to explain sudden changes in the behavior of showers around 10^{15} eV at mountain altitude reported by Nikolsky and his collaborators [103–105]. Their results showed that the number of low energy hadrons (> 1 GeV) increased with N_e as N_e^α where $\alpha \sim 0.2$ for showers up to a size of about 10^5 and 1 beyond. The "absorbability" of showers also showed an unexpected increase and subsequent decrease in the neighborhood of the discontinuity. Peters suggested that these features can be understood if there is a fairly sharp rigidity cut-off for all component at a rigidity of 10^{15} V probably because of the rapid escape of the particles from the containment volume of the galactic disc. Beyond the proton cut-off energy of 10^{15} eV the relative abundance of nuclei increases with energy and the number of hadrons increases more rapidly with shower size than before this energy since $\alpha < 1$ in the lower energy region. Also, the energy per nucleon decreases by a factor of 2 at this energy and showers become more absorbable. According to this model the primary energy spectrum should not continue beyond the Fe cut-off energy of 3×10^{16} eV. But soon it became clear from the experiments at M.I.T. [106] the spectrum extends up to 10^{19} eV without any sign of steepening further. To accommodate this part of this spectrum Peters [89] suggested that a second source starts contributing above 10^{16} eV. He, however, did not identify this second source to be extragalactic.

In the early sixties Linsley et al. [107] found that the fluctuations of low energy muons in showers of primary energy around 10^{17} eV near maximum development are entirely statistical in nature, which led them to conclude that the composition in this energy region is pure, either all protons or all nuclei in the iron group. Distribution of shower ages led them to prefer a pure proton composition. Based on these results Linsley [108] suggested that second component is extragalactic in nature arguing that only heavy nuclei, not protons, can be confined in the galaxy at these high energies.

Since then modifications to Peters' model have been suggested by several authors as better experimental data became available. As discussed in Sec. 6.3.2.1, Chatterjee suggested in mid sixties that the rigidity cut-off occurs at 3×10^{14} V instead of at 10^{15} V and that an extragalactic proton component starts contributing at 2×10^{15} eV.

Analyzing the data on N_e and N_μ spectra at various altitudes, the sea level muon energy spectrum and other experimental data available at that time Hillas [69] showed, using calculations based on scaling model with rising cross sections, that all these data cannot be consistently explained in terms of the rigidity cut-off model if all components, including protons, have the same rigidity cut-off. For example, he pointed out that the energy spectrum should show two knees—one where protons are cut off and the other for iron nuclei—which is not seen experimentally. He also showed that the N_e spectrum at sea level is much more sensitive to the proton component than to the heavy nuclei and the N_μ spectra are dominated by heavy nuclei and thus, in the region of the knee, they determine the energies at which the proton and heavy nuclei spectra bend respectively. He found that the best fit to the data is obtained if all nuclei other than protons had a "bend" at the same energy per nucleon, but the proton spectrum bent at a much higher energy (per nucleon). He thus concluded that the bend in the spectrum is not related to trajectories in a magnetic field, as in the rigidity cut-off model. He then suggested that an originally featureless power law spectrum can acquire such modifications through photonuclear reactions. He argued that in a head-on collision between a nucleus of energy E and mass Am_p (where m_p is the nucleon mass) with a photon of energy ϵ, nuclear photodisintegration occurs when $2\epsilon\,(E/AMc^2) \approx (10 - 20$ MeV), whereas for protons this process does not occur, and considerable energy losses set in due to the commencement of photopion production process when $2\epsilon\,(E/Mc^2) \approx 200$ MeV. This can put the spectrum bend at similar values of E for protons and iron nuclei. He also suggested that it is quite possible that photon fluxes are high enough to cause appreciable energy loss by pair production, in which case the position of the noticeable drop in intensity for protons and for various nuclei is more dependent on the details of the environment. Photons of energy ~ 70 eV would be required for the first process and rather lower energy for the second.

Gawin et al. [100] analyzed the data on the variation of the depth of shower maximum with primary energy available in the early eighties (similar to Fig. 6.6) using a scaling-breaking interaction model of Wdowczyk and Wolfendale [109]and proposed a two component model of composition similar to that of Chatterjee. Linsley and Watson [25] concluded from this data that the mass composition around 10^{16} eV is dominated by heavy primaries whereas at 10^{17} eV primary particles are primarily protons. The high energy muon data of the KGF group [62] also showed the same tendency. The data on gamma ray families at Mount Fuji by the Fuji group [110] indicated that the composition in the lower energy region is dominated by heavy nuclei. Gawin et al. pointed out that the fluxes of hadrons are far too high compared to the expectations from a heavy dominant composition at 10^{15} eV and the fluctuations in air showers at sea level as well as mountain altitude are inconsistent with a heavy dominant composition. They argued that cosmic rays at all energies in this energy region (10^{14}–10^{17} eV) consist of a significant fraction of protons. The widely accepted shock acceleration mechanism of cosmic ray acceleration, either in supernova remnants or interstellar medium, predicts a rather sharp cut-off in the primary spectrum at a certain rigidity, which should be observed around 10^{14} eV. This leads to clear enrichment of heavy nuclei and finally to a cut-off in the cosmic ray spectrum. These effects are not observed. They argued that the only way to preserve the shock acceleration mechanism for low energy cosmic rays is to assume the existence of a second component consisting mainly of protons. They proposed that the integral spectra of various nuclei in the first component are of the form

$$j_A(>E) = \frac{C_A}{1.7} E^{-1.7} e^{-\left(\frac{E}{D_A}\right)^2} \left(1 + \frac{E}{3.33 \times 10^4}\right)^{0.3} \quad \mathrm{m^{-2}\,s^{-1}\,sr^{-1}}, \tag{6.25}$$

Table 6.10 Values of constants in the integral sppectra of various nuclei in the model of Gawin et al (1984).

A	C_A	D_A
p	1.72×10^4	3.33×10^5
α	9.2×10^3	6.67×10^5
CNO	6.2×10^3	2.17×10^6
H+VH	9.2×10^3	4.33×10^6
Fe	6.2×10^3	8.66×10^6

with the values of the constants for different nuclei shown in Table 6.10. The second component consisting of protons is given by

$$\begin{aligned} j_p &= 2.78\left(E^{-1} - 5 \times 10^{-7}\right) \quad \mathrm{m^{-2}\,s^{-1}\,sr^{-1}} \quad \text{for } E < 2 \times 10^6 \text{ GeV} \\ &= 2.78 \times 10^6 E^{-2} \quad \mathrm{m^{-2}\,s^{-1}\,sr^{-1}} \quad \text{for } E > 2 \times 10^6 \text{ GeV} \end{aligned}$$

with E in GeV. The spectra of various nuclei of the two components are shown in Fig. 6.40. While there is general agreement of the calculated cascade curves with

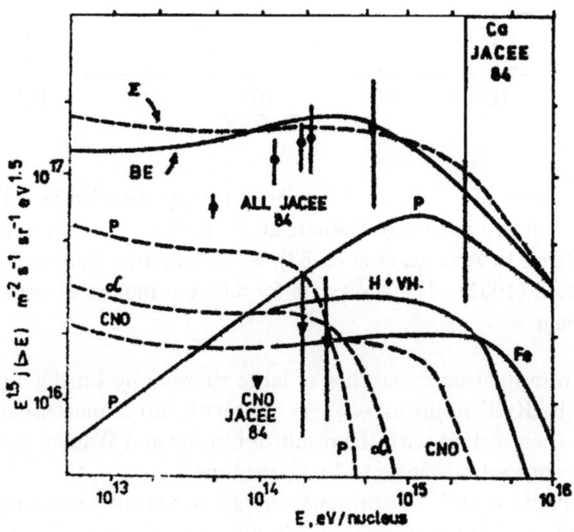

Figure 6.40 Integral energy spectra of various nuclei in the two component of Gawin et al (1984).

experimental data, there is marked disagreement in some parts, especially in the lower third of the atmosphere for smaller primary energies. The authors claim that the model is consistent with observations of secondary cosmic rays.

Using the KGF data on $N_\mu - N_e$ dependence and their Monte Carlo simulations, Fichtel and Linsley [111] calculated the mean of the logarithm of the mass number, $< \ln A >$, as a function the primary energy. It is difficult to obtain the dependence of $< \ln A >$ on primary energy from thus data since nuclei require increasing primary energy with increasing mass number to generate showers of the same size at 920 g cm^{-2} where they are well beyond maximum development. It is not clear how this was done, but they seem to have made some assumptions. Their result is shown in Fig. 6.41. The figure also shows the result of direct measurement by Burnett et al. and

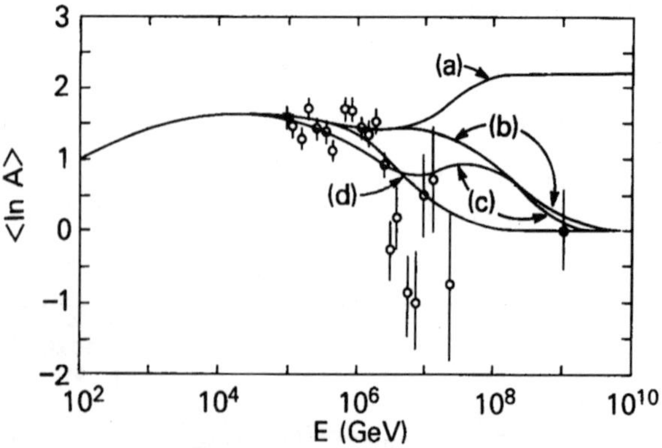

Figure 6.41 Variation of $< \ln A >$ with primary energy compiled by Fichtel and Linsley (1986). ■: direct measurement of Burnett et al. (1983); ○ : from measurement of $N_\mu - N_e$ dependence by Acharya et al (1983); • : Fluctuation studies of large showers by Linsley and Watson (1981). The curves are for different models of primary composition considered by them.

that obtained from fluctuation studies of large showers by Linsley and Watson [31]. It is seen that the KGF result agrees very well with direct measurements at the low energy end and is consistent with the result of Linsley and Watson at the high energy end where the composition seems to be all protons.

Based on this data and, the primary energy spectrum from direct observations as well as air shower measurements as discussed in Chapter 5, and the observed anisotropy of cosmic rays (see Sec. 7.2.4 and Figs. 7.21 and 7.22), which is seen to increase with energy in the range $10^{15} - 10^{19}$ eV, they argued that a single source within the galaxy, either distributed over the galaxy or at its center, cannot explain all the data. These sources with a rigidity cut-off of the type suggested by Peters, while explaining the anisotropy, predict increasing mean mass with energy after the proton cut-off energy which is contrary to observations. They then consider two cases of two source type models, as in some of the models discussed earlier. In both cases the second source is galactic since the anisotropy expected for an extragalactic source is very small.

The predictions of the variation of the mean mass with energy for all these four models are also shown in Fig. 6.41. Curve (a) is for a single type of galactic sources like supernovae and pulsars distributed through out the galaxy and with a rigidity cut-off around 10^6 GV. In this model, heavy nuclei dominate at the high energy end and hence the mean mass is large. Curve (b) is for a similar model but in which for some reason the heavy nuclei decrease very dramatically in relative abundance between about 10^5 GeV and 10^7 GeV and the protons are compensatingly somewhat greater in intensity at high energies, or alternately higher energy particles see more matter during propagation in the galaxy, causing heavy nuclei to fragment. This model appears to be unrealistic. Curve (c) is for a two-source model, one contributing to lower energies and other at higher energies. Both are distributed galactic sources, the lower energy one having composition similar to the one measured by direct experiments and the other mostly protons. Change of spectral slope occurs for both sources at the same energy of 3×10^6 GeV. Curve (d) is for a similar model as in (c), but the two sources have change of slope at different energies; the lower energy one at 5×10^5 GeV and the higher energy one at 5×10^6 GeV. The lower energy source is again a distributed source, but the higher energy one is at the galactic center. This model is similar to the one proposed by Chatterjee [67] in spectral shapes and agrees well with data. The energy spectra of the various components in this model are shown Fig. 6.42. The noteworthy feature of all these models which agree with data is that protons continue right up to and beyond the knee, and nuclei are lost well before the knee.

Pursuing the suggestion of Hillas that photonuclear interactions may be responsible for the changes in the energy spectra of various nuclei, Karakula and Tkaczyk [112] investigated theoretically to see whether the modified spectrum reproduces the observed spectra. They assumed that a point source produces a power law spectrum, $E^{-\gamma}$, for all nuclei. The background photon spectrum around the sources can be either thermal (Planckian) or power law type $n(\epsilon) = (\rho/\rho_0)\,\epsilon^{-\alpha}$ ($\mathrm{cm}^{-3}\ \mathrm{eV}^{-1}$) for $\epsilon \leq \epsilon_\mathrm{m}$ and zero for $\epsilon > \epsilon_\mathrm{m}$, where ρ is the photon density and ϵ its energy. The relative mass composition of nuclei produced by the sources is assumed to be the same as that suggested by experiments in the lower energy region: from Tien Shan (Nikolsky, 1984 [113]), Maryland and NUSEX experiments [114]. For Tien Shan normalization the power index $\gamma = 2.6$ was assumed for all nuclei.

The reciprocal mean free path for a nucleus of mass A for photodisintegration is given by

$$\lambda_A^{-1} = \frac{1}{2\pi}\int_{\epsilon_1}^{2\gamma_A\epsilon_\mathrm{m}} \epsilon^*\sigma_A(\epsilon^*)\,d\epsilon^* \int_{\epsilon^*/2\gamma_A}^{\epsilon_\mathrm{m}} n(\epsilon)\,\frac{d\epsilon}{\epsilon^2} \tag{6.26}$$

where $n(\epsilon)$ is the density of ambient photons with energy ϵ in the observer's system,

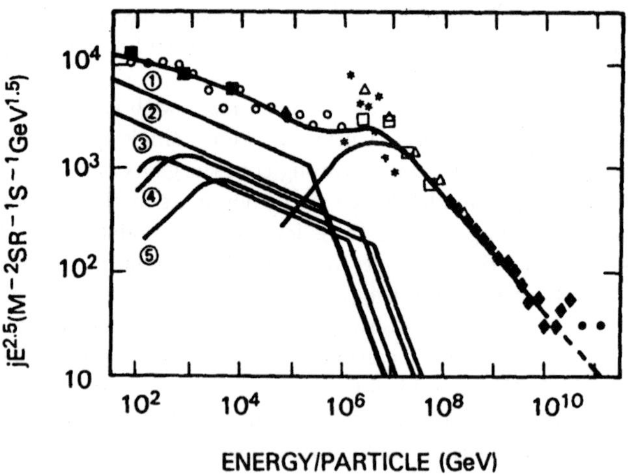

Figure 6.42 Primary energy spectra of the various components of the two-source model of Fichtel and Linsley (1986). The curves marked 1,2,3,4 and 5 in circles are for p, He, CNO, $10 \le Z \le 20$ and $21 \le Z \le 30$ respectively.

ϵ^* is the energy of the photon in the rest system of the nucleon, γ_A is the Lorentz factor of the nucleus, $\sigma_A(\epsilon^*)$ is the photodisintegration cross section.

The photodisintegration cross section for losing one nucleon by a nucleus of mass number A was obtained by approximation of the experimental data and is given by

$$\sigma_A(\epsilon^*) = \sigma_0(A)\frac{(\epsilon^* T)^2}{(\epsilon^{*2} - \epsilon_0^2) + (\epsilon^* T)^2} \text{ mb} \quad \text{for } \epsilon^* \le 30 \text{ MeV}, \tag{6.27}$$

$$= A/8 \text{ mb} \quad \text{for } \epsilon^* > 30 \text{ MeV}, \tag{6.28}$$

where $\sigma_0(A) = 1.45 \times 10^{-27} A$ cm^2, $T = 8$ MeV is the energy bandwidth of the giant resonance, $\epsilon_0 = 42.65A^{-0.21}$ MeV for $A > 4$ and $\epsilon_0 = 0.925A^{2.433}$ MeV for $A \le 4$, the peak energy of the giant resonance. For numerical calculations a threshold energy of $\epsilon_1 = 2$ MeV in the rest system of the nucleus is taken. Fig. 6.43 shows the reciprocal mean free path of various nuclei for photodisintegration and for photomeson production, calculated for the background photon spectrum of power law type with parameters $\epsilon_m = 20$ eV, $\alpha = 1.3$, $\rho/\rho_0 = 1$ (Fig. 6.43a) and Planckian type with $kT = 1.8$ eV (Fig. 6.43b).

During propagation in a space filled by low energy photons nuclei lose nucleons one by one due to photodisintegration process. Proton energy losses due to photomeson production are also included. The energy spectra of nuclei with mass number A at the

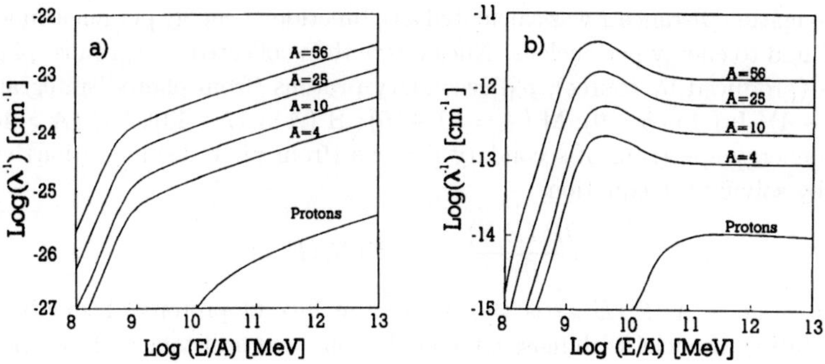

Figure 6.43 The reciprocal mean free path for photodisintegration of nuclei and the reciprocal mean interaction length of protons for photomeson production calculated for a background photon spectrum of power law (a) and Planckian (b) type by Karakula and Tkaczyk (1993).

propagation distance x are obtained by solving the appropriate diffusion equations:

$$\begin{aligned} \frac{\partial N_i(E,x)}{\partial x} &= -a_i(E)\,N_i(E,x) + a_{i-1}(E)\,N_{i-1}(E,x)\,, \quad \text{for } i=1,2,\ldots,54, \\ \frac{\partial N_0(E,x)}{\partial x} &= -a_0(E)\,N_0(E,x)\,, \quad \text{for } i=0. \end{aligned} \tag{6.29}$$

Here $i = 56 - A$, E is the energy per nucleon of the nucleus, $N_i(E,x)$ the differential flux of nuclei with mass number $A = 56 - i$ and energy E at the propagation distance x from the source and $a_i(E)$ is the reciprocal mean free path of nuclei with mass number $A = 56 - i$ for photodisintegration. The general form of the solution of Eqs. 6.29 is given by

$$N_i(E,x) = \sum_{j=0}^{i} b_{ij}(E)\exp\left(-a_j(E)\,x\right), \tag{6.30}$$

where

$$b_{00} = N_0(E,0) \tag{6.31}$$

and other coefficients are given by the recurrence relations

$$\begin{aligned} b_{ik}(E) &= \frac{a_{i-1}(E)\,b_{i-1k}(E)}{a_i(E) - a_k(E)}, \quad \text{for } k<i, \\ b_{ii} &= N_i(E,0) - \sum_{n=0}^{i-1} b_{in}(E)\,, \quad \text{for } k=i. \end{aligned}$$

Solving the diffusion equation was started from iron (i=0) and the spectra of nuclei with mass number $A < 56$ were calculated subsequently. The composition of nuclei at the propagation distance x was calculated as a function of energy per nucleon and then recalculated to energy per nucleus. Nuclei were then collected into groups: p, primary protons (produced by source); p_A, secondary protons (from photodisintegration); α $(A = 2-4)$; L $(A = 5-9)$; M $(A = 10-16)$; H $(A = 17-40)$; VH $(A > 40)$.

The energy spectrum of secondary protons (from photodisintegration) were obtained by solving the equation

$$\frac{\partial P_i(E,x)}{\partial x} = a_i(E)\, N_i(E,x)\,, \tag{6.32}$$

where $i = 56 - A$, $P_i(E,x)$ is the differential flux of protons of energy E from fragmentation of nuclei with mass number $A = 56 - i$ at a distance x from the source, and $a_i(E)$ and $N_i(E,x)$ are as above. The solution of Eq. 6.32 is

$$P_i(E,x) = \sum_{j=0}^{i} \frac{a_i(E)\, b_{ij}(E)}{a_j(E)} \left(1 - \exp\left(-a_j(E)\, x\right)\right). \tag{6.33}$$

The total flux of protons of energy E at the propagation distance x (from photodisintegration of nuclei from iron to deuterium) is given by

$$P(E,x) = \sum_{i=0}^{54} P_i(E,x)\,. \tag{6.34}$$

During propagation in a space filled by background photons high energy protons suffer considerable energy losses due to photomeson production. The energy spectrum of protons at the propagation distance x were obtained by solving the diffusion equation

$$\frac{\partial P(E,x)}{\partial x} = -g(E)\, P(E,x) + g\left(\frac{E}{1-K}\right) P\left(\frac{E}{1-K}, x\right) \frac{1}{(1-K)}, \tag{6.35}$$

where $P(E,x)$ is the differential flux of protons with energy E at the propagation distance x from the source, K is the coefficient of inelasticity for the $p - \gamma$ process, $g(E)$ is the reciprocal mean free path of protons for photomeson production. K was assumed to have the same value of 0.3 for all energies. The solution of this equation, obtained by the method of successive generations is

$$P(E,x) = \sum_{l=0}^{\infty} P_l(E,x)\,, \tag{6.36}$$

where

$$\begin{aligned} P_l(E,x) &= P(E,0)\exp\left(-g(E)\,x\right), \qquad \text{for } l = 0 \\ &= \frac{P\left(\frac{E}{(1-K)^l}, 0\right)}{(1-K)^l} \prod_{j=1}^{l} g\left(\frac{E}{(1-K)^l}\right) \sum_{n=0}^{l} \frac{\exp\left\{-g\left(\frac{E}{(1-K)^n}\right) x\right\}}{\prod\limits_{\substack{m=0 \\ m \neq n}}^{l} \left[g\left(\frac{E}{(1-K)^m}\right) - g\left(\frac{E}{(1-K)^n}\right)\right]}, \\ &\text{for } l > 0 \end{aligned} \tag{6.37}$$

Table 6.11 Values of best fit parameters of the power law and the Planckian background photon spectra in the photodisintegration model of Karakula and Tkaczyk (1993).

	Planck			**Power law**		
	kT (eV)	$\log x$ (cm)	ρx (eV/cm^2)	$\epsilon_{\rm m}$ (eV)	$\log x$ (cm)	ρx (eV/cm^2)
Tien Shan	1.8	14.4	2.25×10^{29}	20	26.4	1.84×10^{29}
NUSEX	0.6	15.6	4.41×10^{28}	8	26.2	6.12×10^{28}
Maryland	1.4	14.5	1.04×10^{29}	8	26.3	7.71×10^{28}

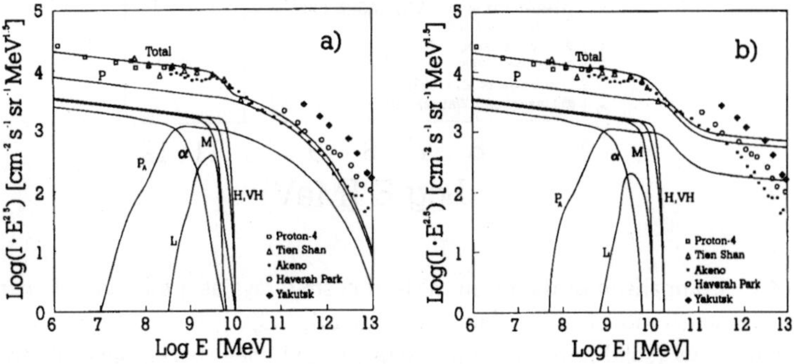

Figure 6.44 Comparison of the cosmic ray energy spectrum with the spectrum obtained by Karakula and Tkaczyk (1993) in their photodisintegration model for Tien Shan normalization and power law (a) and Planckian (b) background photon spectra.

and $P(E, 0)$ is the flux of protons at $x = 0$.

The spectra of all nuclei were calculated by solving the propagation equations and were collected into groups: p, α, L, M, H, VH. The propagation distance x was found from the best fit of the calculated spectrum to the observed one. Table 6.11 shows their best fit parameters for power law ($\alpha = 1.3$) and Planckian background photon spectra. Note that the final solution depends on the product ρx, not on individual values. Fig. 6.44 shows the resultant spectra of individual nuclei as well as the total for the two types of background photon spectra compared with observations. Notice the sharp cut-off for nuclei. Power law photon spectrum seems to agree better with observations than Planckian spectrum, particularly beyond the knee. Fig. 6.45 shows the mean value of $\ln A$ as a function of primary energy in their model compared with observations. Notice the sharp fall of mean mass from the low energy value to one beyond the cut-off energy. The agreement is good for all the variations of the model except that the Maryland normalization seems to be in better agreement in the energy region where the effects of phtodisintegration have not yet set in, probably due to the flatter spectra for heavy nuclei in the Maryland composition.

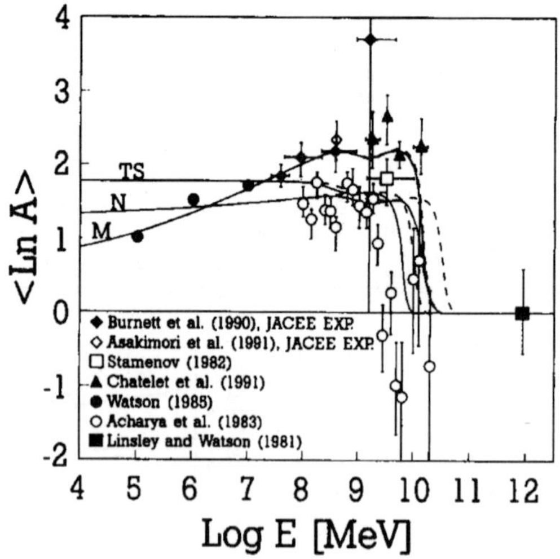

Figure 6.45 Comparison of the mean mass of cosmic rays as a function of energy in the photodisintegration model of Karakula and Tkaczyk (1993), for the three normalisations and the power law (solid lines) and Planckian (dotted lines) background photon spectra mentioned in the text, with observations.

The agreement between the predicted and observed spectra is good up to about 3×10^{17} eV for both power law and Planckian background spectra. Above this energy the power law background spectrum agrees better. The authors comment that it can be made to agree better if protons are lost through an additional process, for example leakage from the storage region. This leakage process seems to be required to explain the observed anisotropy.

Taking into account that the luminosity of a source in the ultraviolet band is $L_{UV} = 4\pi R^2 \rho c$ and $R \leq x$, one can find the inequality $L_{UV} \leq 4\pi R x \rho c \leq L_{Edd}$, where L_{Edd} is the Eddington luminosity ($L_{Edd} = 4\pi G M m_p c/\sigma_T$), R is the radius of the source. For a mass of a source equal to the solar mass and $x\rho = 1.2 \times 10^{29}$ eV/cm^2 (see Table 6.11) the authors estimate the radius of the source to be $\leq 10^9$ cm. Thus there seems to be strong evidence for the compactness of the cosmic ray sources. They note that the linear dimension of the source region R can be smaller than the propagation distance of particles x if magnetic confinement of particles in the source region takes place.

This model worked out in detail by Karakula and Tkaczyk is similar to the model suggested by the KGF group to explain their high energy muon data. Note that any model in which protons continue to be present—even if due to a second source—

beyond the energy where nuclei are removed, either due to photodisintegration in the source or leakage from a confinement region, predicts the mean mass to be nearly constant even in the energy region where the composition is changing due to these processes and becomes one rapidly after the cut-off for iron. Such a rapid change is seen also in the plot of mean mass as a function of shower size since it occurs where iron is cut off and fluctuations in iron initiated showers are small.

It appears that leakage process at least for one of the primary cosmic ray components, say for protons, is needed to explain the observed anisotropy. According to Hillas [115], if the source spectrum per unit time is $Q(E) \propto E^{-\gamma}$ and the residence time in the confinement region $T(E) \propto E^{-\beta}$, then the observed spectrum is $F(E) \propto E^{-(\gamma+\beta)}$. The spectrum thus steepens at an energy where the residence time becomes energy dependent causing the knee. The amplitude of the first harmonic is given by $A \propto t/T(E) \propto E^{\beta}$, where t is the rectilinear propagation time. Thus anisotropy is expected for a confinement and subsequent leakage process. The observed anisotropy (see Figs. 7.21 and 7.22) seems to start increasing with energy where the primary spectrum starts steepening.

In the KGF model and the model of Karakula and Tkaczyk the leakage is due to protons. In the model of Fichtel and Linsley it is due to protons in both the components as well as nuclei. These two types of models can be distinguished experimentally by studying the variation of anisotropy, as a function of primary energy, of showers with different mu-richness. Only mu-poor showers should show anisotropy if the photodisintegration mechanism is operating while anisotropy should set in at increasing shower size with increasing mu-richness if the leakage mechanism is operating for nuclei also. The CASA-MIA and KASCADE experiments and to a lesser degree the KGF gamma ray experiment have large enough muon detectors to carry out such a study.

References

1. A. Tomaszewski and J. Wdowczyk, *Proc. 14th International Cosmic Ray Conference, Munich* **8** (1975) 2899.
2. Simon Swordy, *Proc. XXIII International Cosmic Ray Conference, Calgary, Invited, Rapporteur & Highlight Papers* (1993) 243.
3. N. L. Grigorov et al., *Proc. 12 th International Cosmic Ray Conference, Hobart* **5** (1971) 1946.
4. N. N. Kalmykov et al., *Proc. 17 th International Cosmic Ray Conference, Paris* **6** (1981) 114.
5. R. J. Protheroe and K. E. Turver, *Nuovo Cimento* **51A** (1979) 277.
6. K. Kamata, *Proc. 17th International Cosmic Ray Conference, Paris.* **13** (1981) 305.
7. Yu. A. Fomin, and G. B. Khristiansen, *Sov. Jour. Nucl. Phys.* **14** (1972) 360.
8. V. I. Zatsepin, and A. E. Chudakov, *Sov. Phys. JETP* **15** (1962) 1126.
9. L. G. Dedenko, *Dissertation* FIAN (1968).
10. N. N. Kalmykov et al., *Proc. 14 th International Cosmic Ray Conference, Munich* **8** (1975) 3034.
11. N. N. Kalmykov et al., *Proc. 16th International Cosmic Ray Conference, Kyoto* **9** (1979) 73.
12. T. Hara et al., *Proc. 15 th International Cosmic Ray Conference, Plovdiv* **8** (1977) 308.
13. K. Greisen *Prog. in Cosmic Ray Physics* **3** (1956) 1.
14. H. Messel and D. F. Crawford *Electron Photon Shower distribution function tables* (Pergamon Press) (1970).
15. T. Hara et al., *Proc. 17 th International Cosmic Ray Conference, Paris* **11** (1981) 277.
16. I. P. Ivanenko et al. *Proc. 16 th International Cosmic Ray Conference, Kyoto* **9** (1979) 88.
17. V. V. Makarov et al., *Proc. 17 th International Cosmic Ray Conference, Paris* **11** (1981) 289.
18. V. I. Galkin et al., *Proc. 17 th International Cosmic Ray Conference, Paris* **11** (1981) 285.
19. R. T. Hammond et al., *Nuovo Cimento* **1C** (1978) 315.
20. V. M. Grigorev et al., *Sov. Jour. Nu. Phys.* **27** (1978) 225.
21. I. P. Ivanenko et al., Preprint No. 98, *Phys. Inst., Academy of Sciences of the USSR* (1976).
22. N. Inoue et al., *Proc. 17 th International Cosmic Ray Conference, Paris* **11** (1981) 270.
23. G. J. Thornton et al., *Proc. 16 th International Cosmic Ray Conference, Kyoto* **9**

(1979) 103.

24. J. D. Kuhlman et al., *J. Phys. G* **7** (1981) L13.
25. J. Linsley and A. A. Watson, *Phys. Rev. Lett.* **46** (1981) 459.
26. T. K. Gaisser, *Proc. 17 th International Cosmic Ray Conference, Paris* **6** (1981) 233.
27. G. J. Thornton and R. W. Clay, *J. Phys. G* **4** (1978) L251.
28. G. J. Thornton and R. W. Clay, *J. Phys. G* **5** (1979) L 137.
29. G. J. Thornton and R. W. Clay, *Phys. Rev. Lett.* **43** (1979) 1622 and **45** (1980) 1463 (E).
30. J. Linsley, *Proc. 15 th International Cosmic Ray Conference, Plovdiv* **12** (1977) 89.
31. J. Linsley and A. A. Watson, *Proc. 17 th International Cosmic Ray Conference, Paris* **6** (1981) 238.
32. K. J. Orford and K. E. Turver, *Phys. Rev. Lett.* **44** (1980) 959.
33. D. F. Liebing et al., *Proc. 17 th International Cosmic Ray Conference, Paris* **6** (1981) 90.
34. K. J. Orford and K. E. Turver, *Nature* **264** (1976) 727.
35. M. Chantler et al., *Proc. 16 th International Cosmic Ray Conference, Kyoto* **9** (1979) 42.
36. A. Andam et al., *Proc. 16 th International Cosmic Ray Conference, Kyoto* **9** (1979) 48.
37. K. J. Orford et al., *Proc. 17 th International Cosmic Ray Conference, Paris* **6** (1981) 118.
38. M. P. Chantler et al., *Proc. 17 th International Cosmic Ray Conference, Paris* **6** (1981) 121.
39. A. A. Andam et al., *Proc. 17 th International Cosmic Ray Conference, Paris* **6** (1981) 125.
    ibid. **11** (1981) 281.
40. A. A. Andam et al., *Phys. Rev.* **D26** (1982) 23.
    M. P. Chantler, *Ph. D. Thesis,* (University of Durham, 1982, unpublished).
41. M. P. Chantler et al., *J. Phys. G* **8** (1982) L51.
42. N. N. Kalmykov et al., *Proc. 15 th International Cosmic Ray Conference, Plovdiv* **8** (1977) 244.
43. T. A. Alimov et al., *Proc. 18 th International Cosmic Ray Conference, Bangalore* **11** (1983) 387.
44. I. A. Borezhko et al., *Sov. Journ. Nucl. Phys.* **30** (1979) 215.
45. M. N. Dyakonov et al., *Proc. 17 th International Cosmic Ray Conference, Paris* **6** (1981) 106.
46. M. N. Dyakonov et al., *Proc. 17 th International Cosmic Ray Conference, Paris* **6** (1981) 110.
47. O. S. Diminstein et al., *Proc. 17 th International Cosmic Ray Conference, Paris* **11**

(1981) 242.

48. I. P. Ivanenko et al., *Proc. 16 th International Cosmic Ray Conference, Kyoto* **9** (1979) 83.
49. N. N. Kalmykov et al., *Proc. 18 th International Cosmic Ray Conference, Bangalore* **11** (1983) 383.
50. N. Aliev et al., *Proc. 17 th International Cosmic Ray Conference, Paris* **11** (1981) 262.
51. H. E. Bergessen et al., *Phys. Rev. Lett.* **39** (1977) 847.
52. R. Plaga et al., *Proc. 24 th International Cosmic Ray Conference, Roma* **2** (1995) 693.
53. S. Martinez et al., *Nucl. Instr. and Meth. in Phys. Res.* **A357** (1995) 567.
54. K. Suga et al., *J. Phys. Soc. Japan* **17 Suppl A-III** (1962) 128.
55. F. Kakimoto et al., *Proc. 17 th International Cosmic Ray Conference, Paris* **11** (1981) 254.
56. J. Linsley, *Proc. 18th International Cosmic Ray Conference, Bangalore* **12** (1983) 135.
57. T. K. Gaisser et al., *Phys. Rev.* **D47** (1993) 1919.
58. B. S. Acharya et al., *Proc. 16 th International Cosmic Ray Conference, Kyoto* **8** (1979) 312.
59. B. S. Acharya et al., *Proc. 16 th International Cosmic Ray Conference, Kyoto* **9** (1979) 109.
60. B. S. Acharya et al., *Proc. 16 th International Cosmic Ray Conference, Kyoto* **13** (1979) 287.
61. B. S. Acharya et al., *Proc. 17 th International Cosmic Ray Conference, Paris* **11** (1981) 385.
62. B. S. Acharya et al., *Proc. 18 th International Cosmic Ray Conference, Bangalore* **9** (1983) 191.
63. B. S. Acharya, *Ph. D. Thesis* (University of Bombay, 1983, unpublished).
64. T. H. Burnett et al., *Phys. Rev. Lett.* **51** (1983) 1010.
    ibid. *Proc. 18 th International Cosmic Ray Conference, Bangalore* **2** (1983) 105.
65. M. Simon et al., *Astrophys. J.* **239** (1980) 712.
66. R. Cowsik et al., *Proc. 17 th International Cosmic Ray Conference, Paris* **2** (1981) 120.
67. B. K. Chatterjee, *Ph. D. Thesis* (University of Bombay, 1964, unpublished).
68. B. V. Sreekantan, in *Proc. TIFR Colloquium on Cosmic Ray Studies in Relation to Recent Developments in Astronomy and Astrophysics*, eds. R. R. Daniel, P. J. Lavakare and S. Ramadurai (Tata Institute of Fundamental Research Bombay, 1969) p. 151.
69. A. M. Hillas, *Proc. 16 th International Cosmic Ray Conference, Kyoto* **8** (1979) 7.
70. A. M. Hillas, *Ann. Rev. Astron. Astrophys.* **22** (1984) 425.

71. B. S. Acharya et al., *Proc. 19 th International Cosmic Ray Conference, La Jolla* **2** (1985) 226.
72. M. V. S. Rao, in *Cosmic Pathways-Contemporary Perspectives in Physics and Astrophysics* (ed: R. Cowsik) Tata McGraw-Hill, (1986) 175.
73. T. H. Burnett et al., *Proc. of International Symposium on Cosmic Rays and Particle Physics, Tokyo* (1984) 468.
74. I. N. Kirov et al., *Proc. 17 th International Cosmic Ray Conference, Paris* **2** (1981) 109.
75. Yu. N. Bazhutov et al., *Proc. 19 th International Cosmic Ray Conference, La Jolla* **7** (1985) 151.
76. V. V. Vashkevich et al., *Sov. J. Nucl. Phys.* **47** (1988) 672.
77. The MACRO Collaboration, *Proc. 22 nd International Cosmic Ray Conference*, Dublin **2** (1991) 1;
S. Ahlen et al., *Phys. Rev.* **D46** (1992) 895.
78. C. Forti et al., *Phys. Rev.* **D42** (1990) 3668.
79. J. A. Goodman et al., *Phys. Rev.* **D26** (1982) 1043.
ibid. *Phys. Rev. Lett.* **42** (1979) 854.
80. The MACRO Collaboration, *Proc. 23 rd International Cosmic Ray Conference*, Calgary **2** (1993) 97.
81. M. Aglietta et al., *Phys. Lett.* **B337** (1994) 376.
82. A. E. Chudakov et al., *Proc. 16 th International Cosmic Ray Conference*, Tokyo **10** (1979) 188.
83. V. N. Bakatanov et al., *Proc. 19 th International Cosmic Ray Conference, La Jolla* **8** (1985) 28.
84. S. N. Boziev et al., *Proc. 21st International Cosmic Ray Conference*, Adelaide **9** (1990) 38.
85. V. N. Bakatanov et al., *Proc. 24 th International Cosmic Ray Conference, Roma* **2** (1995) 720.
86. M. V. S. Rao, *Ph. D. Thesis* (University of Bombay, 1966, Unpublished).
87. B. K. Chatterjee et al., *Canadian Journal of Physics* **46** (1968) S131.
88. B. V. Sreekantan, *Proc. 12 th International Cosmic Ray Conference, Hobart* **7** (1971) 2706.
89. B. Peters, *Il Nuovo Cimento* **22** (1961) 800.
90. G. B. Khristiansen et al., *Proc. 12 th International Cosmic Ray Conference, Hobart* **6** (1971) 2097.
91. T. Hara et al., *Proc. 18 th International Cosmic Ray Conference, Bangalore* **11** (1983) 281.
92. V. S. Aseikin et al., *Proc. 12 th International Cosmic Ray Conference, Hobart* **6** (1971) 2152.
93. I. N. Kirov et al., in *Proc. International Seminar on Cosmic Ray Cascades, Sofia*

(ed: T. K. Gaisser and T. Stanev) (1980) p 61.

94. A. I. Mincer et al., *Proc. 18 th International Cosmic Ray Conference, Bangalore* **9** (1983) 383.
95. P. N. Bhat et al., *Phys. Rev.* **D25** (1982) 2820.
96. J. Kempa, *J. Phys. G: Nucl. Part. Phys.* **20** (1994) 215.
97. A. Dudarewicz et al., *Proc. VIII International Symposium on Very High Energy Cosmic ray Interactions, Tokyo* (1994) 616.
98. J. Kempa et al., *Proc. 24 th International Cosmic Ray Conference, Roma* **2** (1995) 681.
99. E. R. Bagge et al., *Proc. 16 th International Cosmic Ray Conference, Kyoto* **13** (1979) 260.
100. J. Gawin et al., *Acta Universitatis Lodziensis, Folia Physica* **7** (1984) 59.
101. B . Peters, *Proc. of the Moscow Cosmic Ray Conference* **III** (1960) 157.
102. B. Peters, in *Cosmic Pathways-Contemporary Perspectives in Physics and Astrophysics* (ed: R. Cowsik) Tata McGraw-Hill, (1986) 201.
103. S. I. Nikolsky, *Proc. of the Oxford Conference on Extensive Air Showers* (1958) 16.
104. S. I. Nikolskij and A. A. Pomanski *Zum Eksp. Teor. Fiz.* **35** (1958) 618.
105. O. Dovzenko et al., *Dokl. Akad. Nauk.* **118** (1958) 899.
106. J. Linsley, L. Scarci and B. Rossi, *Proc. International Conference on Cosmic Rays and the Earth Storm, Kyoto* **III** (1962) 91.
107. J. Linsley and L. Scarci, *Phys. Rev. Lett.* **9** (1962) 123.
108. J. Linsley and L. Scarci, *Phys. Rev. Lett.* **9** (1962) 126.
109. J. Wdowczyk and A. W. Wolfendale, *Il Nuovo Cimento* **54** (1979) 433.
110. M. Akashi et al., *Phys. Rev.* **24** (1981) 2353.
111. Carl E. Fichtel and John Linsley, *Astrophys. J.* **300** (1986) 474.
112. S. Karakula and W. Tkaczyk, *Astroparticle Physics* **1** (1993) 229.
113. S. I. Nikolsky, *Proc. International Symposium on Cosmic rays and Particle Physics, Tokyo* (1984) 507.
114. M. Aglietta et al., *Nucl. Phys. B (Proc. Suppl.)* **14B** (1990) 193.
115. A. M. Hillas, *Composition and Origin of Cosmic Rays* (ed. M. M. Shapiro, D. Reidel Publishing Company, 1983) (1983) 125.

Chapter 7
EXTREMELY HIGH ENERGY COSMIC RAYS

There are several reasons why the study in detail of the extreme end of the cosmic ray spectrum is important.

1. There could be drastic changes in the characteristics of interactions at these energies.
2. If the 3° microwave radiation is universal, as all evidence so far seems to indicate, then there should be a steepening of the primary energy spectrum beyond 10^{19} eV due to the interaction of the primaries with the 3° photons; the threshold for cut-off will be lower in the case of heavy primaries if they are present at these energies.
3. If the sources of these super high energy cosmic rays are within the Super Cluster of galaxies then the steepening may not be seen.
4. Are the primaries at these energies all protons? Could they be all iron group nuclei?
5. Will it become possible to detect the very high energy neutrinos expected to be produced in the environments of the active galactic nuclei (AGNs) either as horizontal air showers or in underground and underwater installations?

It is no exaggeration to say that for practically all the time to come, the only method available for investigations on the spectrum, composition and anisotropy of cosmic rays and also of the characteristics of interactions at energies $>10^{17}$ eV, is the method of extensive air showers, which may as technology progresses incorporate newer and newer methods of detection and analysis.

While the earliest air shower arrays for detection in these energy ranges consisted of just Geiger counter arrays, (the Harwell [1], the Cornell [2] and the Moscow State University [3]) which did not have the ability of shower direction detection, it is the Volcano Ranch array [4] with large area scintillation counters that set the trend for highly quantitative investigations at these energies by determining the shower directions and core locations with high accuracy and also established unambiguously the presence of primary particles with energies at least up to 10^{20} eV.

Table 7.1 Details of the Haverah Park, SUGAR and Yakutsk air shower arrays.

Array	Haverah Park	SUGAR	Yakutsk
Area Covered (km^2)	12	90	25
Type of detectors	1.2 m deep water Cerenkov	scintillator	scintillator & Cerenkov
Area of detector (m^2)	1–34	6	2
No. of detectors	200	94	86 + 43
No. of stations	48	47	43
Threshold energy (eV)	6×10^{16}	10^{17}	10^{17}
Angular accuracy (°)	2.5	2.6	5

The next generation of arrays that operated for several decades are the Haverah Park array [5], the SUGAR array [6] and Yakutsk array [7]. The details of these arrays in terms of their composition, effective areas, angular resolution etc. are given in Table 7.1. We shall discuss the main results from these in the next section.

A new dimension to air shower research was added with the development of the Fly's Eye array by the Utah group [8], using the air scintillation method.

While the idea of using atmospheric scintillation for the study of air showers had been proposed by Suga [9], Chudakov et al. [10], and Greisen [11], with different kinds of setups, the credit for making this a feasible one for the study of large air showers (>10^{18} eV) goes to the Utah group [12]. The main advantage of this method is that the shower core may land several kilometers from the centre and can still be detected and the shower size, core location and direction of arrival determined with high accuracy.

The scintillation is produced by the excitation of the nitrogen molecules in the air through the ionization by the thick column of charged particles in the core regions of the shower as it propagates and develops in the atmosphere. This method enables for the first time a measurement of the ionization density in a single shower at various heights and thus get a record of longitudinal development. In all the other types of arrays, only the lateral distribution of the shower particles at the level of observation is measured.

7.1 The EHE Arrays

7.1.1 The Fly's Eye Detector

The first set of Fly's Eye detectors, Fly's Eye I and Fly's Eye II, were set up by the University of Utah Group on the top of the Little Granite Mountain, Dugway, Utah about 160 km from Salt Lake City [8]. The separation between Fly's Eyes I and II is 3.3 km.

Fly's Eye I, shown in Fig. 7.1 is a cluster of 67 spherical mirror units each

Figure 7.1 Aerial view of the Fly's Eye I detector of the Utah group, located at Dugway.

mirror (Fig. 7.2) being 62″ in diameter coupled to Winston light collectors and photomultipliers packed hexagonally in groups of 12 or 14 light sensing eyes. The units which are mounted on steerable alt-azimuth mounts are serially oriented in such a way as to cover the entire visible sky. While there are in all 880 "eye"s in Fly's Eye I, in the Fly's Eye II there are 120 "eye"s only. The computer controlled electronics is designed to trigger for air showers and record the pulse integrals for each of the "eye"s from which the geometry of the track (Fig. 7.3), the longitudinal development profile (Fig. 7.9) and total shower energy are computed for each recorded event. While Fly's Eye I started operating in 1981, Fly's Eye II became operational in 1986. The Fly's Eye II views half the night sky in the direction of Fly's Eye I.

7.1.2 The Hi-Res

The Hi-Res, which is an improved version of the Fly's Eye, is a joint project of the University of Utah, the University of Illinois and the University of Adelaide, Australia, and has been constructed in the same region of the Five Mile Hill.

While the Fly's Eye had an angular resolution of 5.5°, the Hi-Res has a resolution of 1°. This is achieved by keeping a 16×16 array of hexagonal photomultiplier tubes in the image plane of the 3.8 m^2 mirrors. There are in all 152 units oriented to cover the entire night sky. The 152 units are distributed among three independent sites. Calibration with Laser beam generated events has shown that a reconstruction

Figure 7.2 Single mirror unit of the Fly's Eye detector. A cluster of 14 Winston cone light collectors, PMTs and associated preamplifiers is mounted in the mirror's focal plane.

resolution of 0.3° is feasible. Additional 4 Hi-Res detector units are under construction as of 1994, at a distance of 13 km from Fly's Eye I site. These units will permit stereo track measurements improving the reliability of the measurement of angle and energy of the showers compared to monocular Hi-Res data.

7.1.3 The AGASA Array

The Akeno group carried out a series of experiments with different arrays at different times to determine the primary energy spectrum from 10^{14} eV to 10^{20} eV. The smallest array, the 1 km^2 Array (Array 1) shown in Fig. 5.4, is used to determine the spectrum up to around 10^{17} eV. The other arrays are the Akeno 20 km^2 Array (Array 20) and the 100 km^2 Akeno Giant Air Shower Array (AGASA). The Array 20 [13] is shown in Fig. 7.4. It consists of 19 scintillation detectors, each of area 2.25 m^2, and four detectors, each of area 1 m^2, which are arranged with an inter-detector spacing of about 1 km. Two-way communication between detectors and the central station for data transmission is carried out through optical fibre cables.

The AGASA, shown in Fig. 7.5, is the world's largest air shower array and as of Dec '95, the coverage was 100 km^2 [14]. The old Akeno array of 1 km^2 area

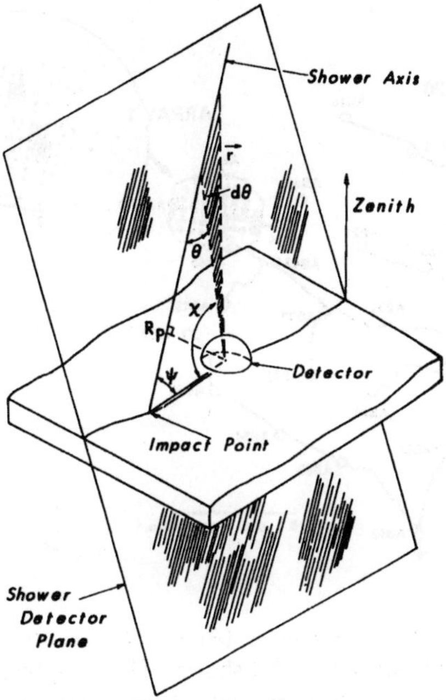

Figure 7.3 Geometry of an EAS trajectory as seen by the Fly's Eye. The shower-detector plane contains both the EAS and the center of the Fly's Eye. The impact parameter, R_p, and the angle ψ are obtained from a fit to the angle χ vs time of observation.

is part of the AGASA, which has 111 scintillation detectors each 2.2 m^2 with a separation of approximately 1 km. 27 muon detectors of varying area, which are shielded proportional counter arrays and having a threshold of 0.5 GeV for muon detection, form part of the array. The lateral distribution of charged particles in a shower of energy $10^{18.25}$ eV measured by the AGASA array is shown in Fig. 7.6. The measurements correspond to the distance range $10^{2.5}$ to $10^{3.5}$ m. In the same figure the ratio of muons (> 1 GeV) to all charged particles is also given.

The largest shower recorded with this array on December 3, 1993 is 2×10^{20} eV from the direction (l = 131°, b = − 41°).

7.2 Results

We shall now discuss the current status of the spectrum, composition and anisotropy of the primaries at energies $> 10^{17}$ eV.

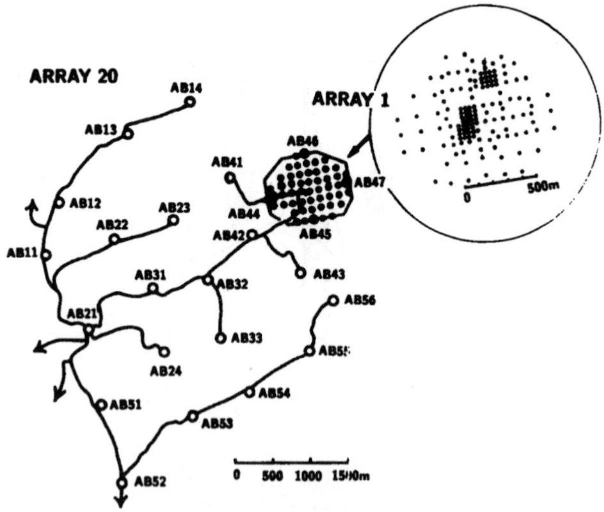

Figure 7.4 The 20 km^2 air shower array (Array 20) operated by the Akeno group (Nagano et al, 1992). The Array 1 is also shown. The detectors are scintillators of area 2.25 m^2 (o) and 1 m^2 (•). The lines are optical fibre cables for data communication. The Array 1 is also shown.

7.2.1 Primary Energy Spectrum

The Haverah Park group operated one of the largest air shower arrays to study the primary energy spectrum at energies greater than 10^{17} eV after the pioneering experiment of Volcano Ranch. The Haverah Park array [15] consists of 200 deep water Cerenkov detectors as shown in Table 7.1. The central region of the array is filled with 30 detectors, each of 1 m^2 area, with a spacing of $\sim$ 150 m ("the infilled" array). The other detectors of different areas are spread from 150 m to 2 km fro the center. The outermost 6 stations are clusters of 4 detectors, each of area 13.5 m^2.

The deep water Cerenkov detector responds to muons and photons also in addition to electrons. A vertically incident muon deposits 270 MeV in the detector. The "particle densities" are measured in terms of this energy loss. The lateral distribution of the particle density ρ is experimentally found to be of the form

$$\rho(r) = kr^{-(\eta + r/r_0)} \tag{7.1}$$

where $r_0 = 4000$ m and

$$\eta = 3.78 - 1.44 \sec\theta \tag{7.2}$$

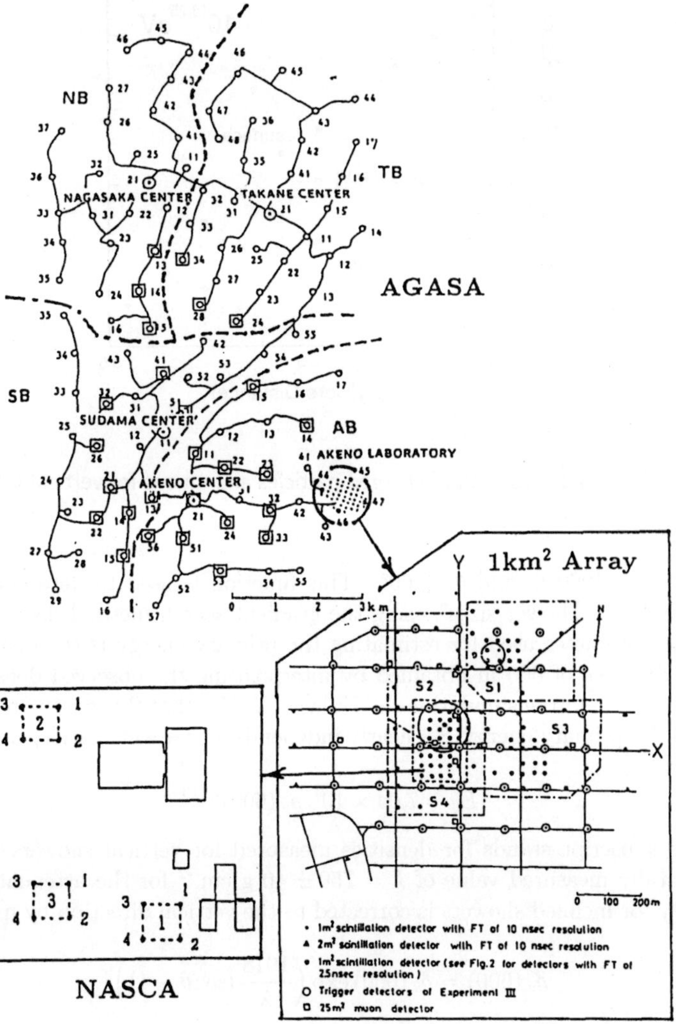

Figure 7.5 AKENO air shower arrays. (i) AGASA (since 1990), (ii) the 1 km^2 array (since 1978), (iii) NASCA (since 1993).

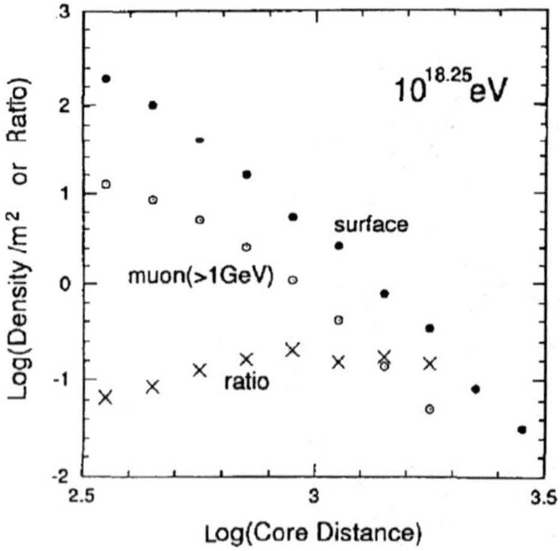

Figure 7.6 Lateral distribution of charged particles and muons for vertical EAS of energy $10^{18.25}$ eV from AGASA.

for $100 < r < 1000$ m and $\theta \le 60°$. This function is used to determine the core location and the "shower sizes" using the gradient search method described in Sec. 3.3.2. The parameter used for estimating the primary energy is the density $\rho\,(600)$ at a core distance of 600 m obtained by interpolating the observed densities. This parameter is found from simulations by Hillas et al. [16] that this parameter is directly related to the primary energy E_0, nearly independent of a wide variety of interaction models, by

$$E_0 = 7.04 \times 10^{17} \rho_v\,(600)^{1.018} \tag{7.3}$$

where the subscript stands for densities measured for vertical showers. Using the experimentally measured value of $\lambda = 760 \pm 40$ g cm^{-2} for the attenuation length, the density for inclined showers is corrected to the vertical direction by means of the relation

$$\rho_v\,(600) = \rho_\theta\,(600) \exp\left(\frac{1018}{\lambda}\,(\sec\theta - 1)\right) \tag{7.4}$$

where 1018 g cm^{-2} is the atmospheric depth of the Haverah Park array. While $\rho\,(600)$ is used for $E_0 > 4 \times 10^{18}$ eV, $\rho\,(500)$, the density at 500 m, is used for lower primary energies.

The differential primary energy spectrum thus obtained is shown in Fig. 7.7 (c), taken from the Rapporteur paper by Teshima at the Calgary Cosmic Ray Conference [17]. Clearly the spectrum cannot be fitted to a power law with a single slope over

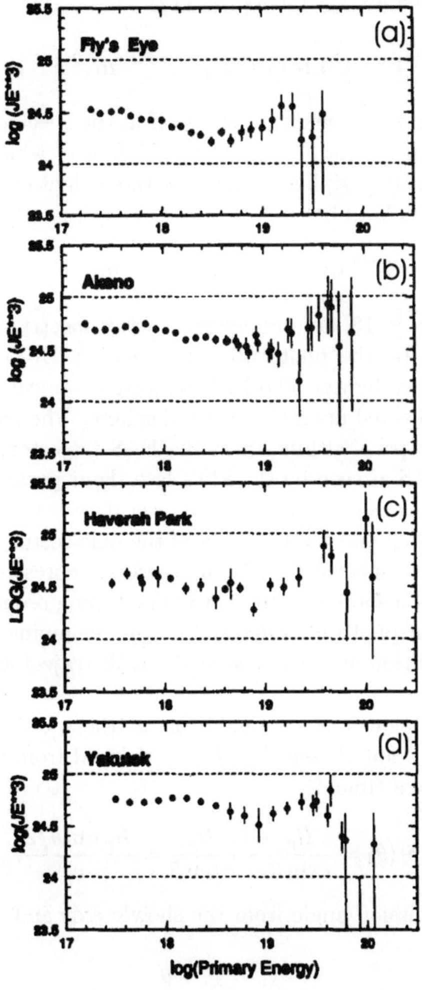

Figure 7.7 Differential primary energy spectra, for $E_0 > 10^{17}$ eV, obtained in the four Northern hemisphere experiments: (a) Fly's Eye (Bird et al, 1993); (b) Akeno (Chiba et al, 1993); (c) Haverah Park (Lawrence et al, 1991); (d) Yakutsk (Efimov et al, 1990).

the entire energy range. The slope changes from -3.0 at $E_0 < 10^{18}$ eV to -3.14 for $10^{18} < E_0 < 10^{19}$ eV. The spectrum becomes flatter above 10^{19} eV with a slope of $2.70^{+0.18}_{-0.17}$. This feature of flattening is known as the "ankle". The spectrum between 5×10^{17} eV and 4×10^{18} eV can be expressed as

$$J(E_0) = (1.31 \pm 0.14)\, 10^{27} E_0^{-3.14^{+0.05}_{-0.06}} \;\; \mathrm{m^{-2}\, s^{-1}\, sr^{-1}\, eV^{-1}} \tag{7.5}$$

with E_0 in eV. There is no evidence for a cut-off in the spectrum at least to 10^{20} eV. They have recorded eight events beyond 10^{20} eV, four of which are at zenith angles $> 45°$ or outside the array periphery. Based on these eight events, they estimate the integral flux above 10^{20} eV to be

$$I\left(> 10^{20}\ \mathrm{eV}\right) = (3 \pm 2)\, 10^{-16} \;\; \mathrm{m^{-2}\, s^{-1}\, sr^{-1}}. \tag{7.6}$$

The Fly's Eye group [8,18] determined the energy spectrum from the fluorescence light pattern recorded by the phototubes in the Fly's Eye detector (see Sec.2.4.5 for details of fluorescence light). The EAS trajectory appears as a sequential track along a great circle projected upon the celestial sphere. The phototube "hit pattern" determines a plane in space within which the EAS trajectory lies. Fig. 7.3 shows the trajectory of an EAS as seen by the Fly's Eye. Four parameters are necessary to completely specify the EAS geometry. The shower-detector plane is specified by two parameters, the azimuthal and zenith angles of the unit vector $\mathbf{n}$ normal to the plane. This vector is found by minimizing $\sum \mathbf{r}_i \cdot \mathbf{n}$ where $\mathbf{r}_i$ represents the ith observation direction vector toward a source on the EAS. The timing sequence of the light pulse arrival times at different phototubes determines the remaining two parameters which characterize the orientation and distance of the EAS trajectory relative to the Fly's Eye.

Since the EAS propagates in a straight line at the speed of light, light reaching the observer from any point P (see Fig. 7.8) is delayed from the arrival time of the passing shower front by a time

$$\delta_t(\theta) = \frac{R_p}{c \sin\theta} - \frac{R_p}{c \tan\theta} = \frac{R_p \tan\theta/2}{c}, \tag{7.7}$$

where θ is the light emission angle from the shower axis and c the velocity of light. Also

$$\chi_0 = \chi_i + \theta, \tag{7.8}$$

where χ_i is the observation angle in the shower-detector plane of the ith phototube and χ_0 represents the direction of the approaching shower. Thus

$$\chi_i(t_i) = \chi_0 - 2\tan^{-1}\left(c\,(t_i - t_0)/R_p\right), \tag{7.9}$$

where t_i is the arrival time at the ith phototube and t_0 an arbitrary reference time. A best fit of the observed $\chi_i(t_i)$ to this function yields the parameters R_p and χ_0

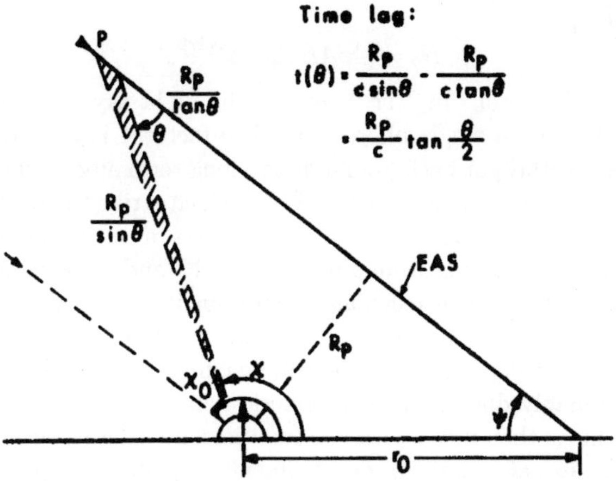

Figure 7.8 Kinematics of EAS timing for the Fly's Eye detector. Light received at the Fly's Eye from a point P on the EAS trajectory lags behind the time of passage of the shower front.

which completes the specification of the EAS trajectory. This technique results in an accuracy of better than 30% in R_p for tracks whose lengths exceed a total subtended angle of 30°. If a shower is seen by both Fly's Eyes I and II, a best shower-detector plane for each detector is determined and the intersection of these planes defines the shower trajectory. The accuracy of the estimated parameters is better in this case.

From Fig. 7.8 it can be seen that in a time Δt, the EAS source as seen by the detector moves a distance

$$\Delta L = \Delta\left(\frac{R_p}{\tan\theta}\right) = c\,\Delta t\,(1+\cos\theta)\,/\sin^2\theta, \tag{7.10}$$

and the number of photons N_{ph} reaching the detector is

$$\begin{aligned} N_{ph} &= \frac{N_e N_\gamma \,\Delta L\, A}{4\pi r^2} e^{-r/\lambda_R} \\ &= N_e N_\gamma \frac{(1+\cos\theta)}{\sin^2\theta}\frac{A}{4\pi r^2} e^{-r/\lambda_R} c\,\Delta t, \end{aligned} \tag{7.11}$$

where N_e is the number of electrons in the EAS generating the light, N_γ is the fluorescent yield in photons/electron/m, $r = R_p/\sin\theta$, A is the optical gathering area of a Fly's Eye mirror and λ_R is the Rayleigh scattering length. The number of photoelectrons N_{pe} is then given by

$$N_{pe} = \varepsilon N_{ph} \tag{7.12}$$

where ε is the photoconversion efficiency of the phototubes. The noise due to night sky background N_b is

$$N_b = (4\varepsilon AB\,\Delta\Omega\,\Delta t)^{1/2}\,, \tag{7.13}$$

where $B = 5 \times 10^5$ photons m^{-2} sr^{-1} μs^{-1} is the night sky background and $\Delta\Omega$ is the solid angle seen by a single phototube. The factor 4 is used to account for an effective increase in starlight background due to long term airglow and Cerenkov light generated by low energy cosmic rays (<1 PeV) which strike the detector head-on.

Having obtained the shower trajectory parameters, as mentioned earlier, the EAS longitudinal size N_e, obtained from Eqs. 7.11, 7.12 and 7.13, as a function of atmospheric slant depth is then calculated using an iterative process to remove the photoelectron contributions due to the direct and scattered Cerenkov beam. Shower size at the calculated depth is then determined from the number of residual photoelectrons presumably due to the atmospheric scintillation process. Each resultant longitudinal profile is then fitted with two functions in order to obtain best estimates of (1) the shower size at maximum development N_m, (2) the depth of shower maximum X_m, and (3) the track length integral $\int N_e\,(x)\,dx$. The two functions used are: (1) an unconstrained (3 free parameters) Gaisser-Hillas shower development function (see Sec. 2.4) of the form

$$N_e\,(x) = N_m \left(\frac{x - x_0}{X_m - x_0}\right)^{(X_m - x_0)/\lambda} e^{(X_m - x)/\lambda} \tag{7.14}$$

where x_0 is the depth of first interaction and $\lambda = 70$ g cm^{-2} and (2) a Gaussian. An example of such a fit is shown in Fig. 7.9. Both functions give nearly the same value for the track length integral $\int N_e dx$. Finally the primary energy E_0 is obtained from either of the functions:

$$E_0 = \frac{\epsilon_0}{X_0}\int N_e\,(x)\,dx, \tag{7.15}$$

where ϵ_0 and X_0 are the critical energy (see Sec. 2.2.2.2) and radiation length (see Sec. 2.2.1.3) respectively. A correction of roughly 10% is applied to the estimated primary energy to account for undetected energy loss via: (1) undetected neutral particles that fail to quickly decay into detectable charged particles before striking the ground, (2) a significant number of high energy muons and (3) nuclear excitation by hadrons in the shower.

The Stereo Fly's Eye energy spectrum based on combination of Fly's Eye I and Fly's Eye II is shown in Fig. 7.10. It is clear that there is a change of spectral index around 4×10^{17} eV and again around 4×10^{18} eV, the slope changing from -3.01 ± 0.06 to -3.27 ± 0.02 to -2.71 ± 0.10. The spectral indices in the various energy regions are shown in Table 7.2.

The usual explanation for this ankle is that primaries of extragalactic origin dominate over the galactic spectrum. If this is true there should be a change in the amplitude and phase in the anisotropy of the arrival direction of the primaries. Also one should observe a steepening or a cut-off due to the interaction of the primaries with

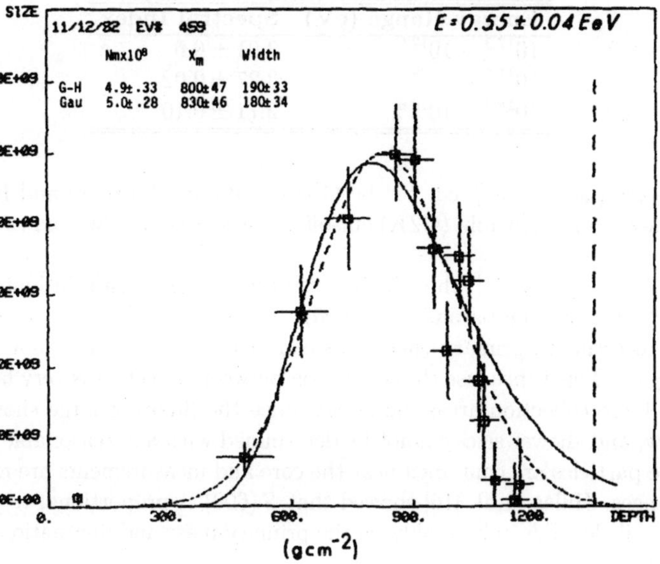

Figure 7.9 An example of the longitudinal developement curve obtained with the Fly's Eye. Observations were fitted to two curves; Gaisser-Hillas (G-H, solid line) and a Gaussian (Gau, dashed line).

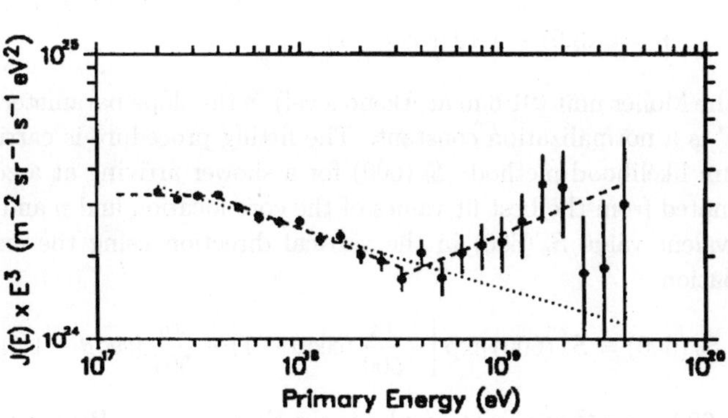

Figure 7.10 Stereo Fly's Eye Energy Spectrum (Bird et al., 1994).

Table 7.2 Spectral Indices from the Fly's Eye Experiment.

Energy Range (eV)	Spectral Index
$10^{17.3} - 10^{17.6}$	3.01 ± 0.6
$10^{17.6} - 10^{18.3}$	3.27 ± 0.02
$10^{18.3} - 10^{19.6}$	2.71 ± 0.10

the microwave photons as predicted by Greisen [19] and Zatsepin and Kuzmin [20], the Greisen-Zatsepin-Kuzmin (GZK) cut-off, if the sources are beyond the local super cluster.

The AGASA group [21], though they use the conventional scintillators for measuring electrons, use the parameter $S\,(600)$, the electron density at a core distance of 600 m, to estimate primary energy instead of the conventional shower size as at lower energies. This is because the separation between detectors is very large in order to achieve large collection areas, necessary since the fluxes of large showers are extremely low, and shower sizes cannot be determined with any reasonable accuracy as most of the particles are contained near the core and measurements are made only at large distances. Hillas et al. [16] showed that $S\,(600)$ is proportional to the primary energy since it depends only weakly on the primary mass and fluctuations in cascade development.

The arrival directions are determined from the relative times of arrival at different detectors using a plane front approximation and core location is determined by fitting the observed densities to the experimentally obtained lateral distribution function, in the shower plane, given by

$$\begin{aligned} \Delta\,(r) &= C\left(\frac{r}{r_0}\right)^{-1.2}\left(1+\frac{r}{r_0}\right)^{-(\eta-1.2)}\left(1+\left(\frac{r}{1000}\right)^2\right)^{-0.6}, \\ \eta &= 3.97 - 1.79\,(\sec\theta - 1) \end{aligned} \tag{7.16}$$

where r_0 is the Moliér unit (91.6 m at Akeno level), η the slope parameter, θ the zenith angle and C is a normalization constant. The fitting procedure is carried out using the maximum likelihood method. $S_\theta\,(600)$ for a shower arriving at a zenith angle θ is then estimated from the best fit values of the core location and η and transformed to the equivalent value $S_v\,(600)$ in the vertical direction using the experimentally observed relation

$$S_\theta\,(600) = S_v\,(600)\exp\left[-\frac{x_0}{500}\,(\sec\theta - 1) - \frac{x_0}{594}\,(\sec\theta - 1)^2\right] \tag{7.17}$$

where $x_0 = 920$ g cm^{-2} is the atmospheric depth at Akeno. Primary energy E_0 is then estimated from the relation

$$E_0 = 2.0 \times 10^{17} S_v\,(600) \quad \text{eV}, \tag{7.18}$$

Table 7.3 Values of the constants in the differential energy spectrum obtained from Array 20 of the Akeno group (Nagano et al, 1992).

Energy Range (eV)	**A**	$\mathbf{E}_c$	γ
$10^{17.9} \sim 10^{18.9}$	$(1.0 \pm 0.1) \times 10^{-29}$	$10^{17.9}$	3.16 ± 0.08
$10^{18.9} \sim 10^{19.9}$	$(6.9 \pm 1.5) \times 10^{-33}$	$10^{18.9}$	2.8 ± 0.3

obtained from Monte Carlo simulations by Dai et al. [22].

While the above relations are used for the data from AGASA, the following relations are used for the data from the Array 20:

$$\rho(r) = N_e C_e \left(\frac{r}{r_0}\right)^{-1.2} \left(1 + \frac{r}{r_0}\right)^{-(\eta-1.2)} \left(1 + \frac{r}{2000}\right)^{-0.5} \tag{7.19}$$

where

$$\eta = (3.80 \pm 0.05) + (0.10 \pm 0.05) \log \left(N_e/10^9\right) \tag{7.20}$$

and

$$S_\theta(600) = S_v(600) \exp\left(\frac{-x_0(\sec\theta - 1)}{500}\right). \tag{7.21}$$

The differential energy spectrum thus obtained from Array 20 [13] is given by

$$J(E_0) = A(E_0/E_c)^{-\gamma} \quad \mathrm{m^{-2}\,s^{-1}\,sr^{-1}\,eV^{-1}}, \tag{7.22}$$

with the values of the constants shown in Table 7.3. The spectrum becomes flatter around $10^{18.9}$ eV, the slope changing from 3.16 to 2.8. We have seen in Sec. 5.2.1 that the spectrum determined by the same group with the 1 km^2 array shows that the slope changes from 2.62 ± 0.12 in the energy range $10^{14.5} \sim 10^{15.67}$ eV to 3.02 ± 0.05 in the range $10^{15.67} \sim 10^{18}$ eV. Thus, after the knee, the spectrum steepens further and then flattens.

The spectrum from the AGASA experiment [21], shown in Fig. 7.11, extends from $10^{18.5}$ eV to 10^{20} eV. The spectrum obviously cannot be fitted to a single power law. It is therefore fitted to a combination of two power laws. The fit is carried out with the two power law indices and the energy E_a where the change of slope occurs—the ankle energy. The index in the lower energy region is constrained to 3.2, the value obtained with the Array 20. The best fit is obtained with $E_a = 10^{19}$ eV and represented as

$$J(E_0) = \left(2.3^{+0.1}_{-0.2}\right) \times 10^{-33} \left(\frac{E_0}{10^{19}}\right)^{-2.3^{+0.5}_{-0.3}} \quad \mathrm{m^{-2}\,s^{-1}\,sr^{-1}\,eV^{-1}}, \tag{7.23}$$
$$\text{for } 10^{19}\ \text{eV} \le E_0 \le 10^{19.9}\ \text{eV}.$$

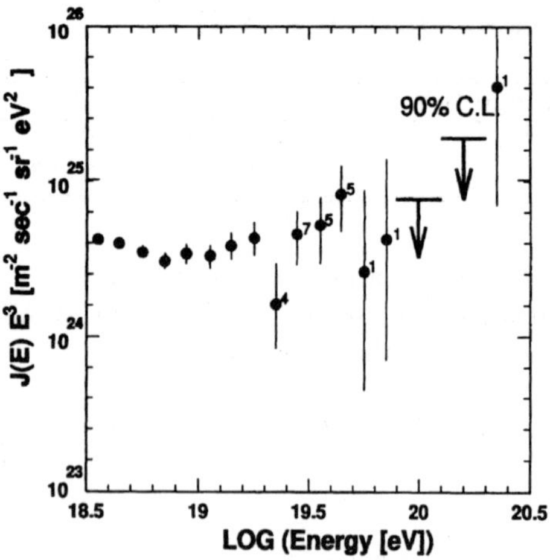

Figure 7.11 The primary energy spectrum obtained from the AGASA experiment (Yoshida et al., 1995). Numbers attached to the points are the number of events in each group.

Extrapolation of the spectrum beyond $10^{19.9}$ eV gives for the operation period 7 showers, while only one has been observed. The combined spectrum from Array 20 and AGASA is shown in Fig. 7.7(b).

The Yakutsk group [23] also use $S\,(600)$ as a primary energy estimator (see Table 7.1 for array details). Since this experiment is at sea level, they use the following conversion factor from $S\,(600)$ to E_0.

$$E_0 = (4.8 \pm 1.0)\, 10^{17} S_v\,(600)^{1.00\pm0.02} \quad \text{eV} \tag{7.24}$$

and

$$S_v\,(600) = S_\theta\,(600) \exp\left(x_0\left(\sec\theta - 1\right)/\lambda_{600}\right) \tag{7.25}$$

where

$$\lambda_{600} = 460 + 32 \log\left(E_0/10^{18}\right) \quad \text{g cm}^{-2} \tag{7.26}$$

Eq. 7.26 is valid up to $\theta < 50°$, beyond which it is 1000–1200 g cm^{-2}—similar to the muon component. Eq. 7.24 is obtained form the calorimetric measurement of Cerenkov light [24]. The energy spectrum from the Yakutsk group is shown in Fig. 7.7(d). It shows similar features as in the other experiments, *viz* steepening and then flattening. The slopes in the different energy regions are shown in Table 7.4. It is clear that the "ankle" is seen in this experiment also.

Thus the primary spectrum in the air shower range is characterized by three steps. The spectrum steepens from 2.62 ± 0.12 below $\sim 10^{15}$ eV to 3.02 ± 0.03

Table 7.4 Values of the logarithmic slopes of the differential primary energy spectrum in different energy intervals from the Yakutsk experiment (Afanasiev et al, 1995).

Energy Range	**Slope**
$10^{17.5} - 10^{18.1}$	-3.03 ± 0.04
$10^{18.0} - 10^{18.8}$	-3.30 ± 0.05
$10^{18.9} - 10^{19.6}$	-2.75 ± 0.2

between $10^{15.7} \sim 10^{17.8}$ eV, steepens further to 3.16 ± 0.3 between $10^{17.8}$ and $10^{18.8}$ eV and flattens beyond to 2.8 ± 0.3. Though all the four experiment have shown similar features of the spectrum and all have evidence for the "ankle", the energies at which the changes of slope occur are somewhat different. The Akeno group have normalized the spectra of the Haverah Park, Yakutsk and the AGASA groups to the Stereo Fly's Eye spectrum by multiplying the energies by 0.9, 0.8 and 0.8 respectively. Considering the different techniques used in the different experiments, systematic errors in the energy estimates of such magnitude are not unreasonable. The resultant spectrum is shown in Fig. 7.12. The spectral slopes of all experiments agree with

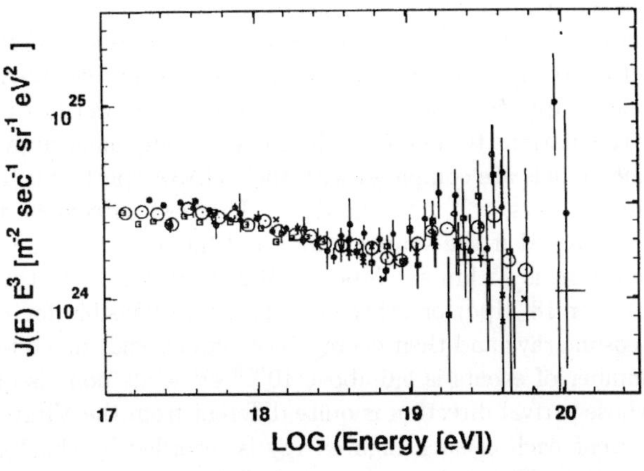

Figure 7.12 Primary energy spectrum at $E_0 > 10^{17}$ eV from the Haverah Park (o), the Yakutsk ($\times$), the Stereo Fly's Eye ($\bullet$), the AGASA ($\square$) experiment and the geometrical average of the four experiments ($\bigcirc$).

each other within 30% in energy and the ankle is clearly seen at $10^{18.9}$ eV and there is a suggestion of cut-off at $10^{19.6}$ eV. Teshima [17] estimates the slopes of this "Grand Unified Spectrum" in different energy regions as shown in Table 7.5.

Table 7.5 Logarithmic slopes of the 'Grand Unified Spectrum of primary cosmic rays in different energy regions (Teshima 1993).

Energy Region	Slope
$\leq 10^{17\pm0.2}$	3.02 ± 0.03
$10^{17.8\pm0.2} - 10^{18.9\pm0.2}$	3.20 ± 0.05
$\geq 10^{18.9\pm0.2}$	2.6 ± 0.1

With better statistics available from all the four experiments combined together, one can now see whether the GZK cut-off is already seen. Teshima notes that the number of events observed above 10^{19} eV in all the experiments is 881. If we extrapolate the "Grand Unified Spectrum" with a slope of 2.6 ± 0.1 we expect 22 ± 5 events at $E_0 > 10^{20}$ eV while only 7 are observed. If the extrapolation is done beyond 2×10^{20} eV, the expected number is 7.3 ± 2.5 whereas only one or two events are observed. Thus, the GZK cut-off may have been already seen!

The implication of this behavior of the energy spectrum is that the galactic component decreases, may be as a power law of $\sim$3.5 and an extragalactic component with a power law of $\sim$2.1 becomes dominant beyond 10^{19} eV. The structure of the spectrum at these energies depends on the distances of the sources. Fig. 7.13 shows the spectra expected for different extragalactic origin models, (a) a single source at different distances (redshift z) and (b) many sources distributed uniformly in the Universe, calculated by Yoshida and Teshima [25], as quoted by the AGASA group [21]. The extragalactic component with an injection spectrum power law index of -2.3 is assumed in the calculations to dominate the galactic component above 10^{19} eV. The predictions of the models are compared with the AGASA spectrum. The model with distributed sources agrees well with the data. The best agreement with the single source model is obtained if the source is at a redshift larger than ~ 0.1 (450 Mpc if the Hubble constant is 75 km s^{-1} Mpc^{-1}). If nearby sources in the Virgo Cluster (redshift 4×10^{-3} or 18 Mpc) or other clusters are responsible for the bulk of the highest energy cosmic rays and their energy spectrum extends much beyond 10^{20} eV, the expected number of events is 5.6 above $10^{19.9}$ eV while none is observed except for one event whose arrival direction is quite different from the Virgo Cluster.

In fact one event each of such high energy is recorded by the Yakutsk and the Fly's Eye groups also. These three events, which are described below (Sec. 7.2.2), however have energies about a factor of two or three larger than the next highest energy events observed in the respective experiments.

7.2.2 The Highest Energy Cosmic Rays (> 10^{20} eV)

So far in the history of cosmic ray research extending over 80 years and of extensive air showers extending over 45 years, only three events, apart from the 5 events reported long ago by the Haverah Park group, have been recorded with energies greater than

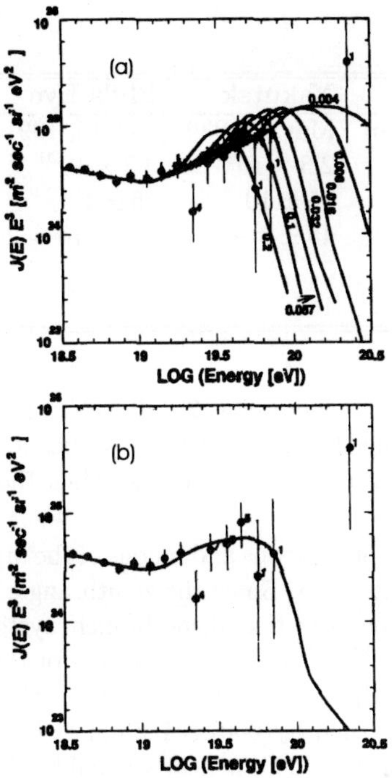

Figure 7.13 Primary energy spectra expected for several models of extragalactic source location calculated by Yoshida and Teshima (1993) taking into account the AGASA energy resolution, compared with the AGASA spectrum: (a) sources are located at the same redshift (shown beside the curves); (b) sources are distributed isotropically in the Universe.

10^{20}eV, one by the Yakutsk group [26] a second by the Fly's Eye group [27] and the third by AGASA [28]. The details regarding these three special events, reported recently, are summarized in Table 7.6, where the Right Ascension α, declination δ, the galactic latitude b and longitude l of their arrival directions are also shown.

The Fly's Eye event is a monocular event observed by Fly's Eye I. The longitudinal development of this event is shown in Fig. 7.14. At the shower maximum the number of particles has reached the value 2×10^{11}. The impact parameter is 12.06 ± 1.15 km and the angle subtended by the track length is more than 40°. These parameters for monocular events imply a good geometry. The shower development for this event leaves no doubt about the energy estimation of 3×10^{20} eV. After a careful examination of systematic errors in energy estimation, the authors conclude that there is no way

Table 7.6 Details of the three highest energy events recorded by the Yakutsk, Fly's Eye and AGASA groups.

Group	**Yakutsk**	**Fly's Eye**	**AGASA**
Date of Observation	May 7, 1989	Oct 15, 1991	Dec 3, 1993
Energy (eV)	2.3×10^{20}	3.2×10^{20}	$(1.7 \sim 2.6) \times 10^{20}$
α (°)	75 ± 10	86 ± 1	18.9
δ (°)	45 ± 4	$44 \pm (10 \sim 20)$	21.1
b (°)	3	8	−41
l (°)	162	167	131

of bringing the energy of this event below 2×10^{20} eV. The depth of shower maximum X_m is 852^{+68}_{-100} g cm^{-2} while the expected mean values for proton and iron initiated showers are 850 and 780 g cm^{-2} respectively. Thus the primary is most likely a proton.

The lateral distribution of electrons and muons of the highest energy event recorded by AGASA is shown in Fig. 7.15. Since the zenith angle of the shower is 22.9° and the core landed near the center of the Akeno branch, systematic error in the estimation of $S\,(600)$ is quite small. The value of $S_{23}\,(600)$ for this event is 892 m^{-2} and its uncertainty, estimated by the authors from Monte Carlo simulations, is +21% and −6.6%. If Eq. 7.17 is used, then $S_v\,(600)$ is 1065 m^{-2}. However if the shower is at its maximum development, then $S_v\,(600)$ is nearly equal to $S_{23}\,(600)$. Therefore $S_v\,(600)$ should be in the range $(892 \sim 1065)^{+21\%}_{-6.6\%} = (833 \sim 1289)$ m^{-2}. The corresponding primary energy range is shown in Table 7.6.

The Yakutsk event arrived at a zenith angle of 59°. The lateral distribution of electrons and muons in this event is shown in Fig. 7.16. The electron and muon densities are comparable. The primary energy is estimated to be 1.2×10^{20} eV, using the conventional $S\,(600)$ to E_0 relation. If the energy is estimated using the muon densities, it comes out to be $\sim 10^{21}$ eV [17] !

It is interesting that all the three events have come within 50° of the antigalactic center. In fact, the space angle between the arrival directions of the Yakutsk and the Fly's Eye events is only $\sim 7°$, within the errors of angle estimates. However, the probability of the two events coming within this angle is estimated by Sommers to be 0.008.

These three events have posed quite a challenge in terms of interpretation regarding the nature of the primaries and the possible sources. Sommers, dealing with the first two cases, has shown that neither nuclei, protons, neutrons nor gamma rays can survive at these energies from a distance of more than 50 Mpc. The puzzle is that there is no likely candidate source within this distance if the particles are coming straight from the source without any deflection by the intergalactic magnetic fields. The direction of the Fly's Eye event is 88° away from M87 and 134° away from Cyg A.

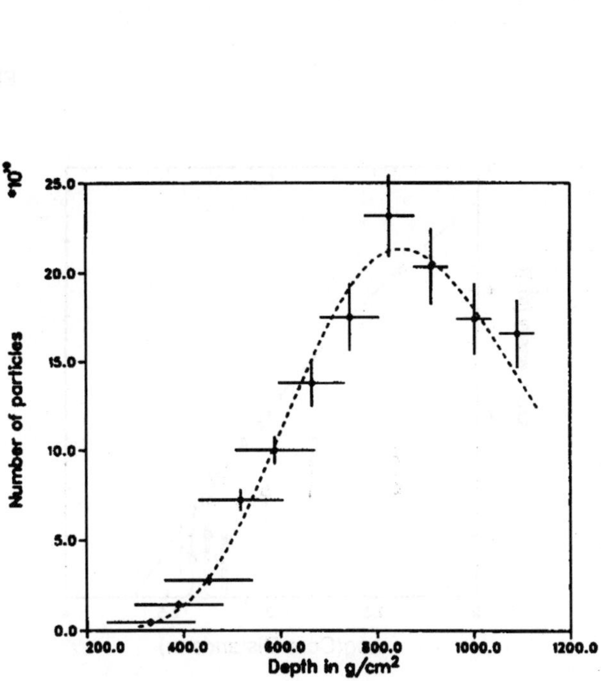

Figure 7.14 The longitudinal developement curve of the highest energy (300 EeV) event recorded by the Fly's Eye.

These cannot be the sources unless our ideas of the 2 micro gauss magnetic field is totally wrong. Hayashida et al. have come to a similar conclusion on the third event which comes from a totally different direction from the first two.

Chi et al. [29] and Efimov and Mikhailov [30] have claimed evidence for clustering in showers of $E > 10^{19}$ eV. The three highest energy showers do seem to fall within these clusters.

The scenarios in which particles can be accelerated to energies beyond 10^{19}eV have been discussed by Yoshida [31] along with the consequences on spectrum and composition :

1. Compact objects in the galactic plane : If these are the sources then there should be a strong anisotropy in the galactic plane which has not been observed.

2. Jokipii and Morfill [32] have proposed the acceleration of particles at the terminal shocks of the galactic wind. In this model, protons can be accelerated to 10^{19}eV and iron nuclei to 10^{20} eV. Of course multiple terminal shocks in the local group can accelerate protons to 10^{20}eV.

3. Hillas [33] has considered the AGNs as the primary source for acceleration of particles. Ip and Axford [34] have shown that diffusive shock acceleration

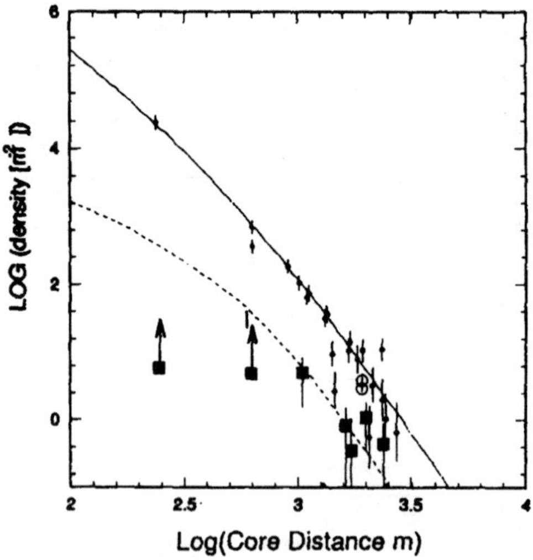

Figure 7.15 The lateral distribution of electrons (•) and muons (■) of the highest energy (200 EeV) event recorded by AGASA.

mechanism operating in extragalactic radio jets can lead to protons of 10^{19}eV while heavy nuclei can be accelerated beyond 10^{20}eV by such a mechanism. This would result in a heavy primary dominant composition beyond 10^{19}eV and a cut-off due to photodisintegration beyond 10^{20}eV.

4. Biermann [35] has suggested hot spots of Fanaroff Riley Class II radio galaxies like Cyg A could be sources of proton acceleration up to 10^{21}eV. In this case, protons will dominate over heavies beyond 10^{19}eV.

5. A new mechanism for the production of super high energy particles without any acceleration process being involved is the decay of super massive 10^{25}eV particles that can be produced through the annihilation or collapse of Topological Defects (TDs) [36]. The same processes also lead to super high energy neutrinos.

7.2.3 Composition Beyond 10^{17} eV

At these very high energies, since the determination of the composition is by a very indirect method, the resolution permits us only to say whether the primaries are predominantly heavies or light (mixed composition) or protons.

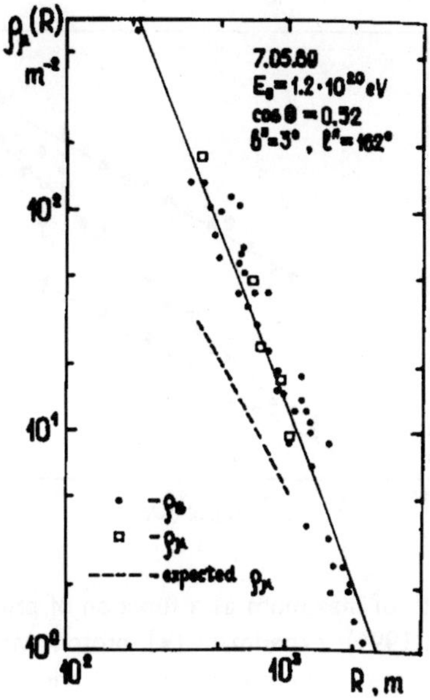

Figure 7.16 Lateral distribution of electrons and muons in the highest energy event observed by the Yakutsk group (Efimov et al., 1990).

The method depends on the fact that while the heavy primary induced showers reach their maxima very high in the atmosphere the proton showers reach their maxima deep down. Also, the fluctuations in the position of shower maximum are much larger for proton induced showers compared to those induced by, say, iron primaries.

The Fly's Eye group [18] has determined X_m as a function of primary energy from the stereo data only using the procedure discussed in Sec. 7.2.1. Their results are shown in Fig. 7.17. Also shown in the figure are the expected dependence for proton and iron showers and a simple energy dependent two component (p and Fe) composition model from Monte Carlo simulations of Gaisser et al. [37]. This model is constructed to fit their primary energy spectrum discussed in Sec. 7.2.1. The elongation rate for the experimental data is 69 ± 1.9 g cm^{-2}. The data is not a good fit for a single straight line. The elongation rate from 0.3 to 10 EeV is 78.9 ± 3 g cm^{-2} while below 0.3 EeV the rate is 50 g cm^{-2}. The elongation rate for any pure elemental composition is 50 g cm^{-2} per decade of energy. This means going from a heavy composition near 3×10^{17} eV to a proton dominated one near 10^{19} eV.

These conclusions are naturally based on the assumption that there are no drastic

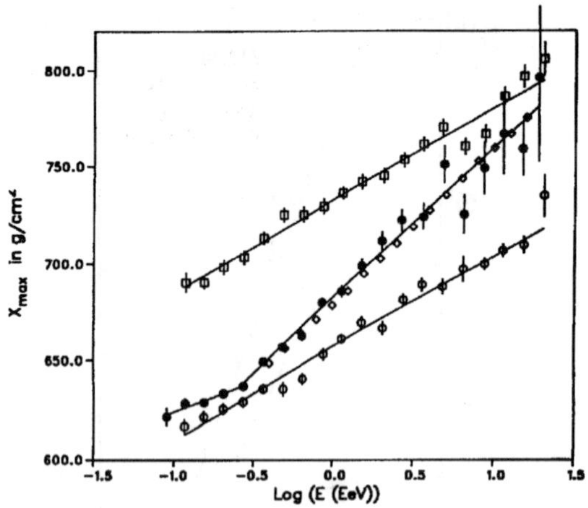

Figure 7.17 Average depth of maximum as a function of primary energy from the Fly's Eye experiment(Bird et al., 1994): experiment (•); proton, Monte carlo (□); iron, Monte carlo (o); two component mixed composition, Monte Carlo (◇).

changes in the characteristics of collisions at super high energies. Hillas [38] summarized the Fly's Eye and the Yakutsk data [39] as shown in Fig. 7.18. The Yakutsk group obtained X_m from the lateral distribution of Cerenkov light as discussed in Sec. 6.2.1. They use the quark-gluon string model for Monte Carlo simulations. It can be seen that this model predicts an elongation rate for both protons (p-qgs) as well as iron nuclei (Fe-qgs) that is different from the calculations of Gaisser et al.'s model (p and Fe). These differences in the predictions of different interaction models have to be resolved before firm conclusions can be drawn regarding the composition in this energy region.

The Haverah Park group [40] determined the elongation rate from measurements of the rise time of pulses in their water Cerenkov detectors. From Monte Carlo simulations they have shown that this parameter is related to X_m. However, instead of depending on Monte Carlo simulations, they adopted a different method suggested by Linsley [41]. He suggested that for an air shower parameter P which does not depend on primary energy but depends on the depth of observation X and X_m the following relation holds:

$$\left(\frac{\partial P}{\partial \log E_0}\right)\bigg|_X = -FD_{10}\left(\frac{\partial P}{\partial X}\right)\bigg|_{E_0} \tag{7.27}$$

where $F = X/X_m$ if the depth dependence of P is of the form $f\,(X/X_m)$ and $F = 1$

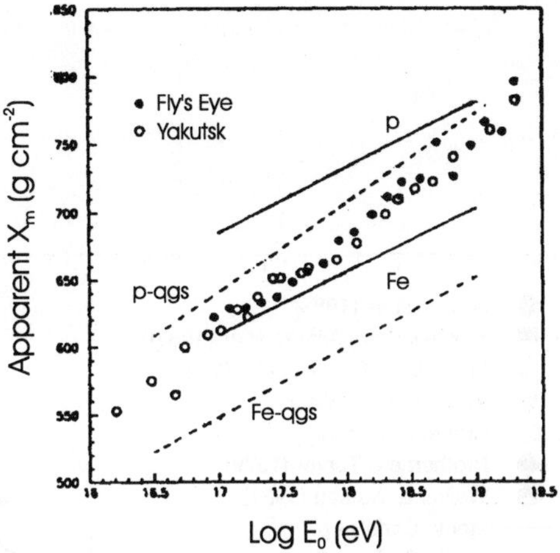

Figure 7.18 Summary by Hillas (1994) of the dependence of X_m on E_0 from the Fly's Eye (Bird et al, 1994) and Yakutsk (Dyakonov et al, 1993) experiments, compared with Monte Carlo simulations of Gaisser et al (1993) (p and Fe) and Dyakonov et al (1993) (p-qgs and Fe-qgs) for protons and iron nuclei.

if the depth dependence is of the form $f\,(X - X_m)$ and $D_{10} = \partial X_m / \partial \log E_0$ is the elongation rate per decade. Thus the energy dependence of the parameter at a fixed depth is related to its depth dependence at a fixed energy through the elongation rate. The Haverah Park group chose $t_{1/2}$ the time for the signal to rise from 10% to 50% of its peak value and showed that the favored form of its depth dependence is $f\,(X - X_m)$ by studying its variation as a function of atmospheric pressure. The elongation rate obtained by them from the experimentally measured values of $\left(\frac{\partial P}{\partial \log E_0}\right)\Big|_X$ and $\left(\frac{\partial P}{\partial X}\right)\Big|_{E_0}$ is 40 ± 20 g cm^{-2} per decade for $E_0 > 7 \times 10^{18}$ eV. Linsley and Watson [46] obtained $X_m = 805 \pm 30$ and 845 ± 80 g cm^{-2} at $E_0 = 10^{19}$ and 10^{20} eV respectively, using this elongation rate by normalizing to $X_m = 730$ g cm^{-2} at $E_0 = 10^{18}$ eV so as to agree with the value obtained from Cerenkov light measurements (see Sec.6.2.1). These values are shown in Fig. 7.19. where the Fly's Eye results as well as those from Cerenkov light measurements in the energy range $10^{15} \sim 10^{20}$ eV are shown. The Haverah Park measurements are the only ones available at such high energies so far. The Monte Carlo simulations are those of Gaisser et al. [37]. It can be seen that while most of the data are consistent with each other, indicating change of composition from a mixed one to all protons at around 10^{16} eV, the Fly's Eye results are at variance with others, particularly in the lower energy region.

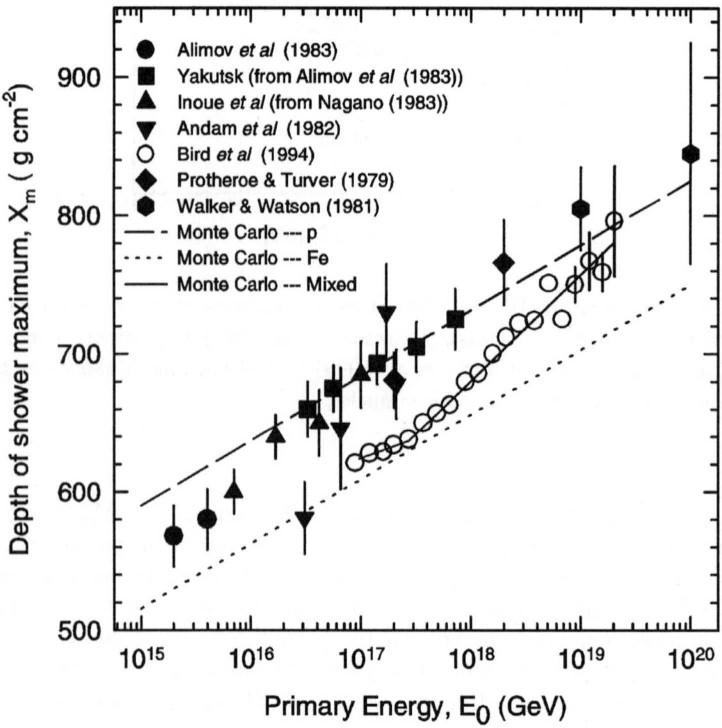

Figure 7.19 Depth of shower maximum as a function of primary energy from Cerenkov and fluorescence light measurements, except those of Walker and Watson (1981) which are from pulse shape measurements in water Cerenkov detectors.

7.2.4 Anisotropy

Studies on the anisotropy in the arrival directions of primary cosmic rays are carried out essentially from the point of view of identifying the sources of cosmic rays. The primaries being predominantly charged particles—protons and heavy nuclei—get deflected in the intergalactic magnetic fields up to very high energies. It is only beyond $\sim 10^{18}$ eV, one can expect the charged particles to arrive essentially from the same direction as the sources. Even at these energies the extragalactic cosmic rays are completely isotropised by the intergalactic and galactic magnetic fields. In recent years, there is considerable interest in looking for very high energy neutrons and γ rays that are expected to be produced by charged particles accelerated to high energies in the special environments of the sources. The neutrons are produced in high energy nuclear interactions and the γ rays through bremsstrahlung. These come directly from the directions of the sources.

The work that was carried out in this field for showers of energy greater than 10^{18} eV has been summarized by Watson [47]. The celestial arrival direction distribution for primaries of energy greater than 4×10^{19} eV is shown in Fig. 7.20.

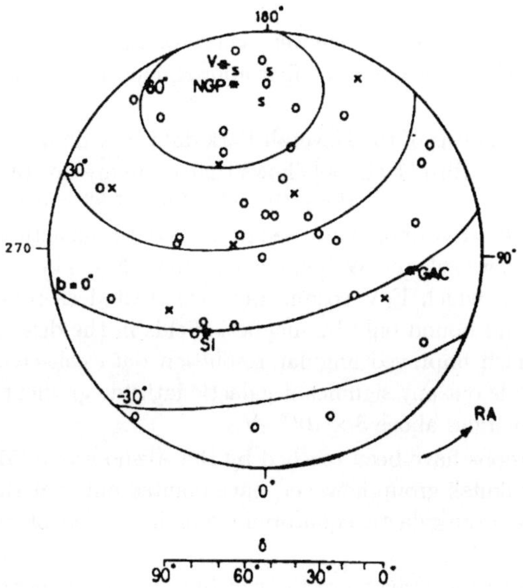

Figure 7.20 Arrival direction distribution of primaries of energy $> 4 \times 10^{19}$ eV summarised by Watson (1985).

There is no evidence for any clustering in the galactic plane which should have been the case if the sources are predominantly galactic. However, there is some weak evidence for an excess from the direction of the local super cluster. It is interesting to note that the three largest showers recorded with the SUGAR array (marked S in the figure) arrive from a direction very close to the galactic north pole.

Another way in which anisotropy was looked at was through a harmonic analysis of the counting rates in sidereal time for showers of different energy. The results from a large number of experiments covering the energy range $10^{13} \sim 10^{19}$ eV have been summarized by Linsley [48].

The important feature seems to be that there are significant, first and second harmonics over the entire range and there is a tendency for the amplitude to increase with energy (see Figs. 7.21 and 7.22). The phase variations with energy show some interesting trend—decreasing from a value of 20 hours below the knee energy to 12 hours at 10^{17} eV with a steep rise to 24 hours at 10^{18} eV and a further tendency to decrease sharply. There are many interpretations of this behavior. Some take it to indicate that the energy spectrum of the primaries is not the same for all directions. Some have talked about a gradient in cosmic ray density within the galactic disk.

Possible extensions of the galactic magnetic field into the halo and contributions from extragalactic sources have also been considered in this context in view of a possible flattening of the spectrum beyond 10^{18} eV. For a clearer understanding it is necessary to broadly separate out showers as proton induced, heavy primary induced and γ-induced which can be done to a first approximation by making use of the muon association criteria.

A more recent analysis of the Haverah Park data by Gilman and Watson [49] and the Fly's Eye data by Bird et al. [50] shows that there are no statistically significant anisotropies at energies greater than 10^{17} eV. Fig. 7.23 shows the galactic plane anisotropy of the Fly's Eye data, expressed in terms of the galactic enhancement factor f_E proposed by Wdowczyk and Wolfendale (see Ref. [29]) The absence of anisotropy rules out models in which EeV cosmic rays are treated as protons accelerated in the galactic disk and bound only by magnetic fields in the disk of the Galaxy. The analysis of the much improved angular resolution data collected with Fly's Eye I and Fly's Eye II rule out any significant galactic latitude gradient and galactic plane enhancement at anergies above 3×10^{17} eV.

Similar conclusions have been reached by the Akeno group [51] and the Yakutsk group [52]. The Yakutsk group however, have pointed out that there is some indication of an increase in the galactic equatorial region in showers of energy $(1 \sim 4) \times 10^{19}$ eV.

Fig. 7.24 showing the arrival directions of 162 showers of energy greater than 10^{19} eV is based on the Akeno data [53]. In the figure the North pole is at the centre. There is no evidence of any clustering in any part of the sky examined contrary to the claim of Chi et al. [29]. The same data has also been analyzed for evidence for any galactic plane enhancement and the answer is negative.

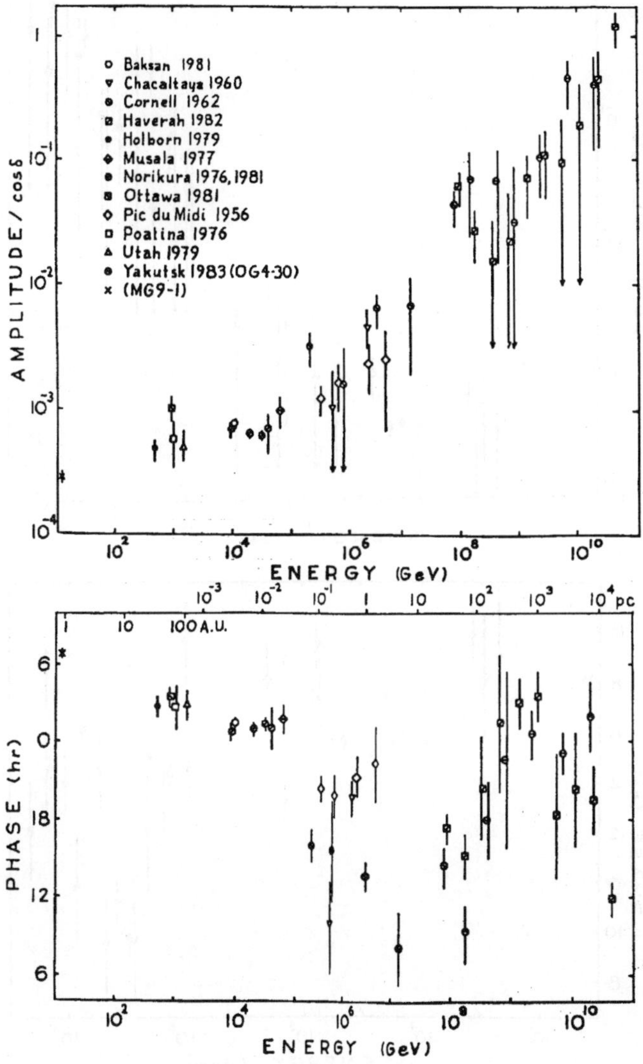

Figure 7.21 Dependence of amplitude and phase of the first harmonic on primary energy summarised by Linsley (1983). For references see Linsley (1983).

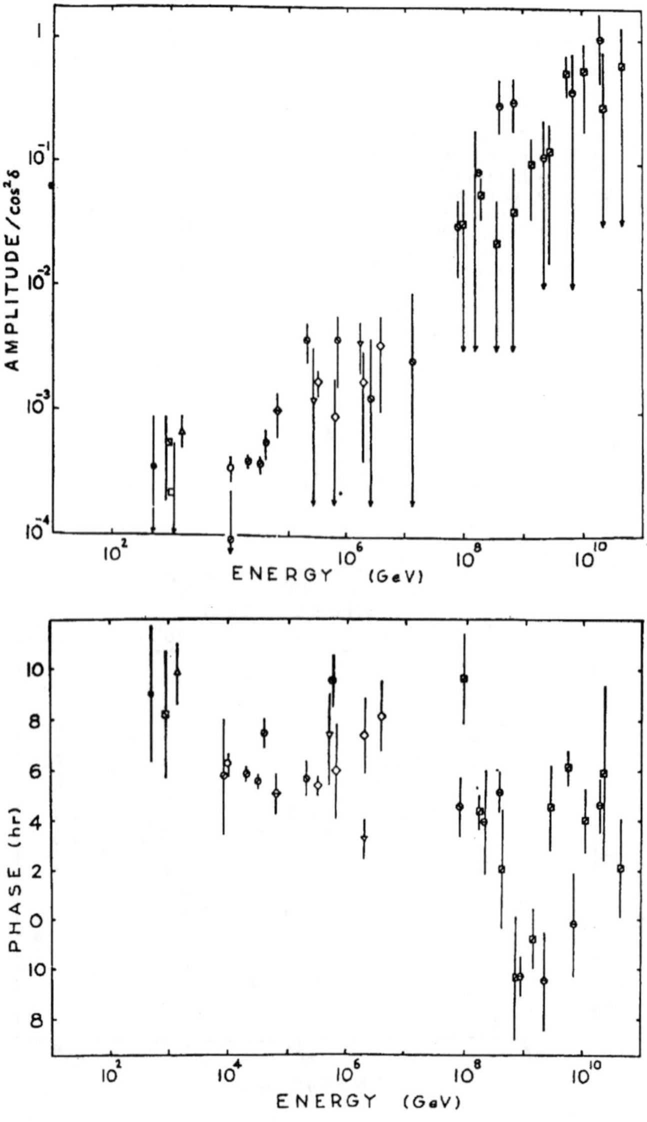

Figure 7.22 Dependence of amplitude and phase of the second harmonic on primary energy summarised by Linsley (1983).

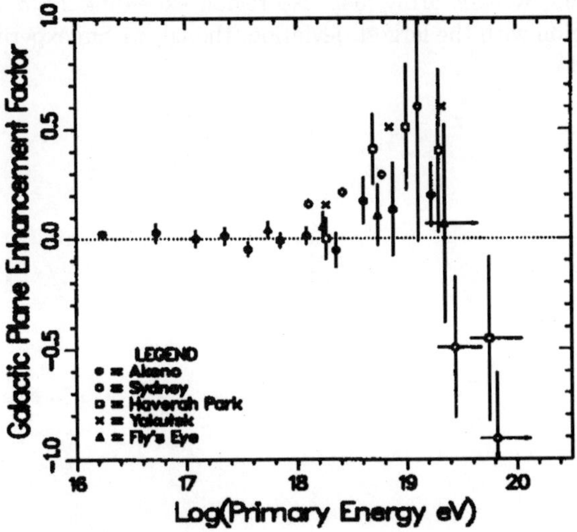

Figure 7.23 The galactic plane anisotropy, from the Fly's Eye experiment, expressed in terms of the galactic enhancement factor as a function of energy.

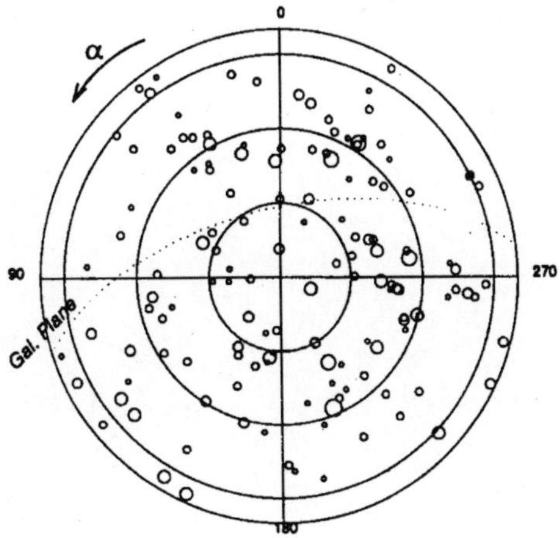

Figure 7.24 Arrival direction distribution of 162 showers of energy above 10^{19} eV from the AGASA experiment.

A search for point sources of energy above $10^{17.5}$ eV has recently been made with the data from the AGASA array [54]. No region exceeding 2.5 σ has been found. However, the region with the largest deviation, though within experimental errors, is the Cyg X-3 region.

References

1. T. E. Cranshaw, W. Galbraith and N. A. Porter, *Phil. Mag.* **2** (1957) 804, 891 and 900.
2. G. Cocconi, V. Tongiorgi and K. Greisen, *Phys. Rev.* **76** (1949) 1020.
3. A. T. Abrosimov et al., *Izvest. Akad. Nauk SSSR, Ser. Fiz.* **19** (1955) 677.
4. J. Linsley, L. Scarci and B. Rossi, *J. Phys. Soc., Japan* **17 Suppl A-III** (1962) 91.
5. A. A. Watson, *Q. J. Roy. Astr. Soc.* **21** (1980) 1.
6. C. J. Bell et al., *J. Phys. A: Math., Nucl. Gen.* **7** (1974) 990.
7. T. A. Egorov et al. , *Proc. 12th International Cosmic Ray Conference, Hobart* **6** (1971) 2059.
8. R. M. Baltrusaitis et al., *Nuclear Instruments and Methods in Physics Research* **A240** (1985) 410.
9. K. Suga, *Proc. V Interamerican Seminar on Cosmic rays, La Paz* (1962)
10. A. E. Chudakov, *Proc. V Interamerican Seminar on Cosmic rays, La Paz* (1962).
11. K. Greisen, *Proc. 9th International Conference on Cosmic Rays, London* **2** (1965) 609.
12. H. E. Bergeson et al. , *Phys. Rev. Lett.* **39** (1977) 847.
13. M. Nagano et al., *J. Phys. G: Nucl. Part. Phys.* **18** (1992) 423.
14. N. Chiba et al., *Nucl. Instr. Methods* **A311** (1992) 338.
15. M. A. Lawrence et al., *J. Phys. G: Nucl. Part. Phys.* **17** (1991) 733.
16. A. M. Hillas et al., *Proc. 12th International Cosmic Ray Conference, Hobart* **3** (1971) 1001.
17. M. Teshima, *Proc. 23rd International Cosmic Ray Conference, Calgary* **Invited Rapporteur & Highlight Papers** (1993) 257.
18. D. J. Bird et al., *Ap. J.* **424** (1994) 491.
19. K. Greisen, *Phys. Rev. Lett.* **16** (1966) 748.
20. G. T. Zatsepin and V. A. Kuzmin, *Sov. Phys. JETP Lett.* **4** (1966) 78.
21. S. Yoshida et al., *Astroparticle Phys.* **3** (1995) 105.
22. H. Y. Dai et al., *J. Phys. G: Nucl. Phys.* **14** (1988) 793.
23. B. N. Afanasiev et al., *Proc. 24th International Cosmic Ray Conference, Roma* **2** (1995) 756.
24. N. N. Efimov et al., *Proc. International Symposium on Astrophysical Aspects of the Most Energetic Cosmic Rays* ed. M. Nagano and F. Takahara (World Scientific, Singapore, 1990) p. 20.
25. S. Yoshida and M. Teshima, *Prog. Theor. Phys.* **89** (1993) 833.
26. B. N. Afanasiev et al., *in Proceedings of the Tokyo Workshop on Techniques for the Study of Extremely High Energy Cosmic rays, Tokyo* (1993) 35.
27. P. Sommers, *in Proceedings of the Tokyo Workshop on Techniques for the Study of Extremely High Energy Cosmic rays, Tokyo* (1993) 23.

28. N. Hayashida et al., *ICRR Preprint No. ICRR-Report-324-94-19* (1994).

29. X. Chi et al., *J. Phys. G: Nucl. Part. Phys.* **18** (1992) 539.

30. N. N. Efimov and A. M. Mikhailov, *Astroparticle Phys.* **2** (1994)

31. Shigeru Yoshida, *in Proceedings of the Tokyo Workshop on Techniques for the Study of Extremely High Energy Cosmic rays, Tokyo* (1993) 167.

32. J. R. Jokipii and G. Morfill, *in Proceedings of the ICRR International Symposium, Astrophysical Aspects of the Most Energetic Cosmic Rays* (World Scintific1991) p. 261.

33. A. M. Hillas, *Ann. Rev. Astron. Astrophys.* **22** (1984) 425.

34. W. H. Ip and W. I. Axford, *in Proceedings of the ICRR International Symposium, Astrophysical Aspects of the Most Energetic Cosmic Rays* (World Scintific1991) p. 273.

35. P. L. Biermann, *in Proceedings of the ICRR International Symposium, Astrophysical Aspects of the Most Energetic Cosmic Rays* (World Scintific1991) p. 301;
P. L. Biermann and P. A. Strittmatter, *Astrophys. Journal* **322** (1987) 643.

36. P. Bhattacharjee, *in Proceedings of the ICRR International Symposium, Astrophysical Aspects of the Most Energetic Cosmic Rays* (World Scintific1991) p. 382;
P. Bhattacharjee, C. T. Hill and D. N. Schramm, *Phys. Rev. Lett.* **69** (1992) 567.

37. T. K. Gaisser et al., *Phys. Rev.* **D47** (1993) 1919.

38. A. M. Hillas, *Proc. of the First International Symposium on Cosmic Ray Physics in Tibet, Lhasa* (1994) 51.

39. M. N. Dyakonov et al., *Proc. 23rd International Cosmic Ray Conference, Calgary* **4** (1993) 303.

40. R. Walker and A. A. Watson, *J. Phys. G: Nucl. Phys.* **7** (1981) 1297.

41. J. Linsley, *Proc. 15th International Cosmic Ray Conference, Plovdiv* **12** (1977) 89.

42. T. A. Alimov et al., *Proc. 18th International Cosmic Ray Conference, Bangalore* **11** (1983) 387.

43. M. Nagano, *Proc. 18th International Cosmic Ray Conference, Bangalore* **12** (1983) 475.

44. A. A. Andam et al., *Phys. Rev.* **D26** (1982) 23.

45. R. J. Protheroe and K. E. Turver, *Nuovo Cimento* **51A** (1979) 277.

46. J. Linsley and A. A. Watson, *Phys. Rev. Lett.* **46** (1981) 459.

47. A. A. Watson, *Proc. 19th International Cosmic Ray Conference, La Jolla* **9** (1985) 111.

48. J. Linsley, *Proc. 18th International Cosmic Ray Conference, Bangalore* **12** (1983) 135.

49. M. S. Gilman and A. A. Watson, *Proc. 23rd International Cosmic Ray Conference, Invited, Rapporteur & Highlight Papers, Calgary* (1993) 447.

50. D. J. Bird et al., *Proc. 23rd International Cosmic Ray Conference, Calgary* **2** (1993) 51.

51. N. Chiba et al. , *Proc. 23rd International Cosmic Ray Conference, Calgary* **2** (1993) 59. (abstract only)
52. A. D. Krasilnikov et al., *Proc. 23rd International Cosmic Ray Conference, Calgary* **2** (1993) 60.
53. Institute for Cosmic Ray Research, University of Tokyo, *Annual Report* (1993-94) 19.
54. T. Doi et al., *Proc. 24th International Cosmic Ray Conference, Roma* **2** (1995) 804.

Chapter 8
GAMMA RAY ASTRONOMY

Ever since cosmic rays were discovered at the turn of this century the quest for their origin was on. Soon after the discovery it was realized that cosmic rays are of extra-solar origin. Being charged particles, cosmic rays are scattered and isotropised by the galactic magnetic field and hence lose their original directions. If a neutral radiation like gamma rays is associated with the presence of energetic particles in a source, they travel in straight lines and their sources can be identified.

Schklovsky [1] pointed out in the early 50's that the high degree of optical polarization from the Crab Nebula (a supernova remnant in the constellation Taurus due to a supernova explosion in the year 1054 AD) could be accounted for by synchrotron process requiring the presence of very high energy ($> 10^{12}$ eV) electrons gyrating round the magnetic fields ($\sim 10^{-4}$ gauss) in the filaments of the nebula. In 1960 Cocconi [2] suggested that since electrons of such high energy cannot arise from nuclear decay processes, a mechanism of acceleration of particles to high energies must be operative in the nebula. If such a mechanism was indeed present, there was no reason why protons could not be accelerated to high energies too. Presence of such high energy protons meant that these particles in their nuclear collisions with the filamentary matter could produce charged and neutral pions. Decay of such high energy neutral pions results in the production of high energy (hundreds of GeV) gamma rays. Thus detection of high energy gamma rays leads to discovery of cosmic ray sources. Apart from supernova remnants pulsars, X-ray binaries, radio galaxies, quasars etc. are also potential sources of cosmic rays and therefore high energy gamma rays.

Though gamma rays are not deflected by the galactic magnetic field, they do interact with matter and radiation. Interaction with interstellar matter leading to pair production is not important since the total amount of matter along the line of sight even in the plane of the galaxy is much less than a radiation length. However, interaction with the background radiation fields is not negligible. The background radiation consists mainly of star light, infrared, radio and the universal microwave background. Of these only pair production due to interaction with the black body radiation is important. Among calculations made by various authors, the absorption mean free path of gamma rays calculated as a function of energy by Wdowczyk et al. [3] is shown in Fig. 8.1. It can be seen that attenuation is not large for galactic objects.

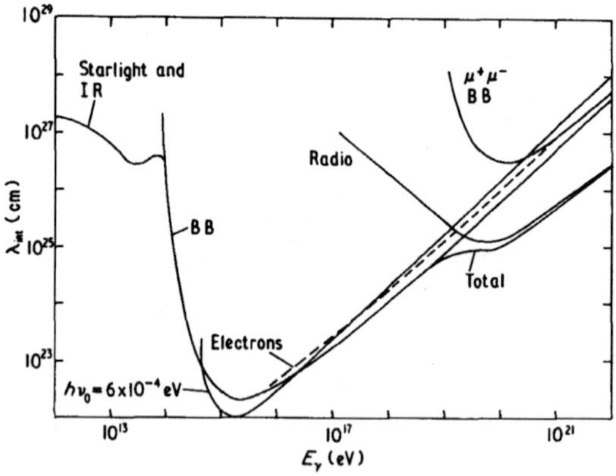

Figure 8.1 Interaction length against photon energy of energetic photons and electrons with photons of various radiation fields (Wdowczyk et al, 1972).

In this chapter, we shall first describe the experimental techniques and some typical experiments used to detect these gamma rays, the results obtained on some of the typical sources of each type mentioned above together with the production mechanisms of gamma rays.

There are some excellent reviews and books [4–7], apart from Rapporteur talks at International Cosmic Ray Conferences, on gamma ray astronomy which the reader may also refer to.

8.1 Experimental Techniques

Gamma rays of energy $> 10^{12}$ eV (TeV) are detected with ground based detectors through the cascades they produce in the Earth's atmosphere, as their fluxes are too low to be detected by instruments of limited area flown in balloons or satellites. However cosmic ray showers produced by nucleons and heavy nuclei constitute an irreducible background against which the gamma ray showers have to be detected. Since cosmic rays are highly isotropic and uniform in time, despite their statistical fluctuations gamma rays from point sources above a certain intensity can be detected as peaks in spatial distributions, as in the case of supernova remnants etc., or as temporal distributions in the case of pulsars, X-ray binaries etc. The sensitivity with which gamma ray sources can be detected depends on the collection area and the angular resolution of the experiment. For example, the number of gamma rays, N_γ, detected over an area of A cm^2 in a time of operation of T s is given by

$$N_\gamma = ATF_\gamma \tag{8.1}$$

where F_γ is the flux of gamma rays from the source. The number of background cosmic rays detected is given by

$$N_{CR} = AT\pi\theta^2 F_{CR} \tag{8.2}$$

where θ is the half angle over which events are accepted and F_{CR} is the flux of cosmic rays. The signal to noise ratio, S/N, is then given by

$$S/N = \frac{A^{1/2}T^{1/2}\theta^{-1}F_\gamma}{F_{CR}^{1/2}}. \tag{8.3}$$

Thus, large areas of collection, long times of operation and small angular apertures lead to high sensitivity. In addition, certain very special techniques are used to select gamma ray showers preferentially to improve the sensitivity further. We shall discuss these techniques later.

The technique used to detect gamma rays depends on the energy of the gamma rays. From Eq. 2.14, one can see that the number of electrons in a gamma ray shower of primary energy 1 TeV at an atmospheric depth of 800 g cm^{-2}, where experiments are generally carried out, is 55. A shower of this size can not be detected with conventional EAS arrays using charged particle detectors. However, the Cerenkov light emitted by the electrons in the shower is large enough to be detected. The Cerenkov photon density up to a core distance of 150 m in such a shower is ~ 200 m^{-2} and can easily be detected. This technique is called the Atmospheric Cerenkov Technique (ACT) and is used for gamma ray energies between 100 GeV and 100 TeV. The number of particles in a gamma ray shower of primary energy 100 TeV is about 3×10^4 and so conventional air shower detectors (EAS Technique) is used for energies above this value.

8.1.1 The Atmospheric Cerenkov Technique

The Atmospheric Cerenkov Technique uses the pool of Cerenkov photons produced by the cascade electrons in traversing the atmosphere. Because of this, observations have to be carried only on dark, moonless, cloudless nights and the duty cycle is only about 10%. The pool of light is collected by a system of search light mirrors mounted on orientation platforms. These could be pointed in the direction of any candidate source and followed for several hours; the photomultipliers at the foci of the mirrors record the intensity of Cerenkov radiation. The requirement of fast coincidence (~ 10 ns) of pulses from several mirrors, aligned parallel to each other, reduces the background counting rates essentially due to star light. Since the telescope has to be pointed at the source, only one source can be observed at a time.

Because of the Coulomb scattering of the electrons and the opening angle of emission, Cerenkov light has an angular spread of a few degrees. The photomultipliers are, therefore, covered with a mask to limit the angular aperture to a half angle of 0.5°–1.5°. A typical array of mirrors, used by the Tata group [8], is shown in Fig.

Figure 8.2 A typical closely packed array of parabolic mirrors used by the Tata group (Bhat et al, 1980) to search for TeV gamma ray point sources.

8.2. The array, consisting of eight parabolic mirrors of 1.5 m diameter and twelve of 0.9 m diameter, was operated initially at Ootacamund, South India, and later at Pachmarhi, Central India, for almost three decades.

Two methods are commonly used to observe TeV gamma ray sources. In one method (drift scan), the mirrors are oriented towards a direction that is in the path of the source, as it traverses across in the sky. The counting rate increases significantly when the object enters the field of view, if the source is a gamma ray emitter. In the second method, the source is tracked as it moves across the sky over a certain angular region typically for a couple of hours. An adjacent source free region in the sky is then tracked over the same zenith angular region for the same time to estimate the background. Any significant excess of rate from the source above the estimated background indicates gamma ray emission from the source.

An improvement in the S/N ratio can be obtained by selecting showers over a cone around the source that is narrower than that determined by the mask. To achieve this the arrival directions of the showers are determined using the fast timing technique described in Sec. 3.1.1.3. Angular accuracies of about $0.1°$ can be achieved using this method since the shower front of the gamma ray cascade is much thinner than in the case of cosmic ray showers. By selecting showers within about $0.2° - 0.3°$, even though the mask is much larger, the S/N ratio is improved and the sensitivity increased by a large factor.

The S/N ratio can be improved further by making use of the differences between gamma ray and proton showers. In gamma ray showers, the angular spread of Cerenkov photons is due only to the Cerenkov angle of emission and the multiple Coulomb scattering of electrons that emit the light, where as in the case of proton showers the opening angles of the π^0 mesons, due to the transverse momentum they acquire at production in nuclear interactions, also contributes to the spread. Thus, the angular spread of the photons, and therefore the size of the Cerenkov image in the angular space, is larger for proton showers compared to gamma ray showers [9]. This effect also leads to flatter lateral distribution of Cerenkov light in gamma ray showers compared to proton showers, with a characteristic hump at a distance of about 140 m from the core. This hump is absent in proton showers [10]. Rao and Sinha [10] showed that the S/N ratio can be improved by at least a factor of two by locating this hump as a ring of high photon densities in a suitably spaced array of mirrors.

The angular spread method, also known as Cerenkov imaging technique, was developed by the Whipple group. Their imaging camera is shown in Fig. 8.3. It is a

Figure 8.3 The 10 m diameter reflector operated by the Whipple collaboration to search for TeV gamma ray sources using the Cerenkov imaging technique.

10 m diameter reflector at the focal plane of which a camera head is mounted. The camera head, shown in Fig. 8.4, consists of 91 pixels of 0.25° (2.5 cm diameter photo-

tubes) surrounded by an outer ring of 18 pixels of 0.5° (5 cm diameter phototubes), arranged in a hexagonal pattern, giving a full field of view of 3.75° diameter. The

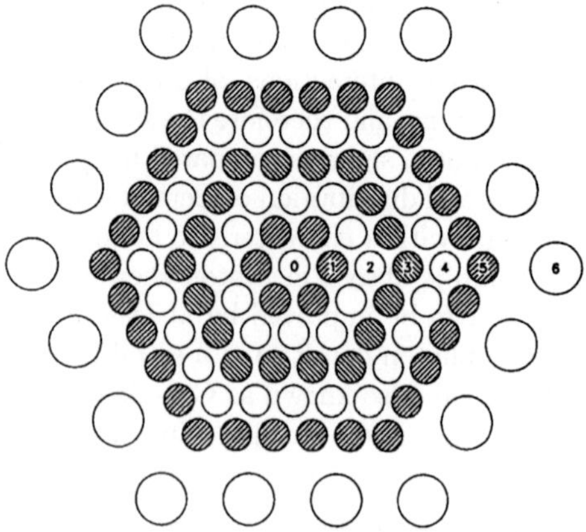

Figure 8.4 The hexagonal array of photomultipliers located at the focal plane of the Whipple telescope. The numbers refer to different zones used to designate the position of the Cerenkov image.

effective energy threshold is about 0.3 TeV and collection area for gamma ray showers is about 3×10^4 m^2. In addition to being compact in angular size, the Cerenkov images due to gamma ray showers from a point source are roughly elliptical with the major axis pointing towards the center which is the direction of the source. These two properties are combined in a single parameter, "azwidth", which is the width of the image in the direction perpendicular to the radial axis through the image centroid. By making a suitable cut on this parameter, on the basis of Monte Carlo simulations, they showed that gamma ray showers could be selected with at least 50% efficiency while rejecting 97% of the background cosmic ray showers.

Yet another method, which exploits the differences in the lateral distribution of Cerenkov light in gamma ray and proton showers, was suggested by Hillas and Patterson [11] and elaborated by Rao and Sinha [10]. While Hillas and Patterson calculated the lateral distributions in both gamma ray and proton showers, using Monte Carlo simulations, and showed that the characteristic hump is absent in proton showers, Rao and Sinha carried out Monte Carlo simulations for gamma ray showers only and showed analytically that the hump will not be present in proton showers. They argued that the hump in gamma ray showers is due to contribution from high energy electrons for which the cumulative RMS Coulomb scattering angle is smaller than the Cerenkov emission angle at the depth where Cerenkov emission takes place.

The energy at which the two angles are equal determines the prominence of the hump. In proton showers, this energy is about a factor of two larger, therefore their relative numbers are smaller, leading to the absence of the hump. The lateral distributions calculated by Hillas and Patterson for gamma ray and proton showers are shown in Fig. 8.5. The difference between the two types of showers is clear when the

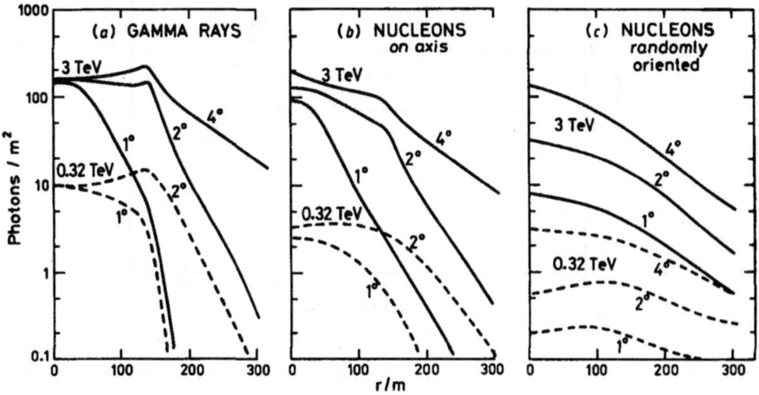

Figure 8.5 The lateral distribution of Cerenkov light in gamma ray showers and randomly oriented on-axis proton showers calculated by Hillas and Patterson. The numbers on the curves represent fields of view used in the calculations. Notice that the hump is present in gamma rays.

aperture of the mirrors is large enough ($\sim 4°$). Fig. 8.6 shows the calculations of Rao and Sinha for gamma rays of energy 100 GeV at different altitudes. It can be seen that the prominence of the hump reduces with increasing altitude as one approaches the shower maximum. Thus this method can be used at atmospheric depths well beyond the shower maximum with wide angle detectors. Furthermore, fluctuations in the Cerenkov photon density are very much larger in proton showers compared to gamma ray showers.

These features are exploited by the Tata group [12] to improve the S/N ratio. Their experimental setup, shown in Fig. 8.7, consists of 13 banks of reflectors, each containing two or more mirrors of total area 2.5 m^2, spread over an area of 80 m $\times$ 100 m. They define a flatness parameter, which is essentially the ratio of the variance to the mean of the photon densities recorded in all the mirrors, and select gamma ray-like showers by making a cut on this parameter. This procedure selects events with not only a flat lateral distribution, but also with smaller fluctuations in the photon densities. Preliminary analysis indicates that gamma ray signal is enhanced when events are selected with this criteria.

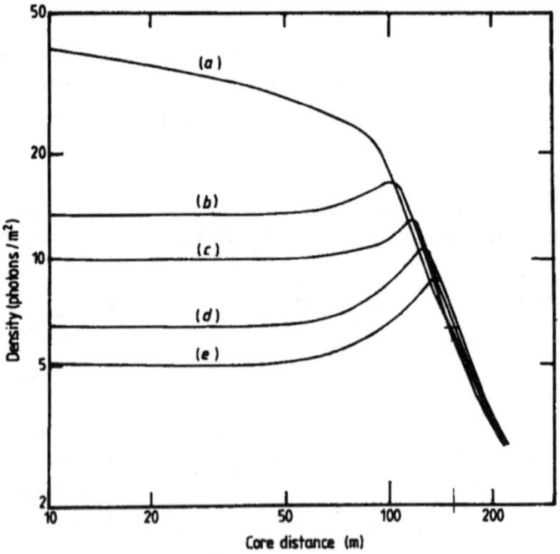

Figure 8.6 Lateral distribution of Cerenkov light in gamma ray showers of energy 100 GeV at various atmospheric depths calculated by Sinha and Rao: (a) 540 g cm^{-2}; (b) 700 g cm^{-2}; (c) 800 g cm^{-2}; (d) 920 g cm^{-2}; (e) 1030 g cm^{-2}.

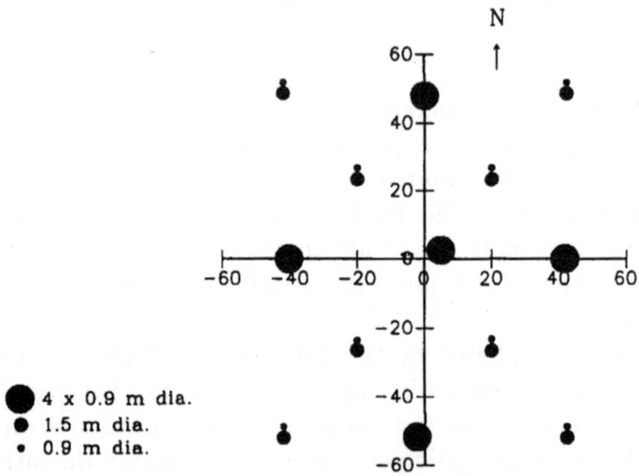

Figure 8.7 The widely distributed array of parabolic mirrors being operated by the Tata group at Pachmarhi to search for TeV gamma ray sources exploiting the differences in the lateral distribution of Cerenkov light in gamma ray and cosmic ray showers.

Table 8.1 Cerenkov telescopes operated by various groups.

Observatory (Group)	**Latitude** (°)	**Depth** (**g cm**$^{-2}$)	**Remarks**
Akeno (ICRR, Tokyo)	36N	930	20 mirrors
Crimea (Crimean Astrophys. Obs.)	45N	890	mirror array
Dugway (Durham)	40N	860	telescope array
Dugway (Univ. of Utah)	40N	860	3 mirror array
Gulmarg (Bhabha Atomic Res. Ctr.)	33N	765	6 mirrors
Haleakala (Hawaaii-Wiscinsin Coll.)	21N	715	6 mirrors
Malta (Dublin, Harwell)	36N	1010	4 mirrors
Narrabri (S.A.O. Sydney)	31S	1010	2 mirrors
Narrabri (Durham)	31S	1020	2 mirrors
Ootacamund (Tata Institute)	11N	790	mirror array
Potchefstroom (Univ. Potchefstroom)	27S	840	4 telescope array
Sandia (Riverside, Iowa)	35N	980	2 mirrors
Targasone (Saclay, Dublin, Bordeaux, Turin)	43N	840	solar array (10 m)
Tien Shan (Lebedev Inst., Crimean A.O.)	43N	690	3 mirrors
Whipple Obsy. (S.A.O., Dublin, Iowa, Hawaii)	32N	780	2 mirrors (10 m, 11 m)
WhiteCliffs (Aelaide Univ.)	31S	1010	3 mirror array
Woomera (Univ. Adelaide)	31S	1010	3 mirrors
ASGAT	42N		7 mirror array
Pachmarhi (Tata Institue)	22N	900	multi-mirror array
BlackBirch Range (JANZOS Collab.)	42N	930	3 mirrors

Many Cerenkov telescopes have been operating all over the globe. Table 8.1 lists some of the past and currently operating telescopes with some details.

8.1.2 EAS Technique

Since the EAS technique depends on detection of electrons, it can be used round the clock and several sources can be observed at the same time. As can be seen from Eq. 8.3, large area of collection and good angular accuracy are required for an array to detect gamma ray sources efficiently against the cosmic ray background. In addition, differences in features of gamma ray and proton showers, such as the lateral distribution of electrons and the muon and hadron contents, can be exploited to enhance the sensitivity.

The lateral distribution of electrons is naively expected to be flatter in gamma ray showers than in proton showers. Detailed Monte Carlo simulations by several authors [13] have, however, shown that there is practically no difference in the electron lateral distribution. Muons arise in gamma ray showers due to pion production in photonuclear interactions with air nuclei, the cross section for which is about 100 times smaller than p-air cross section. The muon content of gamma ray showers, however, is expected to be less than 10% of that in proton showers. Thus, selection of showers with poor muon content will enhance the S/N ratio. Similarly events with small hadron content also will be enriched in gamma rays. This criterion is more difficult to use. To establish the absence of hadrons one has to be close to the core of the showers since the hadrons have a steep lateral distribution.

Typically collection areas of $10^4 - 10^5$ m^2 are achieved and angular accuracy better than 1° obtained using a large number of fast timing detectors. The angular accuracy is approximately proportional to the inverse of square root of the number of timing detectors. The first dedicated array for gamma ray source search, shown in Fig. 8.8 was operated by the Tata group [14] at Kolar Gold Fields (12.95° N, 73.3°, 920 g cm^{-2}), India. The array consists of 127 plastic scintillation detectors, each of 1 m^2 area and 7 muon detectors, of threshold energy 1 GeV, each having an area of 28.8 m^2. The scintillators are arranged in a hexagonal pattern with a spacing of 20 m between neighboring detectors. The detectors extend to 120 m from the center, in six hexagonal rings, covering an area of 3.7×10^4 m^2. The innermost 61 detectors are instrumented for both fast timing and density measurements while the rest measure only density. The threshold density for triggering the timing detector is kept at 0.3 particles m^{-2} to minimize effects of rise time fluctuations. A stable 5 MHz Oscilloquartz crystal provides real time information with an uncertainty of about 0.5 ms. Six of the muon detectors are located at the vertices of a hexagon of side 60 m and the seventh at its center, which coincides with the center of the array. The muon detector is shown in 3.7. An on-line LSI-11 microprocessor records the events and also continuously monitors the performance of all the detectors [15]. The angular accuracy of this array is 1.5° for showers of size 5×10^4 and improves with shower size.

Another dedicated array being operated by the CYGNUS collaboration [16] in Los Alamos, U.S.A. (35.9° N, 106.3° W, 800 g cm^{-2}), is shown in Fig. 8.9. It consists of 108 scintillation detectors, each of area 0.825 m^2, spread over an area of 2.2×10^4 m^2 with a detector spacing of 7 m at the center of the array increasing to 20 m at the edge. Each detector is viewed by a single fast photomultiplier to measure particle density as well as arrival time. The area of the muon detector is 36 m^2. The array was later expanded to include a total of 204 scintillation detectors over an area of 8.5×10^4 m^2 and second muon detector of area 70 m^2. The angular accuracy of the CYGNUS array was 1.1° for showers of size 3×10^4 and improved with shower size. A lead shield of one radiation length thickness was added on top of each scintillator in the expanded array, which enabled them to lower the gamma ray energy threshold

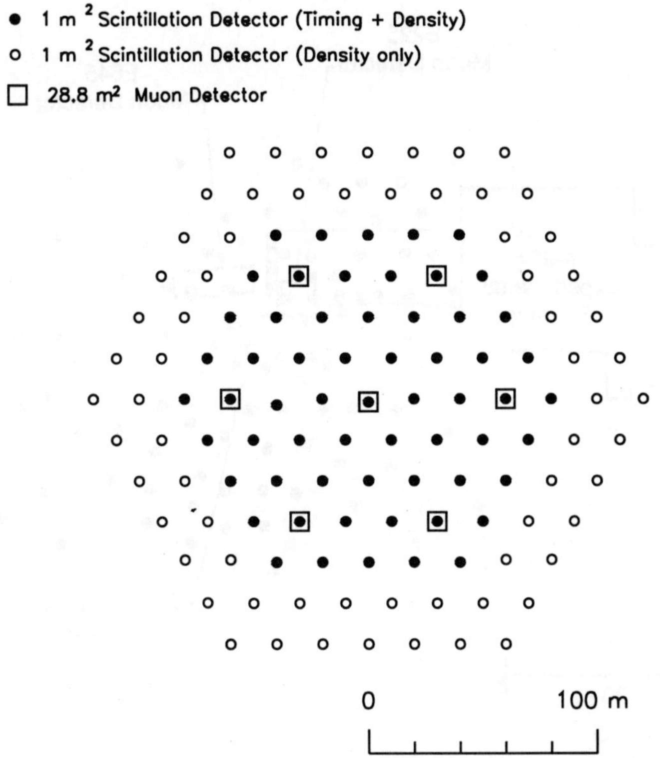

Figure 8.8 The air shower array operated by the Tata group at Kolar Gold Fields to search for PeV gamma ray sources.

by about 25% and improve the angular resolution by 26% on the average.

The largest array, shown in Fig. 8.10, being operated for PeV (10^{15} eV) gamma ray source search is that of the CASA-MIA collaboration [17]. It is located at Dugway, Utah (40.2° N, 112.8°W, 870 g cm^{-2}), U.S.A., around Fly's Eye II detector. It consists of 1089 stations on a 15 m square grid with an enclosed area of 2.3×10^5 m^2. Each station consists of 4 scintillation counters, each of size 0.36 m^2 and covered by a lead sheet one radiation length thick. In addition, there are 1024 scintillation counters, each of area 2.5 m^2, buried 3 m below the surface in 16 patches of 64 counters each. The angular resolution of the array is 0.75° for showers of size 10^4.

Some of the PeV gamma ray arrays operating in the world are listed in Table 8.2.

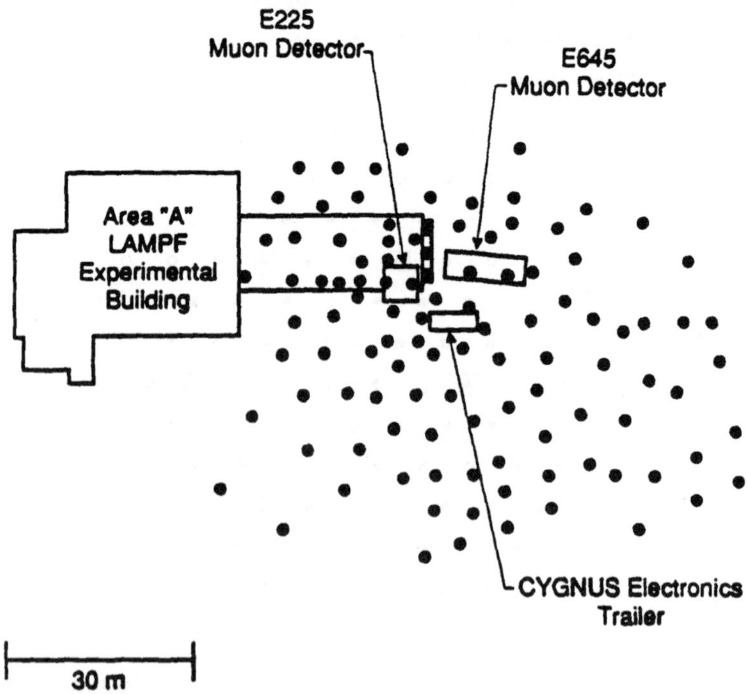

Figure 8.9 The gamma ray air shower array of the CYGNUS group.

8.2 Gamma Ray Source Search

The very first object that was studied was the Crab Nebula. Soon after Cocconi's suggestion [2] that Crab might be emitting TeV gamma rays, Chudakov et al. [18] operated a Cerenkov telescope and observed Crab in the TeV energy region. Having found no signal, they set an upper limit of 5×10^{-11} photons cm^{-2} s^{-1} for gamma rays of energy > 5 TeV, almost three orders of magnitude smaller than that suggested by Cocconi. Similar limits were set on a number of supernova remnants. In the PeV energy region, search for gamma ray sources were carried out in the early days using conventional air shower arrays which, however, were not well suited for point source search. No statistically significant source was reported except Crab which was seen at a marginally significant level. The announcement by the Kiel group [19] of detection of the X-ray binary source, Cygnus X-3, as a PeV gamma ray emitter gave a sudden impetus to the field of PeV gamma ray astronomy. Many dedicated air shower arrays, listed in Table 8.2 have been built with the requisite features such as large collection

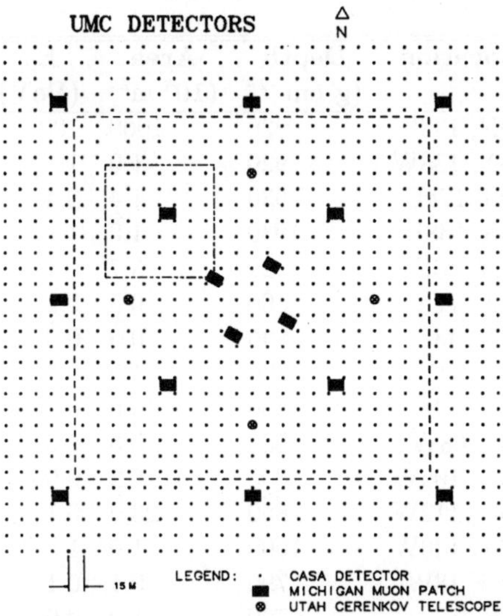

Figure 8.10 The largest gamma ray air shower array of the CASA-MIA collaboration for PeV gamma ray source search.

areas, good angular resolution and large area muon detectors, for gamma ray source studies. We shall discuss some of the results obtained from these observations.

8.2.1 Crab Nebula

The supernova remnant, Crab, is one of the strongly suspected sources of cosmic rays and most extensively studied in the entire electromagnetic spectrum. For a detailed review on the Crab, see Rao and Sreekantan [20]. We shall discuss the results from Crab in some detail.

8.2.1.1 TeV Gamma Rays

The first positive detection of TeV energy gamma rays from the Nebula was reported by Fazio et al. [21] (Smithsonian group) in 1972 after three years of observation. From 150 hours of observations they detected a signal at 3 σ level and estimated a flux of $(5.7 \pm 1.8) \times 10^{-11}$ photons cm^{-2} s^{-1} at energies above 0.14 TeV. They also presented some evidence for time variation correlated with glitches in the pulsar. Mukanov et al. [22] using the drift-scan method, in which the source was allowed to drift across the field of view of the telescope, at Tien Shan obtained a flux of $(5.7 \pm 1.3) \times 10^{-11}$ photons cm^{-2} s^{-1} at energy > 2 TeV.

Table 8.2 PeV gamma ray arrays operated by various groups.

Array	**Location**	**Depth** ($\mathbf{g\ cm^{-2}}$)	**Area** ($\mathbf{10^4\ m^2}$)	$\mathbf{E_{th}}$ **(PeV)**	**Muon Area** ($\mathbf{m^2}$)
Akeno	35N, 138E	920	1000	100	225
SPICA	35N, 138W	920	1.0	1	25
JANZOS	42S, 174E	930	1.4	0.4	—
Buckland	35S, 138E	1030	1.0	1	—
KGF	13N, 78E	920	3.7	0.1	210
Ooty	11N, 77E	800	0.5	0.1	—
Baksan	43N, 43E	840	0.5	0.3	—
GREX	54N, 1W	1030	1.0	0.5	40
EAS-TOP	42N, 13E	800	10.0	0.05	144
BASJE	16S, 68W	530	0.5	0.2	60
CYGNUS	36N, 106W	800	6.2	0.05	106
CASA-MIA	40N, 112W	850	23.0	0.1	2500
Ohya	37N, 140W	1000	1.0	1	400
HEGRA	29N, 18W	790	0.1	0.1	—
SPASE	90S, 0	760	1.0	0.05	—
Tibet	30N, 91E	606	0.8	0.01	—

A very important development has taken place recently in detecting steady emission of TeV gamma rays from the Crab Nebula. The Whipple group [23–26], using the imaging technique (see Sec. 8.1.1), has succeeded in detecting a significant flux with their 10 m imaging camera. They detected the source at a level of 9 σ during 1986–88, which was the most significant signal from Crab at that time. These observations yielded a flux of 1.8×10^{-11} photons cm^{-2} s^{-1} at $E > 0.7$ TeV, with a factor of 1.5 uncertainty in both flux and energy estimates. They also found that the emission is constant within statistics over this period and indicated the possibility that the Crab Nebula might be a "standard candle" for TeV gamma ray astronomy.

The Whipple collaboration have in recent years further improved their camera with a smaller pixel size, a lower energy threshold and a still better discrimination against cosmic ray showers and have obtained a 20 σ signal during 1988–89, which gave a flux of 7×10^{-11} photons cm^{-2} s^{-1} at $E > 0.4$ TeV. With such a strong signal and the measured pulse height spectrum, they have been able to estimate the source energy spectrum [26]. The differential energy spectrum is found to be $N(E)dE = 2.5 \times 10^{-10}(E/0.4)^{-2.4\pm0.3}dE$ photons cm^{-2} s^{-1} TeV^{-1} with E in TeV. The virtue of obtaining the energy spectrum from a single experiment cannot be overstated. It obviates the uncertainties in energy thresholds and fluxes of widely different experiments carried out in different locations and at different times. They found that the monthly average fluxes are constant over this entire period, confirming that Crab Nebula is indeed a "standard candle" in the TeV energy region.

Several other groups [27–29], using a similar technique, have also detected steady emission and confirmed the Nebula to be a steady TeV source. The University of Michigan group [27] applying a selection criteria, similar to the earlier crude criteria of the Whipple group, on their data collected during Nov-Dec 1988 obtained a 5.8 σ signal and gave a flux of 1.8×10^{-10} photons cm^{-2} s^{-1} at $E > 0.2$ TeV. The University of California Collaboration [28,29], using a slightly different technique of pulse shape discrimination and drift-scan method, have also detected the signal at a level of 4.2 σ and reported a flux of $(2.5 \pm 1.3) \times 10^{-11}$ photons cm^{-2} s^{-1} at $E > 0.6$ TeV.

8.2.1.2 PeV Gamma Rays

In the PeV energy region, where the EAS method is used, the estimate of background is easier because the background is automatically monitored all the time. The Lodz group [30, 31] found a 5.4 σ excess at $E > 10$ PeV from the direction of Crab, summing up the results of a seven year period from 1975 to 1982. Due to poor angular resolution the excess was seen over a large angular bin around Crab (37.5° in Right Ascension and 10° in Declination); so it is not clear if all the excess is from Crab. The deduced flux was 2×10^{-13} photons cm^{-2} s^{-1} at $E > 10$ PeV. The muon content of the showers from Crab was not significantly smaller than that in the cosmic ray background showers. However, the errors on the muon measurements were such that no conclusions could be drawn about the nature of the primaries. For example the muon content, relative to that in cosmic ray showers, in the data sample collected during 1968–75 was found to be [31] 0.60 ± 0.12, which was consistent with both gamma ray and proton primary hypotheses within 3 σ. The total data sample collected during 1968–82 seems to show somewhat higher muon content. The Tien Shan group [32] also detected the signal from Crab during 1974–82 with their EAS array at the Tien Shan mountain station, but only in muon poor showers of somewhat lower primary energy. They also used a large angular bin of 15°× 15° around Crab. The fluxes reported by them are $(2.8\pm0.8)\times10^{-13}$ and $(1.9\pm0.7)\times10^{-13}$ photons cm^{-2} s^{-1}at energies greater than 350 and 550 TeV respectively. The only other positive detection from Crab was during an episode of about six hours on December 9, 1980, reported by the Fly's eye group [33]. While they detected a 3 σ signal at $E > 1$ PeV on that day, they did not see any excess in the runs in February of the following year. The reported fluxes during the December 9 episode were $(8.2 \pm 3.1) \times 10^{-12}$, $(2.1 \pm 0.7) \times 10^{-12}$ and $(3.3 \pm 1.3) \times 10^{-13}$ photons cm^{-2} s^{-1} at $E > 0.3$, 1 and 3 PeV respectively.

The Akeno group [34], however, did not find any excess at $E > 2$ PeV within $10° \times 10°$ bin around Crab either in normal or mu-poor showers in their data collected during 1978–81. Their upper limit is an order of magnitude lower than the Lodz flux. The Haverah Park group [35] also did not find any excess in the same energy region during 1979–84. Their 95% CL upper limits are 1.5×10^{-13} and 1.5×10^{-15} photons cm^{-2} s^{-1} at $E > 1$ and 10 PeV respectively. Their upper limit at 10 PeV is two orders of magnitude smaller than the finite flux reported by the Lodz group. No information on the muon content is available from this experiment.

The Durham group [36] also looked for excess in a $15° \times 15°$ bin around Crab with their widely spread Cerenkov array at Dugway operated during 1977–80. Failing to find any excess, they set 3 σ upper limits of 1.7×10^{-13} and 6.6×10^{-14} photons cm^{-2} s^{-1} at energies > 9 and 30 PeV respectively. Their limit at the lower energy is, however, consistent with the Lodz flux.

In all these PeV observations the absolute or relative time keeping was not accurate enough to subject the data for pulsar analysis. So, one cannot rule out the possibility that at least part of the emission in the reported positive detections is pulsed.

In recent years, many groups have been operating EAS arrays optimized for PeV gamma ray observations with good angular resolution and equipped with large area muon detectors. None of them has succeeded in detecting a finite steady flux from the Nebula.

8.2.1.3 Comparison of Observations

The fluxes and the limits from the various observations over the energy range 10^{-1} to 10^5 TeV are plotted as a function of energy in Fig. 8.11 and the energy spectra predicted by different theoretical models are also shown. The spectrum observed by the Whipple collaboration is shown as the solid box, with the dashed lines indicating extrapolation to higher energies. The disagreement with the earlier observations of the Smithsonian group and the Tien Shan group may be due to large uncertainties in these estimates, which are truly not reflected in the error bars.

In the PeV energy region, all the upper limits are consistent with the extrapolation of the Whipple spectrum shown by dashed lines. The Fly's eye points correspond to episodic emission lasting a few hours and is several orders of magnitude higher. The Tien Shan fluxes at a few hundred TeV, based on mu-poor selection are also higher than the extrapolated spectrum. The upper limits for mu-poor showers given by the Utah-Michigan collaboration, Corbato et al. [37], marked M2 in the figure, are however in complete disagreement with the Tien Shan results. There is no time overlap between the two experiments. Long term variability of the source on the scale of several years has to be invoked in order to reconcile these results. The most disturbing discrepancy in the PeV region is between the Lodz and the Haverah Park experiments which are contemporaneous. The finite flux of the Lodz group is about 100 time larger than the upper limit of the Haverah Park group. One way to understand the wide discrepancy is to assume that the Lodz flux is not from the Crab alone, but also forms part of general emission from the galactic plane. The Lodz group found from the same data base, excess mu-poor showers within $\pm 17.5°$ of the galactic plane. They constitute $1 - 2\%$ of the cosmic ray flux and so the gamma ray flux from the galactic plane is about 2.3×10^{-14} photons cm^{-2} s^{-1} sr^{-1} at $E > 10$

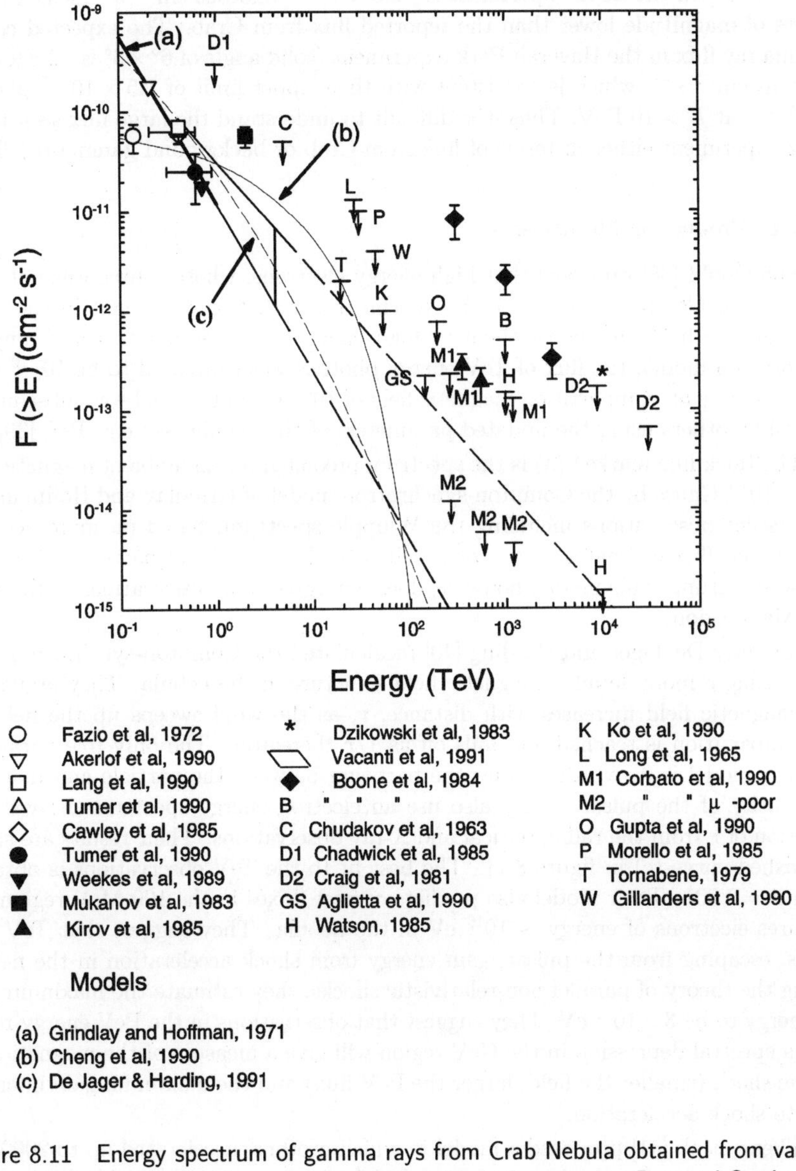

Figure 8.11 Energy spectrum of gamma rays from Crab Nebula obtained from various observations compared with model predictions. For references see Rao and Sreekantan (1992).

PeV. Therefore the cosmic gamma ray flux expected in the solid angle around Crab ($37.5° \times 10°$) in the Lodz experiment is 2.6×10^{-15} photons cm^{-2} s^{-1}, which is two orders of magnitude lower than the reported flux from Crab. The expected cosmic gamma ray flux in the Haverah Park experiment (solid angle of $6° \times 6°$ is 2.5×10^{-16} photons cm^{-2} s^{-1}, which is consistent with their upper limit of 1.5×10^{-15} photons cm^{-2} s^{-1} at $E > 10$ PeV. Thus it is difficult to understand the large flux seen in the Lodz experiment either in terms of flux from Crab or background gamma ray flux.

8.2.1.4 Production Mechanisms

In 1965 Gould [38] proposed that high energy electrons, whose source was not specified, would Compton scatter on the synchrotron radiation they emit in the nebular magnetic field and boost them to high energies. According to this Compton-synchrotron model, the flux of TeV energy photons was estimated to be 10^{-10} cm^{-2} s^{-1}, assuming an equipartition magnetic field of 10^{-4} G. This model was subsequently refined by others using the updated parameters of the nebula (see e.g. Ref. [39]).

The thick line marked (a) is the spectrum predicted for an ambient magnetic field of 3×10^{-4} Gauss by the Compton-synchrotron model of Grindlay and Hoffman [39]. The recent observations including the Whipple spectrum, based on more accurate energy and flux estimates, agree very well with this model—in particular the slope of the spectrum. This model, however, does not agree with observations in the lower 100 MeV region.

Recently De Jager and Harding [40] recalculated the Compton-synchrotron spectrum using a more detailed magnetic field structure in the nebula. They argue that the magnetic field increases with distance, r, as the wind sweeps up the field until equipartition is reached and falls off as $1/r$ thereafter. The only free parameter in their model is σ, which decides the partition between the particle and magnetic luminosity of the pulsar. They also use an electron energy spectrum derived as a function of r from the radio, optical and X-ray observations. Their results are shown as dashed curve (c) in figure 8.11. The best fit to the TeV observations is obtained with $\sigma = 0.001$. Their model also predicts correct fluxes in the 100 MeV region, but requires electrons of energy $\sim 10^{16}$ eV in the nebula. They suggest that TeV electrons, escaping from the pulsar, gain energy from shock acceleration in the nebula. Using the theory of parallel non-relativistic shocks, they estimate the maximum gain in energy to be 3×10^{15} eV. They suggest that observations in the PeV energy region and a spectral depression in the GeV region will give a measure of the magnetic field at the shock (smaller the field, larger the PeV flux) and the maximum gain in energy due to shock acceleration.

Cheng et al. [41] proposed a model in which protons, accelerated up to 2000 TeV in the outer magnetospheric gap of the pulsar (as in their earlier model [42]) are stored for a long time in the nebula due to the ambient magnetic field and these protons interact with the ambient matter in the nebula and produce π^0 mesons, which then decay into gamma rays. The spectrum predicted by this model is shown as curve (b) in the figure. While this model predicts the right magnitude of fluxes, the spectral shape does not agree. In yet another model proposed by Kwok and Cheng [43] and referred to by Vacanti et al. [26], the relativistic electrons accelerated in the outer magnetosphere enter the Nebula and Compton-scatter on the infra red photons they themselves produce by synchrotron mechanism and boost the photons to very high energies; these photons further cascade and produce TeV gamma rays which escape and are the ones that are detected in the experiments. Perhaps a combination of all these processes is responsible for the observed fluxes and spectral characteristics.

The emission due to the Compton-synchrotron process does not extend to PeV energies because of severe energy losses of the electrons during acceleration. In the hadronic model of Cheng et al. [41] the spectrum extends beyond 10 TeV, but shows a sharp cut off around 100 TeV, even though according to the model protons could be accelerated up to 2000 TeV in the pulsar outer magnetosphere and then confined in the Nebula. This sharp cut off in the gamma ray spectrum may be because they have used the scale-breaking model of Wdowczyk and Wolfendale [44] for hadronic interactions. In this interaction model scaling is violated in the fragmentation region also, because of which the π^0s cannot acquire high enough energies. There is evidence from recent accelerator data [45], however, that scaling is preserved in the fragmentation region at PeV energies. If this feature is included in the calculations of Cheng et al., then perhaps it is possible to get higher fluxes in the PeV energy region. Also, if protons can be re-accelerated in the nebula due to shock mechanism, as electrons are in the model of De Jager and Harding, higher fluxes of PeV gamma rays can be expected.

The latest situation on the spectrum from the Nebula, compiled by Borione et al. [46], is shown in Fig. 8.12. While some new observations in the TeV energy region have been added and upper limits in the PeV region improved, the general picture is nearly same as in Fig. 8.11. The upper limits in The PeV region are still above the model predictions and one has to wait for further improvements in this energy region.

8.2.2 Crab Pulsar

8.2.2.1 TeV Gamma Rays

Soon after the discovery of pulsars, several groups looked for TeV pulsed emission from the Crab pulsar. While some of them reported only upper limits the first positive detection came from Grindlay [47] of the Smithsonian group. He used a special technique, for the first time in the TeV region, to distinguish between gamma ray and cosmic ray showers. He used an assembly of three mirrors to collect the Cerenkov light produced by the shower particles. Two of the mirrors, located 70 m apart, were pointed but with a slight convergence between them along the direction of the Crab such that they collected the Cerenkov light produced by the particles at the shower maximum high in the atmosphere. The third mirror, also oriented in the direction

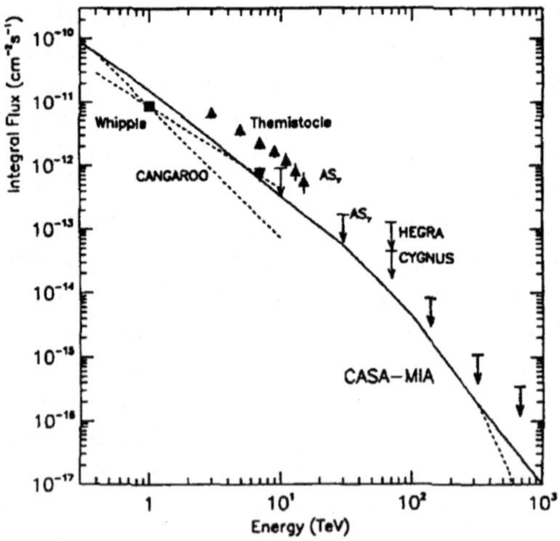

Figure 8.12 Energy spectrum form Crab nebula compiled by Borione et al. (1995). The three highest energy upper limits are from Borione et al. (1995). The curve is from de Jager and Harding's model with $\sigma = 0.003$ and maximum electron energies 10^{16} eV (solid) and 10^{15} eV (dashed). Other points are from the Themistocle, CANGAROO, AS$_\gamma$, HEGRA and CYGNUS experiments (see Borione et al., 1995 for references).

of Crab, had a larger tilt and pointed to a region lower in the atmosphere to record the Cerenkov light emitted by muons which penetrated to deeper levels. To select gamma ray showers the third mirror was used in anticoincidence since gamma ray showers were expected to have very low muon content. Using this technique, Grindlay detected 3.5 σ and 5.4 σ signals in the main and interpulse regions respectively. The flux estimated from these observations was $(1.25 \pm 0.6) \times 10^{-11}$ photons cm^{-2} s^{-1} at $E > 0.68$ TeV. This is the first time that a signal has been detected from the pulsar when gamma ray-like events are selected.

In the TeV energy region several groups have improved in recent years the sensitivity of their telescopes, by increasing the number of mirrors, and installing accurate clocks that enable analysis of the data for pulsed emission. The availability of contemporaneous ephemeris from radio observations by the Jodrell Bank Radio Observatory has been a vital input for pulsar analysis. This is necessary, especially in the case of Crab, since the period increases continuously with time and occasional glitches that produce sudden changes occur. While many of these observations have led to negative results regarding steady pulsed emission, evidence for episodic pulsed emission lasting over several minutes to several days has been accumulating slowly. Some of these episodes have shown emission at the radio main pulse and interpulse positions and a few at other phases also. The duration of the pulse has not been constant. Gupta et al. [48] observed two peaks (Fig. 8.13(f)) with a probability of 1.4×10^{-4} in the phasogram in the data collected at Ootacamund during 9 runs in February 1977 (but not in the runs in March), which they attribute to the main and interpulses on the basis of their separation; they did not have information on the absolute phase. The estimated flux was $(1.19 \pm 0.33) \times 10^{-11}$ photons cm^{-2} s^{-1} at $E > 0.5$ TeV. They subsequently [49] revised their energy and flux estimates to 6.4 TeV and $(8 \pm 2) \times 10^{-12}$ photons cm^{-2} s^{-1} due to the re-calibration of the light collection efficiency of the mirrors.

More recent observations have shown transient emission of several minutes duration, essentially because of the special efforts made to search for them. The burst of 15 min duration observed by the Durham group [50] in October 1981 showed a broad pulse of width of about 0.3 in phase (Fig. 8.13(a)). Absolute phase was, however, not available. The estimated flux was $(2 \pm 0.3) \times 10^{-10}$ photons cm^{-2} s^{-1} at $E > 3$ TeV during the burst. The burst reported by the Tata group [51] was also of 15 min duration and has shown a narrow pulse at the position of the radio main pulse (Fig. 8.13(b)). They reported a flux of $(2.5 \pm 0.6) \times 10^{-10}$ photons cm^{-2} s^{-1} at $E > 1.2$ TeV during the burst. An important feature of this observation is that two telescopes tracked the Crab from locations separated by 11 km and both of them showed the signal while a third telescope adjacent to one of them but looking 8° away did not show any signal. This confirms that the signal is from the Crab and not due to any terrestrial phenomenon like lightning. Another burst of similar duration and flux in the main pulse has been seen by the Tata group from their Pachmarhi Observatory on 2 January 1989.

A very interesting feature has been reported by Vishwanath [52] from an analysis of the data base of the Tata group on the Crab in the TeV energy range. He divided all the data collected during the period 1979–85 into mini runs of one minute duration each and selected those which showed large χ^2 for the phase distribution. Since χ^2 is only a measure of deviation from uniform distribution, excess events in individual mini runs can be at any phase and the resultant phasogram for the high χ^2 runs should show a uniform distribution if the parent distribution is uniform. However, he found that the distribution showed not only peaks at the main and interpulses but also some small excess in between the two pulses compared to the rest of the phase region (Fig. 8.13(c)). Based on these observations, he obtained a mini burst flux of $(1.6 \pm 0.3) \times 10^{-10}$ photons cm^{-2} s^{-1} at $E > 0.6$ TeV during 1982–83 and $(2.1 \pm 1.1) \times 10^{-11}$ photons cm^{-2} s^{-1} at $E > 2.2$ TeV during 1984–85. Furthermore, data on two particular nights in 1985 showed a similar phase distribution (Fig. 8.13(i)) in the two telescopes, separated by 11 km.

The only instance of pulsed emission lasting for a long period of time was reported by the Durham group [53] during 103 hrs of observation from September 1982 to November 1983. The emission was confined to a narrow phase region at the radio main

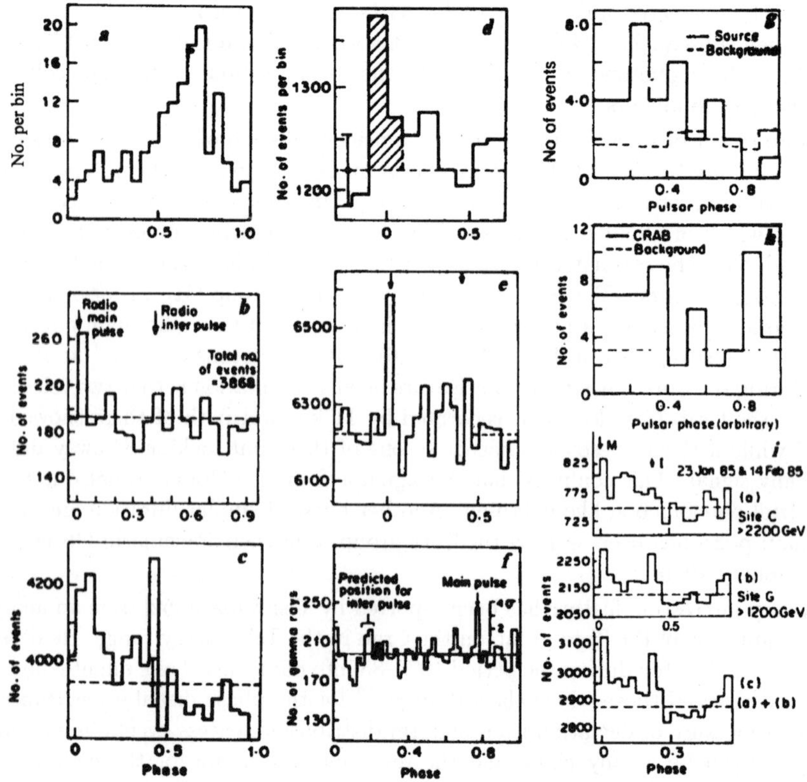

Figure 8.13 Light curves observed from the Crab pulsar in various observations. a: Gibson et al (1982); b: Bhat etal (1986); c & i: Vishwanath (1987); d: Tumer et al (1985); e: Dowthwaite et al (1984); f: Gupta et al (1978); g: Acharya et al (1990); h: Acharya et al (1991).

pulse with some weak evidence at the interpulse (Fig. 8.13(e)). Additional support for the genuineness of the signal was that the excess events came preferentially from the center of the field of view identified from the relative arrival time information from well separated detectors. They quote a probability of 6×10^{-7} for the observation to be due to chance. The detected signal constitutes $(0.233 \pm 0.054)\%$ of the cosmic ray flux, the weakest among all reported. The resulting flux is $(7.9 \pm 1.8) \times 10^{-12}$ photons cm^{-2} s^{-1} at $E > 1$ TeV. There was, however, no evidence for transient emission during this period. The University of California collaboration [54] also found emission at the main pulse in 22 hrs of observation during the two overlapping months of September and October, 1982. Their pulse width (Fig. 8.13(d)), however is broader than that in the Durham observations. They found evidence that the signal enhances from 2 σ to 3.2 σ when they select showers with a broader lateral distribution of Cerenkov light expected for γ ray showers compared to cosmic ray proton induced showers. The flux reported is $(2.5 \pm 0.8) \times 10^{-11}$ photons cm^{-2} s^{-1} at $E > 0.2$ TeV.

The Whipple collaboration [55] also reported one episode of 29 minutes duration at UT 04.02 on January 11, 1991 during which pulsed emission is seen at the position of the main pulse with an overall chance probability of $< 1\%$. This result was obtained with a more sophisticated selection of gamma ray-like events, called "supercuts", recently developed by them, and therefore suggests that the TeV radiation from the pulsar probably consists of gamma rays.

The light curves obtained in these various observations are shown in Fig. 8.13. We shall discuss these results along with those in PeV energy region later.

To summarize, pulsed emission in the TeV region is mostly transient in nature though on one occasion some weak emission over a long period has been detected. The light curve seems to be highly variable, often only the main pulse is seen, occasionally both the pulses are seen and at other times a broad pulse encompassing the main and interpulses. The fluxes reported are in the region of a few times 10^{-11} photons cm^{-2} s^{-1}.

8.2.2.2 PeV Gamma Rays

In the PeV region there were no reports of pulsed emission for a long time, not even upper limits, the main reason being that the clocks operating with the EAS arrays were not accurate enough to do pulsar analysis. Only recently this lacuna has been rectified in the EAS arrays. The arrays have been modified to provide good angular resolution and also equipped with large area muon detectors. The only report of pulsed emission lasting for a long period has come from the Ooty group [56]. Their phasogram, for the three year period June 1984 to May 1987, shows a 3.9 σ excess at the interpulse position. They estimate an overall chance probability of 1.6×10^{-3} of observing such an excess at any position in the phasogram. The reported flux is $(4.1 \pm 1.2) \times 10^{-13}$ photons cm^{-2} s^{-1} at $E > 0.2$ PeV. They claim that the signal enhances somewhat if they select older showers (shower age > 1.4). Monte Carlo simulations by several groups, however, do not show any detectable difference in the age parameter between gamma ray and proton showers, as mentioned in Sec. 8.1.2. There are no other reports of positive detection during the period 1984–87. The Los Alamos group do not see any significant excess at any phase in their data collected during April 1986 to August 1989, which overlaps part of the period of the Ooty observations. They set a 90% confidence level upper limit of 6.9×10^{-14} photons cm^{-2} s^{-1} at $E > 50$ TeV, which corresponds to an upper limit of 2.1×10^{-14} at $E > 200$ TeV, factor of 20 smaller than the Ooty flux. The reasons for these discrepancies are not clear.

An important development of far reaching consequence to PeV gamma ray astronomy has recently taken place. The Baksan group [57] reported a PeV energy burst from the direction of the Crab Nebula during the period 14.00–19.00 UT on 23 February 1989 with their extensive air shower (EAS) array at Baksan (atmospheric depth, 840 g cm^{-2}; longitude, 43°E; latitude, 43°N). They recorded 57 events of energy > 0.2 PeV within 2°.5 of the Crab direction, corresponding to an excess of 3.9 σ (according to the likelihood ratio method of Li and Ma [58]) over the expected background of 31.1 events. The source transited at 16.5 UT. Since they searched for 333 days during 1985 and 1989 the overall chance probability for such an excess to occur was 0.02. Even though a re-analysis of their data resulted in a reduced significance in the same angular bin, it is important to note that the excess still remains at a 3.1 σ (Li and Ma) significance level. This observation triggered the Tata group [59] to examine their data for that day only, i.e. 23rd February, from the EAS array operating at Kolar Gold Fields (KGF; atm. depth 920 g cm^{-2}; long. 78.3°E; lat. 12.9°N). They detected the burst during the time interval 13.25–16.0 UT. The Crab transit at KGF was 14.1 UT. KGF recorded 35 events of energy > 0.1 PeV against an expected background of 17.8, within an acceptance angle of 4° radius, corresponding to an excess of 3.4 σ (Li and Ma). This is the first time that what may be regarded as a simultaneous observation of PeV gamma rays has been reported from any source.

The event times were recorded in the KGF experiment with an absolute timing accuracy of 1 ms. They could therefore analyze the data for pulsed emission and found that the first half of the period, which spans both the main and interpulses seen at radio frequencies, contained 26 events against an expectation of 8.9 (Fig. 8.13(g)). Thus all the excess events in the DC mode seem to be concentrated in the first half of the pulsar period. The significance of this result is 0.01. An analysis, by the Tata group [60], of the arrival times of the Baksan events (supplied by the Baksan group) also showed excess events in one half of the phase (Fig. 8.13(h)) though absolute phases were not available in this case. The significance of this result is 0.05 if the main pulse position is identified to be at phase 0.8 in this phasogram. This is the first time that pulsed emission of PeV gamma rays has been detected from Crab.

This Crab burst on 23rd February 1989 has subsequently been confirmed by two more groups. The Gran Sasso collaboration [61], using the EAS-TOP array (long. 13.6°E; lat. 42.5°N), where the source transit was at 18.4 UT, detected 38 events above 0.2 PeV within an acceptance angle of 1.6° radius against an expected background of 25.5 at a significant level of 2.1 σ (Li and Ma). There is some indication that the ON source events are somewhat older than normal cosmic ray showers. They also find an asymmetric light curve with 9 out of 13 older showers contained in one half of the period. Absolute phases were not available in this case also. An independent lower energy trigger in their experiment also showed an excess of 1.2 σ (Li and Ma).

The variation of the Crab zenith angle and the 15 minute counting rates from the three experiments are shown as function of UT in Fig. 8.14. It is seen that the burst has lasted several hours starting at 13:15 UT or earlier and ending at 19:00 UT.

Recently Chudakov [62] has reported the results of the Tien Shan array which also show an excess of 2.4 σ (Li and Ma) from the Crab during its meridian transit at Tien Shan, which is only 3° east of KGF in longitude. The Ooty EAS experiment is in a location to detect the burst and was in operation, but there is no report from that group as yet. Thus at least four different experiments have recorded the burst in the proper time sequence, confirming the reality of the burst. The probability that all the four groups have recorded the respective excesses from the same source on the same day purely due to chance is 6×10^{-6}. If we further consider that the KGF group has recorded a non-uniform phase distribution also, then the overall probability is 1.3×10^{-7}. Even if we take into account the reduced significance of the Baksan result and also the fact that they have extended their observations from 333 to 1256 days, this overall probability works out to be 3.5×10^{-6}.

Weekes opines in his review article [6] that the reduced excess in the Baksan result made the combined effect only marginally statistically significant and concludes that there is no strong evidence yet for emission from the Crab. The quantitative estimate of the combined probability, given above, does not corroborate his opinion and therefore his conclusion. The burst remains statistically very significant in spite of the reanalysis of the Baksan data because (i) the excess in the Baksan data on 23 February remains by itself, even after reanalysis, at a significance level of 3.1 σ. (ii) four different experiments operating at widely separated locations on the globe have recorded excess events on that day from Crab in the expected time sequence and (iii) the combined probability that all the four experiments have seen the excess on that day from Crab purely due to background fluctuations is 3.5×10^{-6}.

The Akeno group has not seen any excess during 10–12 UT and the HEGRA group also did not record any excess with their Canary Islands array, where the Crab transit time is 20.6 UT. These reports help to define roughly the starting and stopping times of the burst as indicated earlier. The temporal structure of the burst reconstructed by the Tata group using the observations of the KGF, Baksan and EAS-TOP groups, is shown in Fig 8.15.

If the radiation responsible for the showers from Crab during the burst are gamma rays, then the muon content of these showers must be small compared to the normal cosmic ray showers as discussed earlier. It is estimated from Monte Carlo simulations

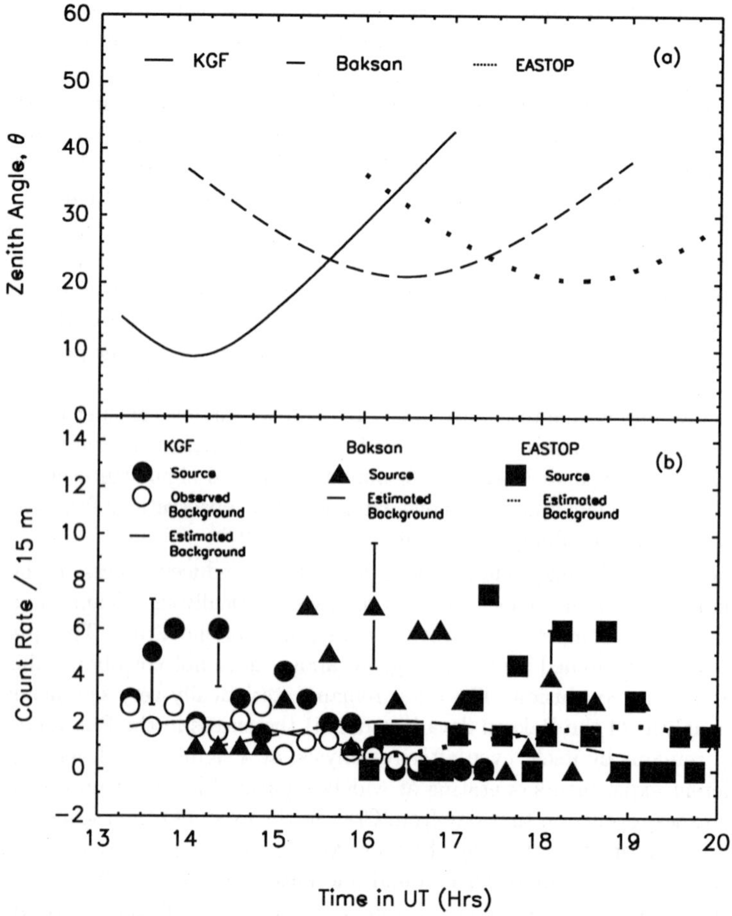

Figure 8.14 (a) Variation of the zenith angle of Crab with UT on 23 February 1989 for the KGF, Baksan and EASTOP arrays. (b) Counting rate in 15-min intervals of events arriving from the Crab direction, plotted against Universal Time for the KGF, Baksan and EASTOP data.

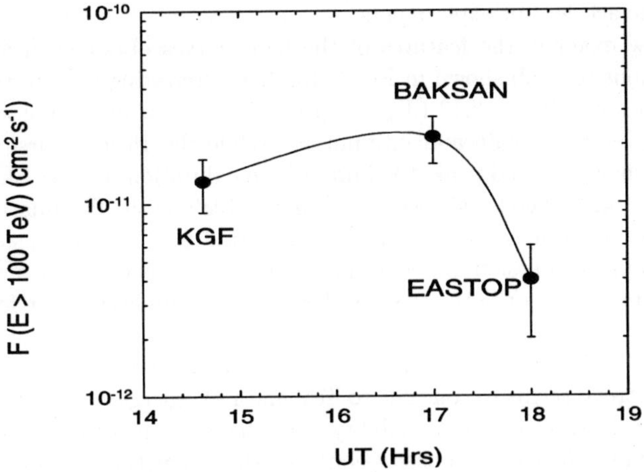

Figure 8.15 Variation of the flux of excess events of energy $>$ 100 TeV from Crab pulsar with UT during the burst on 23 February 1989 from the KGF, Baksan and EASTOP experiments.

that gamma ray showers should contain less than 10% of the muons in proton initiated showers. Of the four experiments cited above, only the KGF array had muon detectors of total area of about 200 m^2 and could measure the muon content. The KGF experiment showed that showers from the direction of Crab contained nearly the same number of muons as in normal cosmic ray background showers. The ratio of the muon content in the excess showers from the Crab to that in background showers was estimated to be 0.93 ± 0.34. Based on this, it could be stated that at the 95% confidence level the muon content in the excess showers from the Crab was greater than 37% of that in normal cosmic ray showers. It is of importance to note here that the Kiel group, who first discovered Cygnus X-3 as a PeV gamma ray source also found that the muon content of showers from that object also was large (see Sec. 8.2.4.1). The Los Alamos group also report a similar finding for showers in the burst they detected from the X-ray binary source Hercules X-1 (see Sec. 8.2.4.2. These observations regarding the muon content of the showers raise several important questions. Are gamma ray interactions different at PeV energies compared to lower energies as suggested by some authors [63]? Are the primaries of these showers neutrinos which interact with almost hadronic cross section as expected in some composite models [64]? Or is Crab emitting a hitherto unknown type of strongly interacting stable neutral particle? If the third alternative is the case then the mass of the neutral particle should be less than 40 MeV c^{-2} in order to maintain phase locking over half the Crab pulsar period. Such a low mass particle should have been detected in the accelerator experiments.

Let us now examine the features of the light curves obtained in the TeV and PeV observations of Crab shown in Fig. 8.13. It is interesting to note that episodes lasting several hours (Figs. 8.13 (d), (g), (h) and (i)) show in general a broad peak encompassing both the main and interpulses. While the 15 min burst reported by Gibson et al. shows a broad peak, the burst of same duration seen by Bhat et al. has a narrow main pulse. The burst seen on 2 January 1989 by the Pachmarhi group also shows a narrow main pulse. The mini bursts reported by Vishwanath suggest, apart from the two pulses, emission in between also. There are also occasions when narrow main pulse or interpulse or both are seen. Thus there is ample evidence for variability of the light curve apart from the emission itself. Manchester and Taylor [65] noted a long time back that the emission in between the pulses increases with energy of the photons. This trend seems to continue to TeV and PeV energies, though quantitative estimates are not yet available. Variability of the light curve is seen at lower energies also. For example, the COS-B observations [66] show that the intensity of the main and interpulses change with time. In fact, the ratio of the intensity in the interpulse to that in the main pulse decreased with time during 1973–80. Dowthwaite et al. [53] also found that the ratio in the TeV energy range during 1982–83 is small and consistent with the nearest COS-B observations. Ozel [67] suggested that this variation in the ratio is due to the free precession, and possibly nutation, of the pulsar with a period of 10.75 years. His fit of a sinusoidal curve to the COS-B and SAS-2 data is shown in Fig. 8.16 along with the result of Dowthwaite et al. If this is the case, then combining

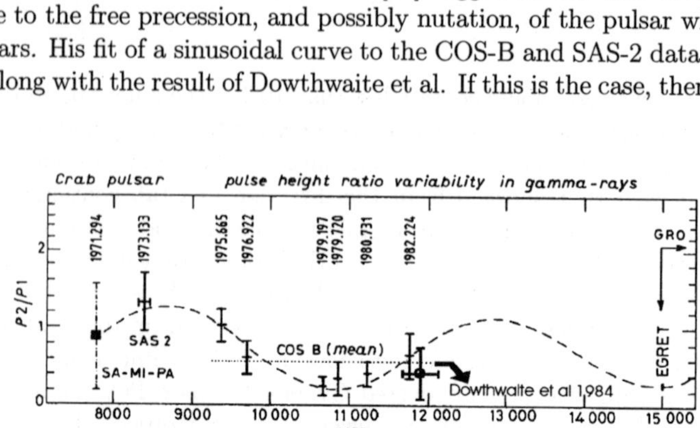

Figure 8.16 Dependence on time of the ratio of the intensities in the interpulse and the main pulse of Cos B observations in the GeV energy region fitted to a sinusoidal curve by Ozel (1991). The observation of Dowthwaite et al. (1984) in the TeV energy region is also shown.

observations over a long period of time is not a good procedure since it will not reveal the variations and also the signal will be diluted as more background is allowed in when one pulse or the other is weak. Obviously, measurement of the features of the light curve in the TeV and PeV regions with high sensitivity detectors is of utmost importance.

It is clear that steady pulsed emission from Crab, if at all present, is very weak and perhaps just at the threshold of sensitivity of the present day detectors. The pulsed emission can be detected only when it is enhanced for short durations or when the background fluctuates on the lower side as in the case of the mini bursts reported by Vishwanath. In his analysis, 263 out of a total of 10,489 mini runs show the signal; so the time averaged flux over the entire period of observation works out to be $(4.0 \pm 0.8) \times 10^{-12}$ photons cm^{-2} s^{-1} at $E > 0.6$ TeV, which is compatible with the time averaged flux of $(7.9 \pm 1.8) \times 10^{-12}$ photons cm^{-2} s^{-1} at $E > 1$ TeV reported for the period 1982–83 by Dowthwaite et al., considering the uncertainties in the estimates of energies and fluxes. Though Dowthwaite et al. mention that there is no evidence in the data for sporadic activity on time scale of minutes as strong as that reported by Gibson et al., it is not clear whether the detected emission is due to a number of mini bursts spread over the entire observation time as in the case of the Ooty observations. It is, therefore, very important to search for such mini burst activity.

Some gamma ray astronomers are somewhat skeptical about the episodic activity in the TeV and PeV region arguing that the episodes are not reproducible and several of them are at marginal significance. In the case of Crab pulsar, however, there are several features which suggest that at least some of the observations are not due to background fluctuations. These are (i) the Smithsonian group found the signal only when they selected gamma ray-like events, (ii) the Tata group observed the episodes on at least two occasions simultaneously at two different locations 11 km apart; they did not see any excess in the telescope looking away from Crab, (iii) the Durham group's observation shows that the signal is concentrated near the direction of the source, (iv) the University of California collaboration found the signal during part of the Durham group's observation, (v) the same group found that the signal is enhanced when gamma ray-like events are selected, (vi) bursts of similar duration (15 m) and intensity were observed by the Durham and the Tata [51] groups, and (vii) the PeV energy burst on 23 February 1989 was seen by four different groups at widely different locations on the globe. Thus TeV and PeV emission from the pulsar appears to be sporadic in nature and future efforts should be concentrated on detecting these with coordinated observations by several groups.

The fluxes and upper limits reported in the various observations are shown in Fig. 8.17 where they are compared with theoretical models.

8.2.2.3 Production Mechanisms

Theoretical models for TeV and PeV gamma ray emission from pulsars are not yet well developed, perhaps because the observations are not yet on a firm statistical footing. In the lower energy region, however, some models have been put forward to explain

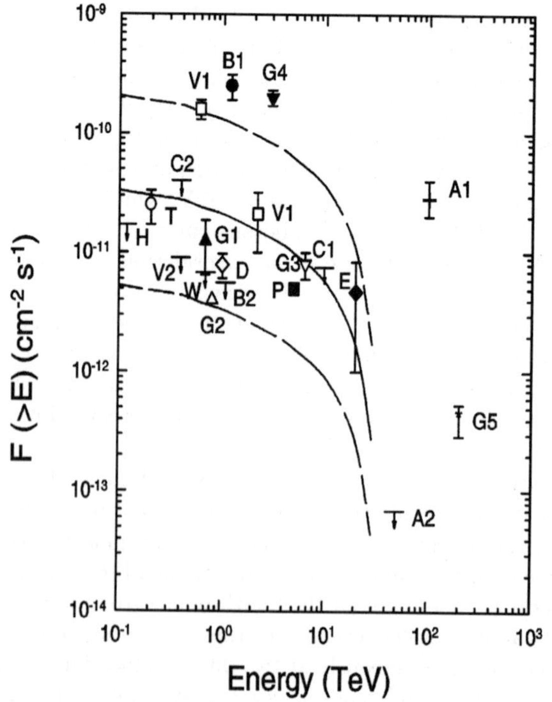

A1 B.S.Acharya et al, 1990
A2 Alexandreas et al, 1991
B1 P.N.Bhat et al, 1986
B2 P.N.Bhat et al, 1990
C1 B.K.Chatterjee et al, 1971
C2 M.F.Cawley et al, 1985
D J.C.Dowthwaite et al, 1984
E R.Erickson et al, 1976
G1 J.E.Grindlay, 1972
G2 J.E.Gringlay et al, 1976
G3 S.K.Gupta et al, 1978
G4 I.A.Gibson et al, 1982
G5 S.K.Gupta et al, 1990
P N.A.Porter et al, 1976
T O.T.Tumer et al, 1985
V1 P.R.Vishwanath, 1987
V2 G.Vacanti et al, 1991
W T.C.Weekes et al, 1989

Figure 8.17 Energy spectrum of gamma rays from the Crab pulsar. The solid curve is the prediction of the model of Bogovalov and Kotov (1990). The dashed curves are the result of fluctuations in the product of $T\lambda$ by a factor of 2.5 on either side estimated by us. The points marked B1, G4 and V1, which are due to transients, are interpreted as due to temporal variations in this parameter. For references see Rao and Sreekantan (1992).

the light curve as well as the energy spectrum. Cheng et al. [42] proposed a model in which electrons and positrons are accelerated in the outer gap in the magnetosphere of an obliquely rotating pulsar, where large electric fields can be sustained. These particles generate photons due to various electromagnetic processes, which in turn create $e^{\pm}$ pairs. The pulsed low energy gamma rays are a result of synchrotron radiation from these $e^{\pm}$ pairs. Up to a few GeV the predictions of the model on light curve and spectra agree with observations on Crab as well as Vela pulsars. A small fraction of the $e^{\pm}$ is lost by inverse-Compton scattering on the soft photons, which are boosted to a few TeV. In Crab, these photons are absorbed by the soft photons and create $e^{\pm}$, which synchrotron radiate with a characteristic energy of $\sim$ TeV. However, the authors argue that these gamma rays may not be able to escape further absorption in the magnetosphere itself, except when density fluctuations occur in the secondary flows. They predict only a single pulse in the light curve because the inward-moving TeV gamma rays pass through strong magnetic fields and are absorbed by Sturrock pair creation. The observations, however, show that both pulses as well as a broad pulse are seen often. Also, there is no prediction of the energy spectrum in the TeV region.

Bogovalov [68] has suggested that in an aligned rotator, a rotational discontinuity could be formed in the flow of the plasma near the light cylinder, and particles are accelerated in this discontinuous region to TeV energies deriving energy from the azimuthal magnetic field generated by the rotating neutron star. These particles then produce the TeV gamma rays by the inverse-Compton scattering on the thermal photons emitted by the neutron star. The gamma rays can easily escape and be detected as they are produced near the light cylinder. Bogovalov and Kotov [69] have calculated the energy spectrum of the gamma rays in this model. Their spectrum is shown in Fig. 8.17. In their model, the flux is proportional to $T^2\lambda^2$, where T and λ are the surface temperature of the neutron star in °K and the plasma density in units of the Goldreich density respectively. Their result for $T = 10^6$ °K and $\lambda = 10^3$ is shown as the full curve in the figure. The dashed lines are drawn by us for a factor of 2.5 variation in $T\lambda$, commensurate with the upper limit of 2.5×10^6 °K for the surface temperature derived by Harnden and Seward [70]. It is seen that almost all the observations agree with this model, except the burst events of references [50–52] in the TeV region marked B1, G4 and V1 respectively. These are suggested to be due to sudden increases in the surface temperature, perhaps during starquakes. The two points from Vishwanath [52] marked V1 at 0.6 TeV and 2.2 TeV correspond to different periods of time separated by 2 years. The widely different intensities may be due to different values of $T\lambda$. Bogovalov and Kotov, however, have not calculated the light curve. The long term flux of the Ooty group [56] and the 7-hour burst seen by the KGF and other groups are not reproduced in this model since there is a sharp cut off at a few tens of TeV. To explain these, one may have to consider acceleration of protons to PeV energies and subsequent photo-nuclear interactions with the ambient photons and production of π°s which then decay into gamma rays. Cheng et al. [42]

consider briefly such a possibility within the light cylinder, but conclude that the gamma rays cannot escape creation of Sturrock pairs in the local magnetic field. The proton acceleration process may, however, just be possible in the Bogovalov-Kotov model.

8.2.3 Other Pulsars

Some of the other radio pulsars that showed emission of TeV gamma rays over extended periods of time are the Vela pulsar (PSR 0833-45), PSR 0355+54 and PSR 1706-44.

The Vela pulsar is located at a distance of 500 pc near the center of the large supernova remnant Vela X. It has a periodicity of about 89 ms. The emission from this pulsar is similar to that from the Crab pulsar at various wave lengths, though there are some marked differences. The radio light curve of Vela pulsar shows only one pulse and no significant emission is detected in the X-ray band. While the Vela pulsar also shows two peaks in optical and medium (GeV) energy gamma ray bands, their phases do not coincide, unlike Crab. Grindlay et al. [71] observed the pulsar in the TeV energy region during 1972–73 and detected 4.2 σ and 3.0 σ peaks in the pulsar phasogram. The Tata group [72] found, during the five year period 1979–85, a 4 σ signal in the phasogram (Fig. 8.18) with a chance probability of 2.7×10^{-5} when they selected lower energy events of 5–10 TeV. The observed energy spectrum (Fig. 8.19)

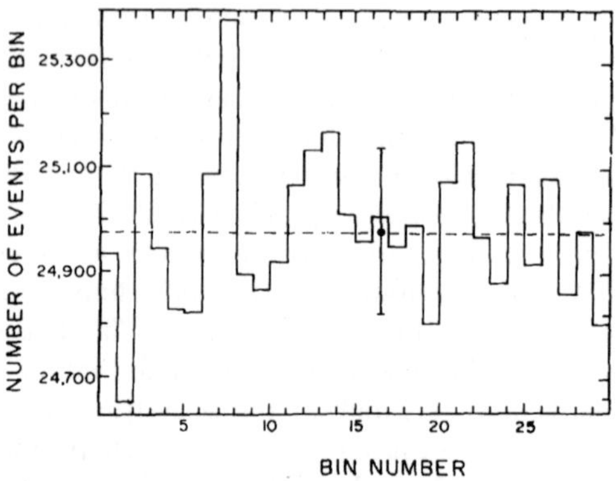

Figure 8.18 Phasogram of the Vela pulsar observed by Bhat et al. (1987) during the five year period 1979–85.

had a slope of $-(2.5 \pm 0.3)$ and the fluxes are several orders of magnitude lower than the extrapolation of the COS-B spectrum suggesting that the production mechanisms in the two energy regions may be different. The Durham group [73] observed this source

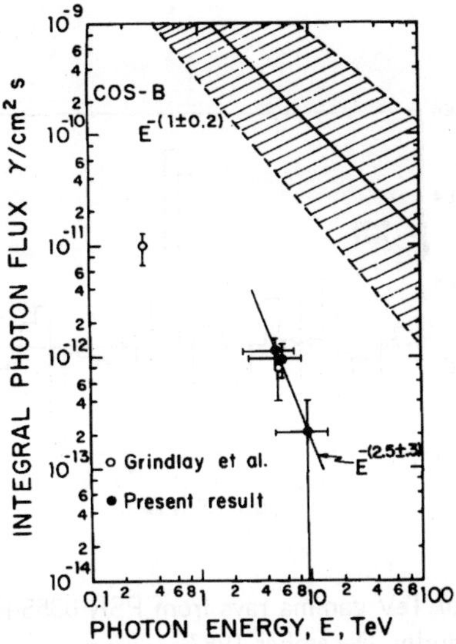

Figure 8.19 Integral energy spectrum of TeV gamma rays from Vela pulsar observed by Bhat et al. (1987) and Grindlay et al (1975). The hatched band is the extrapolation of the COS-B spectrum.

for 40 hours during Jan–Feb 1987 with their Narabri telescope in Australia. Finding no excess, they set a 3 σ upper limit of $(6 \pm 1.5) \times 10^{-11}$ cm^{-2} s^{-1} at $E > 300$ GeV, which is consistent with the lower energy value of Grindlay et al.

PSR 0355+54 is a short period (156 ms) weak radio pulsar, with a period derivative of 4.4×10^{-15} s s^{-1}, at a distance of ~1.5 kpc. It is characterized by extremely low timing-noise and occasional large glitches. A glitch in 1986 was one of the largest ever recorded from any pulsar, with $\frac{\Delta P}{P} \approx 4 \times 10^{-6}$. It is suggested that this pulsar might possibly be a member of an evolutionary family of objects including Crab, Vela, unidentified COS-B sources, Geminga-like objects and gamma ray burst sources. It was observed in the TeV energy range by the Pachmarhi group [74] for 7 days during December 1987. They found a 4.3 σ peak, with an over all chance probability of 8.7×10^{-4}, at a phase of 0.53 with respect to the radio phase in the 24.9 hours of data. This corresponds to a gamma ray flux of $(7.9 \pm 2.0) \times 10^{-12}$ cm^{-2} s^{-1} at $E > 1.3$ TeV. Their phasogram is shown in Fig. 8.20. The observed duty cycle is 3%, similar to Crab and Vela pulsars. The source depicts a steady behavior as shown in Fig. 8.21, where the signal excess, in standard deviations, is plotted for 31 independent runs comprising the whole data set.

PSR 1706-44, which is at a distance of 1.5 kpc, was observed as a gamma ray source

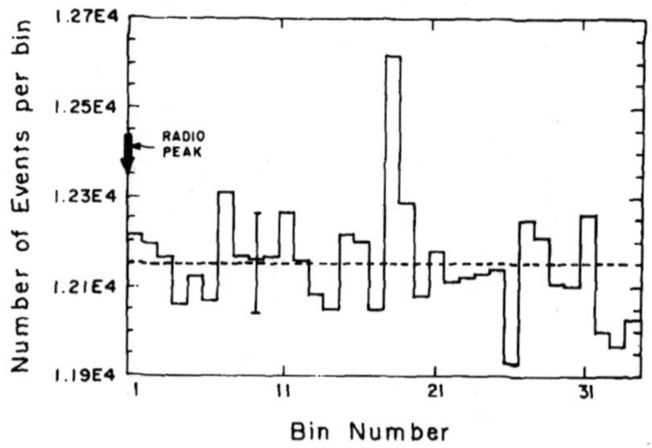

Figure 8.20 Phasogram of TeV gamma rays from PSR 0355+54 observed by Bhat et al. (1990) for 24.9 hours during December 1987.

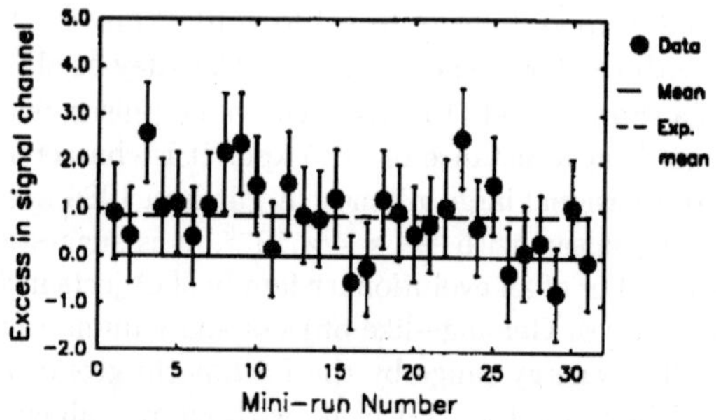

Figure 8.21 Signal strength, in units of standard deviation, from PSR 0355+54 plotted against run number for 31 mini-runs observed by Bhat et al. (1990). It shows a persitent signal with a mean of 0.864 while the expectation for random phasograms is 0.0.

by the COS-B satellite and identified with the radio pulsar through the discovery of 102 ms pulsed signal with EGRET. The CANGAROO collaboration [75] observed this object, during the months of July and August in the years 1992 and 1993, with their 3.8 m diameter imaging camera of 220 pixels, each with a field of view of $0.12° \times 0.12°$, located at Woomera in Australia. The total on-source and off-source observation times were 84 and 60 hours respectively. When they made cut on the parameter "alpha", which essentially selects fully contained on-source events, they found a 13 σ excess. The three periods July-August 1982, July 1983 and August 1983 each showed 7 σ excess, suggesting steady emission. The corresponding flux is $(1.1 \pm 0.2) \times 10^{-11}$ cm^{-2} s^{-1} at $E > (1 \pm 0.2)$ TeV. The estimated luminosity of the source, assuming isotropic emission, is $\sim 3 \times 10^{35}$ erg s^{-1}, which is about 10^{-3} of the total spin down energy of the neutron star, but smaller than the GeV energy gamma ray luminosity and greater than the X-ray luminosity. After Crab, this is the second most convincing and steady galactic TeV gamma ray source detected.

8.2.4 X-ray Binaries

8.2.4.1 Cygnus X-3

Cygnus X-3 is an X-ray binary system at a distance of 11.4 kpc. It is a controversial source in the GeV energy region as it was seen by two groups ([76, 77]) but not by the COS-B group. This source has been seen in the infrared and radio regions also. It is, however, unfavorably located for optical observations. It has a 4.8 hour periodicity believed to be due to corotation of a neutron star and a star of several solar masses. Apart from quiescent emission, intense flares in the radio region have been observed. It is one of the very few sources in which jets have been observed indicating the presence of high energy particles. This source has been extensively studied in TeV and PeV energy regions.

TeV Gamma Rays The Crimean Astrophysical Observatory (CAO) group [78,79] observed Cygnus X-3 for a number of years and found 5.4 σ and 3 σ excess at phases 0.2 and 0.8, of the 4.8 hr period, corresponding to fluxes of 1.8×10^{-10} and 9×10^{-11} cm^{-2} s^{-1} respectively at $E > 2$ TeV. They also found that the signal is modulated with a 34.1 d period, also seen in X-rays. This was later confirmed by several other groups [80–83]. The phasogram obtained by Danaher et al. [81] is shown in Fig. 8.22. While the CAO group saw the signal at two phases, most other groups saw it around only one phase—around 0.5–0.8. The luminosity of the source at energies > 2 TeV is around 10^{37} ergs s^{-1}, if one assumes isotropic emission. Later observations indicate that the flux may be variable as the emission is seen only on some occasions.

The fast pulsar, suspected to be present in the Cygnus X-3 system, was searched for by many groups. The Durham group [84] reported detection of a 12.59 ms periodicity at the binary phase of ~ 0.6. They also found that the pulsar period is slowing down at the rate of $(1.9 \pm 0.3) \times 10^{-14}$ s s^{-1}. Careful search by several other

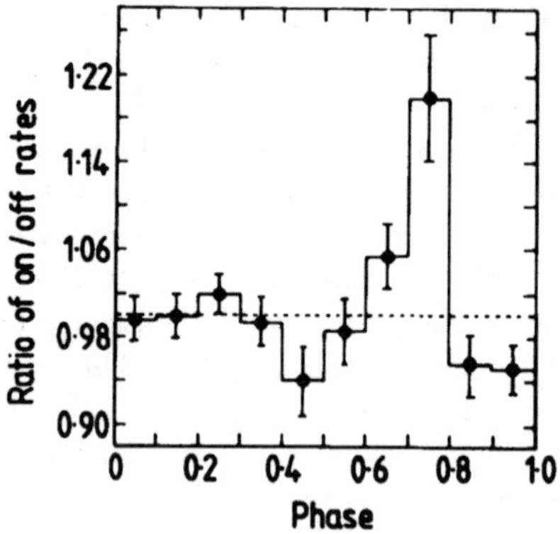

Figure 8.22 Orbital phasogram of TeV gamma rays from Cygnus X-3 obtained by Danaher et al. (1981).

groups, however, did not reveal this periodicity. We have to await conformation by other groups before the existence of a pulsar is considered to be established.

PeV Gamma Rays In 1983, the Kiel group [19] announced the discovery of Cygnus X-3 as a PeV gamma ray source. In the data collected during 1976–79 with their air shower array operated at Kiel (sea level), they found excess events at the level of 4.4 σ among older showers (> 1.1) of energy > 1 PeV from the direction of Cygnus X-3. In addition, the events are modulated with the 4.8 hr periodicity with the peak occurring at a phase of 0.35. Their phasograms with the period divided into different numbers of bins is shown in Fig. 8.23. The excess events are concentrated in a very narrow phase interval. The probability of this being purely due to chance is 2×10^{-4} for the case of 150 bins. The corresponding flux is $(7.4 \pm 3.2) \times 10^{-14}$ cm^{-2} s^{-1} at $E > 2 \times 10^{15}$ eV. Cawley and Weekes [85] estimated, taking into account absorption due to the black body radiation, the luminosity of the source in this energy region to be 2×10^{37} ergs s^{-1}, comparable to that in X-rays. Combining data from lower energies, the Kiel group deduce the energy spectrum to be

$$N(> E_\gamma) = (2.5 \pm 0.9) \times 10^{-7} E_\gamma^{-1.128\pm0.013} \qquad \mathrm{cm}^{-2}\ \mathrm{s}^{-1}, \tag{8.4}$$

with E_γ measured in GeV.

The Kiel group also measured the muon content of showers arriving from the direction of Cygnus X-3 and found that it is 0.77 ± 0.09 of that in normal cosmic ray showers, compared to less than 10% expected for gamma ray showers. So, the radiation from Cygnus X-3 is not gamma ray-like unless new processes are invoked as in the case of Crab.

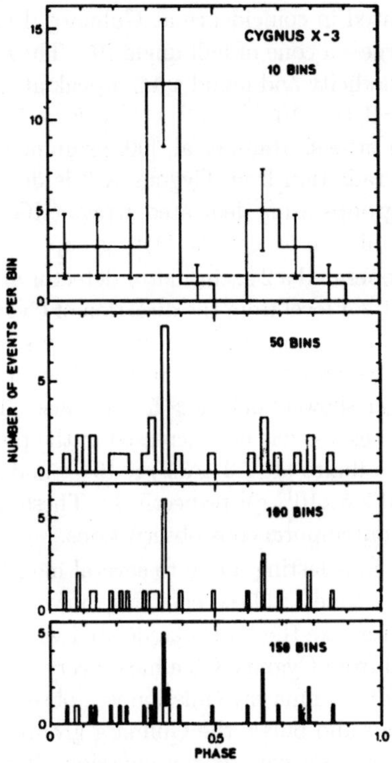

Figure 8.23 Orbital phasograms of events from the direction of Cygnus X-3 obtained by the Kiel group.

The Kiel result was confirmed by the Haverah Park group [86]. They observed a 1.7 σ excess from the direction of Cygnus X-3 in their data collected with the Haverah Park air shower array during 1978–82. The peak in the phasogram in their case, however, occurs in the phase interval 0.225–0.250. They see 73 events in this phase bin against an expected average of 39. The chance probability for such an excess to occur in any phase bin is $\sim 10^{-3}$. Their flux is $(1.5 \pm 0.3) \times 10^{-14}$ cm^{-2} s^{-1} at $E_\gamma > 3 \times 10^{15}$ eV, which is about a factor of ~ 3 smaller than the Kiel flux. Both the groups find that the energy spectrum steepens at a few times 10^{16} eV.

The difference in the phase where the excess occurs in the two experiments is because of different ephemera used by the two groups—the Kiel group used the ephemeris of Parsignault et al. while the Haverah Park group used that given by van der Klis and Bonnet Bidaud.

Bhat et al. [89] used the atmospheric Cerenkov technique with two photomultipliers facing upwards operated in coincidence at Gulmarg, India, during 1976–77. The field of view was very large—a cone of half angle 70°. They analyzed the event times to look for the 4.8 hr periodicity and found a 4.5 σ peak at the phase 0.55–0.60. They deduced a flux of $(1.6 \pm 0.4) \times 10^{-12}$ cm^{-2} s^{-1} at $E > 5 \times 10^{14}$ eV. This is about a factor of 5 larger than others. Rana et al. [90] examined the various observations and concluded that the radiation from Cygnus X-3 is decreasing with time. Later observations by several groups using dedicated arrays (KGF, CASA-MIA, CYGNUS etc.) did not show any emission except the Ooty group.

The Ooty group [91] operated a 24 scintillator detector array at Ooty during 1984–87 and found an excess at 2 σ level from the direction of Cygnus X-3 in showers of size larger than 5×10^4, which increased to 3.4 σ when they selected older showers. The phasogram, however, shows only 2 σ excess in the phase interval 0.60–0.75. Similar features are seen for larger showers of size $> 5 \times 10^5$ also. Thus, unlike the Kiel and Haverah Park results, the excess has not increased in the phasogram. Based on these observations, they deduce fluxes of $(7.2 \pm 3.2) \times 10^{-13}$ and $(1.0 \pm 0.4) \times 10^{-13}$ cm^{-2} s^{-1} at $E > 2.5 \times 10^{14}$ and 2.5×10^{15} eV respectively. These fluxes are at least a factor of 10 larger than other contemporaneous observations.

Several episodic emissions lasting a day to several months have been reported by various groups at various times. These episodes, however, are of marginal significance. Of particular interest are the bursts associated with radio flares. Strong radio flares have been observed from Cygnus X-3 almost every year except during 1985–88. During a flare in 1972, TeV gamma ray emission was observed by Neshpor et al. [78]. During the October 1985 radio burst, the Gulmarg group reported TeV gamma ray flux and the Baksan group PeV gamma ray emission. The radio burst reached its maximum on 10 October. While the TeV gamma rays were detected on 10 and 12 October, the PeV flux was detected 3–5 days later. While the connection between radio and high energy gamma ray emission is not yet clear, Berezensky [92] suggested that where as TeV gamma rays can escape interaction with the radio and infra-red radiation during the burst and can be detected promptly, PeV radiation is absorbed within the source during the maximum of the radio flare and hence delayed.

Cygnus X-3 has been detected in the EeV ($\sim 10^{18}$ eV) energy region also. The Fly's Eye group [93] (see also Ref. [94]) detected during the period 1981–89, with their Fly's Eye detector, a 5.8 σ excess at $E > 2$ EeV with a chance probability of 2×10^{-5}. They also see an excess in the phase interval 0.05–0.35. The combined probability for the phase excess and the DC excess is $\sim 1.4 \times 10^{-5}$. The detected flux is 4×10^{-18} cm^{-2} s^{-1} for $E > 2$ EeV.

The Akeno group [95] recorded a 3.5 σ excess with a chance probability of $\leq 10^{-3}$ during the period 1984–89 with their 20 km^2 air shower array. No modulation with the 4.8 hr period has been seen. Their estimated flux is $(1.8 \pm 0.7) \times 10^{-17}$ cm^{-2} s^{-1} for $E > 0.5$ EeV.

The Haverah Park group, however, do not see any excess during the period 1974–

87 in this energy range and place an upper limit of 4×10^{-18} cm^{-2} s^{-1} on the flux at $E > 0.5$ EeV at the 95% confidence level. This discrepancy is not understood.

At these very high energies, neutrons can survive the 11 kpc distance from the source and the radiation need not be gamma rays.

Muons from Cygnus X-3 If the UHE gamma rays from Cygnus X-3 are due to decay of π^0s produced in interactions, with matter in the source, of protons and nuclei accelerated to high energies, then the charged pions should also produce neutrinos through various decay processes. These neutrinos can be detected in detectors operated underground through muons produced in their interactions in the overburden. The IMB group searched for neutrinos from Cygnus X-3 in their nucleon decay detector, operated underground in the Cleveland salt mines, and found a marginally significant signal. This prompted other underground installations to search for muon signals from Cygnus X-3. The Soudan [96] and the NUSEX [97] (see also Ref. [98]) reported high energy muon fluxes modulated with the 4.8 hr periodicity. The muon signal varied with zenith angle (and hence with the amount of matter above the detector) in such a way that neutrinos could be ruled out as the source of muons. The muons would, then, have to be produced in the Earth's atmosphere or the rock above the detector (muons even of TeV energies can not survive the long distance from Cygnus X-3). The Kamioka nucleon decay experiment in Japan, however, did not see any signal.

While a small number of muons are indeed expected from gamma ray showers, the fluxes of such muons are several orders magnitude smaller than the gamma ray fluxes from Cygnus X-3. Fig. 8.24, taken from Ref. [98], shows the flux of muons as a function of energy reported by these groups as well as the observed gamma ray fluxes (γ) and the expected atmospheric muon fluxes (μ) due to gamma rays. Since the reported muon fluxes are nearly same as gamma ray fluxes, a new neutral particle or new physics at TeV energies has to be invoked to explain these results. Search at accelerators for such new phenomena yielded negative results. Thus, underground muons from Cygnus X-3 remains an enigma. Perhaps the reported detections are just statistical fluctuations.

Production Mechanisms Because of the intense radiation fields in the vicinity of X-ray binaries like Cygnus X-3, electrons are unlikely to be accelerated to high energies since they rapidly lose energy through synchrotron and inverse-Compton processes. Protons and nuclei can, however, be accelerated as such energy losses are small because of the higher mass. Once produced, these particles interact with target matter in the companion star and produce π^0s which subsequently decay into gamma rays. The amount of matter should be small enough for these gamma rays to escape without interaction. Several models incorporating these features have been proposed. These models, shown in Fig. 8.25 differ mainly in the phase at which gamma ray emission occurs.

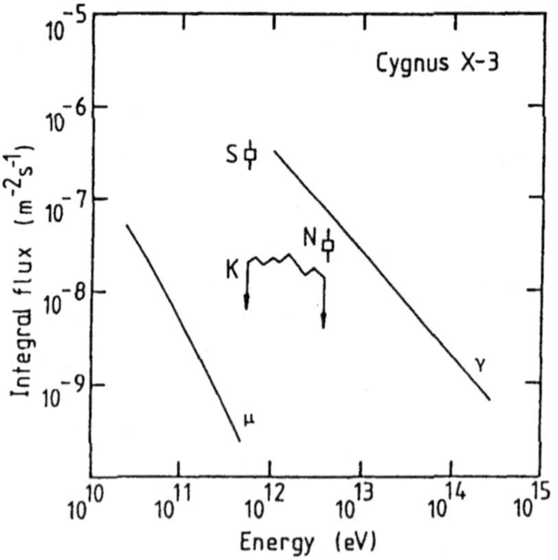

Figure 8.24 Integral flux of muons from Cygnus X-3 as a function of energy observed by Marshak et al. (S) and Battisoni et al. (N). The upper limits are from the Kamioka experiment (K). The lines γ and μ denote the observed gamma ray flux and the estimted muon flux.

The simplest scenario, shown in Fig. 8.25(a), in which the stellar atmosphere forms the target matter, predicts emission at phases 0.1 and 0.9, which is not in agreement with observations. Hillas [99] proposed that a high energy proton beam from the neutron star disrupts the stellar surface producing a fountain of stellar material. Orbital motion would distort the flow providing target material for the production of gamma rays at phases 0.2 and 0.75 (Fig. 8.25(b)), nearly where emission is observed. In this model, nearly monoenergetic protons of energy $\sim 10^{17}$ eV at $\sim 10^{39}$ ergs s^{-1} from the neutron star generate electromagnetic cascades in the target material through production and decay of π^0 mesons. If there is an embedded magnetic field in the gas surrounding the companion star, the cascade develops by alternate pair production and very rapid synchrotron radiation in which case the amount of target material required is very much less. The resulting gamma rays in the cascade escape and are detected. This model can account for all gamma rays above a few GeV. The energy spectrum obtained in this model is compared with observations in Fig. 8.26. The curves are for 1,2 and 4 bremsstrahlung radiation lengths when magnetic field is present. The dotted curve is for the non-magnetic cascade. The steepening of the spectrum at $\sim 10^{16}$ eV is due to attenuation of gamma rays by the magnetic pair pro-

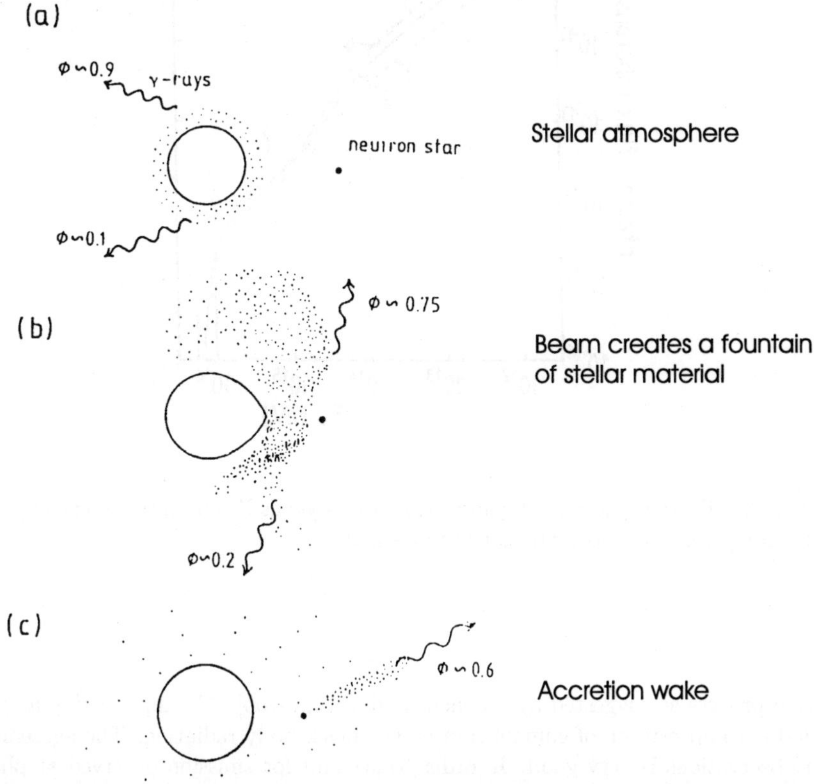

Figure 8.25 Configurations of the Cygnus X-3 system proposed by different authors to explain the light curve of UHE gamma rays; (a) Berezensky (1979); (b) Hillas (1984); (c) Hillas (1985).

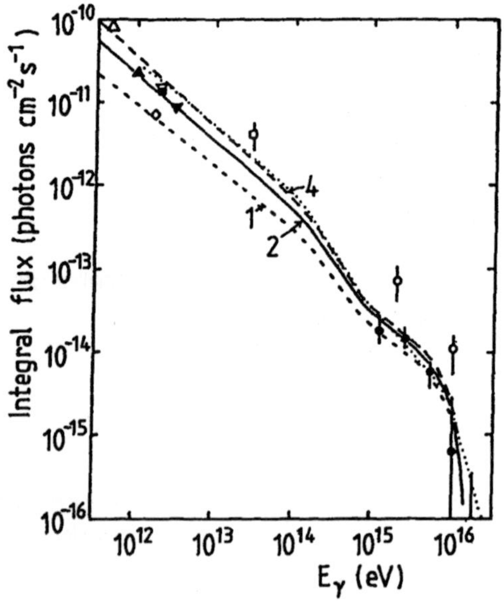

Figure 8.26 Energy spectrum of gamma rays from Cygnus X-3 from the model of Hillas (1984) compared with observations by various groups.

duction process as suggested by Stephens and Verma [100]. The dip starting at 10^{14} eV is due to absorption of gamma rays by the black body radiation. The agreement with observations is very good. In order to account for emission observed at phase 0.6, Hillas [101] suggested that an accretion wake is formed at the phase 0.625 (Fig. 8.25(c)) if the stellar wind velocity and the neutron star orbital velocity are nearly same. This accretion wake provides the material for gamma ray production at this phase.

In this model, most of the protons (or nuclei) escape into the galaxy and form the cosmic rays. Since the source is monoenergetic, one has to rely on evolution of beam energy with time to reproduce the galactic cosmic ray spectrum as suggested by Gunn and Ostriker [102]. To account for the energy density of cosmic rays, only a few sources like Cygnus X-3 are needed. Wdowczyk and Wolfendale [103] suggested that about 30 such sources might supply the power to maintain galactic cosmic rays.

8.2.4.2 Hercules X-1

Hercules X-1 is a pulsating and eclipsing X-ray binary system containing a neutron star and a low mass companion star, Hz-Herculis. It is located at a distance of ~ 5 kpc. In the optical and X-ray regions, the object shows modulations on the time scales of 1.24 s, 1.7 d and 35 d. These periodicities are thought to be the pulsar period, the orbital period and the precession period of the neutron star or the accretion disk respectively. The object has been extensively studied in TeV and PeV energy regions. Most of the reported emissions from this object in these energy regions are episodic in nature. Apart from Crab, this is the only object which has been simultaneously detected by two separate groups and also from which the most significant episodic emission has been reported.

TeV Gamma Rays The first detection of episodic emission from this object was reported by the Durham group [104]. They detected the signal as a $> 3\ \sigma$ excess in the counting rate lasting for 3 min on April 17, 1983 and also a similar signal when analyzed for modulation at the X-ray period of 1.24 s. The emission occurred at a phase of 0.76 in the 1.7 d period and 35 d prior to a detected turn-on in X-rays. They quote a flux of 1.2×10^{-9} cm^{-2} s^{-1} at $E > 1$ TeV. On April 4, 1984 both the Durham group with their telescope at Dugway, Utah, U.S.A. and the Whipple group have seen emission simultaneously. Further episodes were detected by the Pachmarhi group on April 11, 1986, the Whipple group on June 11, 1986 and the Haleakala group on May 13, 1986.

The Pachmarhi observation [105] was a 42 σ excess in counting rate and it is the most significant detection so far. The emission occurred at $\Phi_{1.7\,\mathrm{d}} = 0.19$ and $\Phi_{35\mathrm{d}} = 0.31$. Their flux is $(1.80 \pm 0.04) \times 10^{-8}$ cm^{-2} s^{-1} at $E > 0.4$ TeV and if the source spectrum is $E^{-3}dE$ the luminosity of the source is 1.8×10^{37} ergs s^{-1}. They could not do pulsar phase analysis because of gaps in recording. They did not see any more episodes in 1987 and 1988.

The Whipple group's [106] observation was an excess of 1.9 σ in counting rate and a periodicity of 1.2358 ± 0.0003 s with a chance probability of 9×10^{-3}. Their episode occurred at $\Phi_{1.7\,\mathrm{d}} = 0.70$ and $\Phi_{35\mathrm{d}} = 0.04$ and the quoted flux is 2×10^{-10} cm^{-2} s^{-1} at $E > 1$ TeV. When they applied the "azwidth" cut to the data, however, the signal disappeared. Thus the signal is not gamma ray-like.

The Haleakala group's [107] episode consisted of a 1.4 σ excess in counting rate and the derived pulsar period was 1.2359 ± 0.0002 s with a chance probability of 6.8×10^{-3}. The detected flux was 5×10^{-10} cm^{-2} s^{-1} at $E > 0.2$ TeV. The emission occurred at $\Phi_{1.7\,\mathrm{d}} = 0.81$ and $\Phi_{35\mathrm{d}} = 0.22$.

Thus, the episodes are not very significant except for the Pachmarhi result, which, however, has no pulsar analysis. In recent years there have been no reports of such episodes except for a few marginal ones.

PeV Gamma Rays In the PeV energy region, only two groups reported significant episodes. The Los Alamos group [108] observed two bursts, each lasting 30 minutes, on July 24, 1986. They observed 7 events with a background of 0.53 events in one episode while the source and background events in the second episode were 10 and 2.6 respectively. The periodicity in this case was 1.23575 ± 0.0003 s with a chance probability of 2×10^{-5}. They reported a flux of 2×10^{-11} cm^{-2} s^{-1} at $E > 200$ TeV. The episode occurred at $\Phi_{1.7\,\mathrm{d}} = 0.8 - 0.9$. The muon content of the events from Her X-1 was, however, is nearly same as normal cosmic ray showers ruling out gamma rays. If a neutral particle is responsible for the signal, it should have a mass less than 60 MeV/c^2 in order to maintain phase locking of the signal at the observation site.

The Ooty group [109] reported three episodes, one each on July 1, August 8–9 and November 21, 1986. Since the data were spread over 143 days, they searched for phase-locked signal and obtained a very precise period of 1.2357701 s consistent with the other observations during 1986. The events on all the days clustered at $\Phi_{1.7\,\mathrm{d}} \sim 0.65$ and $\Phi_{35\mathrm{d}} \sim 0.65$. The corresponding flux was $(3.0 \pm 0.7) \times 10^{-11}$ cm^{-2} s^{-1} at $E > 10^{14}$ eV.

It is interesting to note that though all the above episodes occurred on widely separated days the observed period is nearly same, but smaller than the X-ray period of 1.2378 s—blue shifted by about 0.16%.

Production Mechanisms Eichler and Vestrand [110] proposed a model in which gamma rays are generated by particles that interact with the precessing accretion disk. Particles are accelerated within the neutron star's magnetosphere by shocks in an accretion flow. Gamma ray emission occurs preferentially at the onset and decline of the high-intensity X-ray states, as is observed in some cases. Gamma rays of energy up to $2 - 3 \times 10^{15}$ eV can be obtained in this model. This model can not explain the observed blue-shifted period.

In order to explain the blue-shifted period, Cheng and Ruderman [111] proposed that gamma rays are produced in rapidly moving thin target "windows" within the accretion disk (see Fig. 8.27). These windows are drifting towards the neutron star at periods determined by their instantaneous Keplerian orbits. The duration of emission reflects the life time of these target regions.

Slane and Fry [112] suggested that clumps of matter from the accretion disk, which arise due to instabilities at the disk or magnetosphere interface, may give rise to anomalous periodicities as a result of the interaction of a broad beam of particles from the pulsar with them. The gamma rays along the line of sight to the neutron star will prevail as long as the target remains in the beam and will exhibit the periodicity of the Keplerian orbit of the target rather than that of the neutron star.

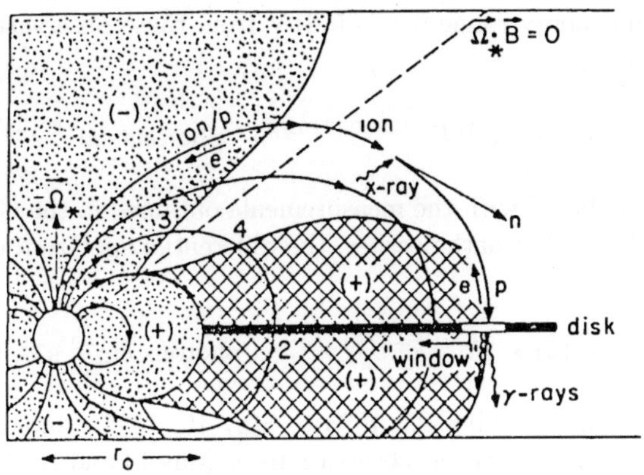

Figure 8.27 Model of Her X-1 by Cheng and Ruderman (1989) to explain the blue shifted period in TeV gamma rays.

8.2.5 Extragalactic Sources

Active galactic nuclei (AGNs) are recognized as a potential class of cosmic ray sources with spectra extending to $> 10^{20}$ eV. They contain enough target material in the form of gas, radiation and magnetic fields for efficient gamma ray production. Their enormous luminosity (ranging from 10^{43} to more than 10^{48} erg s^{-1}) and compact size ($10^{13} \sim 10^{15}$ cm) suggests that the ultimate energy source is accretion on to a massive black hole. Recently several AGNs have been detected as very powerful and variable GeV gamma ray emitters by the detector EGRET on board the Compton Gamma Ray Observatory (CGRO) satellite. The nearest among them is Markarian 421, which is at a red shift $z = 0.031$. TeV gamma rays can reach us from a red shift of up to 0.1 without much attenuation due to interaction with the estimated inter-galactic infrared radiation.

8.2.5.1 Markarian 421

Markarian (Mkn) 421 is a BL Lac object and belongs to the blazar class of AGNs which have associated relativistic jets, emanating from the central power house, with the jet axis oriented near the observer's line of sight. It was extensively observed at radio, UV/optical and X-ray frequencies. Its parent galaxy has been identified as a giant elliptical. The Whipple collaboration [113] observed this source for 7.5 hrs during March 24 to June 2, 1992 and found a 6.3 σ excess after selecting gamma ray-like events applying supercuts on the Cerenkov images. The corresponding gamma ray flux is 1.5×10^{-11} cm^{-2} s^{-1} at $E > 0.5$ TeV. Assuming isotropic emission and taking the source distance of 124 Mpc into account, they estimate the TeV gamma ray luminosity to be $\sim 10^{43}$ erg s^{-1}. They also extracted the energy spectrum of gamma rays in the energy range 0.4 $\sim$ 5 TeV from these observations [114], which can be expressed as

$$\frac{dN}{dE} = (1.04 \pm 0.14) \times 10^{-11} E^{-2.25 \pm 0.19 \pm 0.3} \qquad \text{photons cm}^{-2}\ \text{s}^{-1}\ \text{TeV}^{-1}. \tag{8.5}$$

with E in TeV. Combined with the measurements obtained in the GeV energy range by the EGRET about 10 months earlier, the differential spectrum can be described as

$$\frac{dN}{dE} = (1.02 \pm 0.14) \times 10^{-11} E^{-2.06 \pm 0.04} \qquad \text{photons cm}^{-2}\ \text{s}^{-1}\ \text{TeV}^{-1} \tag{8.6}$$

Observations from December 23, 1993 to May 10, 1994 showed an average flux which is only half of this. However, observations on May 14 and 15, 1994 showed [115] an increase in flux above this quiescent level by a factor of $\sim$ 10 with an e-folding time of $\sim$ 2 days. In addition, a second period of enhanced emission, though of lesser significance of 3 σ lasting for 4 days with the emission peaking on June 7, 1994, was also observed. These observations put constraints on the size of the TeV gamma ray emission region. The peak flux observed on May 15, 1992 is nearly simultaneous with a 20 fold increase in X-ray emission seen by the ASCA group [116]. This is the first time that variability in the TeV energy region is seen in an extragalactic object.

Several other groups carried out observations on this source at higher energies. Finding no excess, they set upper limits on the flux which are shown in Fig. 8.28, taken from the Rapporteur paper of Haines at the Calgary Cosmic Ray Conference [117]. The figure also shows the expected integral fluxes multiplied by the energy due to absorption by the microwave background (solid line) and an estimated infrared component (dash dot line), assuming that the source spectrum continues with an integral power law index of -1. Depending on the magnitude of the infrared flux, Markarian 421 may or may not be visible at these higher energies.

8.2.5.2 Markarian 501

Markarian 501 is another nearby AGN with a redshift of 0.034. This object has not been detected by EGRET in the GeV energy region. The Whipple group [118] carried out observations on this object in two modes—the tracking mode and the ON/OFF mode. In the tracking mode, which is the discovery mode, the object is continuously tracked without taking data off source on a control region. Background is determined from the events which are not from the direction of the source. In the ON/OFF mode, the source and a control background regions are tracked alternately for 28 minutes each. They have made several improvements such as higher mirror reflectivity, use of light cones to collect more light and lower discriminator threshold, which reduced the threshold to 200 GeV. Improvements in the selection of gamma ray events are also introduced making use of the data on Crab and Monte Carlo simulations.

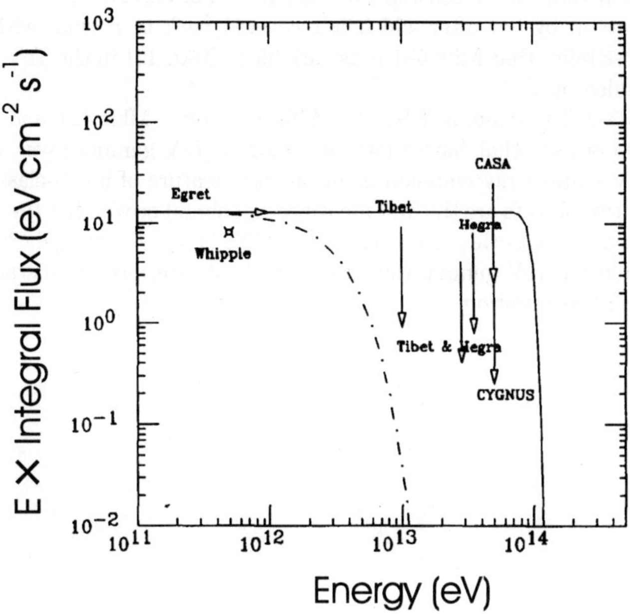

Figure 8.28 Integral gamma ray flux times energy as a function of energy from the AGN, Markarian 421. The expected spectra due to interactions of gamma rays with the microwave (—) and an estimated infrared background radiation fields(– · –) are also shown.

In the tracking mode, a total of 31 hours of data were taken in March, 1995. They observed 699 events in the gamma ray domain of 0° to 10° of the "alpha" parameter while the expected number of background events is 405, constituting an excess of 11.7 σ using Li and Ma statistics. The authors opine that the real significance may be 9 σ because of systematics. They subsequently carried out observations in the ON/OFF mode during April 23 to July 30, 1995 taking 90 pairs of data of 28 minutes each. After rejecting 7 pairs of data due to relatively bad conditions and selecting gamma ray-like events by applying "supercuts", they found an excess of 334 events against an expected background of 632 amounting to 8.4 σ excess, thus confirming detection of Mkn 501 as a TeV gamma ray emitter. The corresponding integral flux is $(8.1 \pm 1.4) \times 10^{-12}$ cm^{-2} s^{-1} at an energy threshold of 300 GeV. Taking the upper limit of the integral flux at 100 MeV to be $< 1.5 \times 10^{-7}$ cm^{-2} s^{-1} from the EGRET observations, the exponent of the integral spectrum between 100 MeV and 300 GeV must be harder than -1.2, which is quite consistent with the value of -1.0 observed for Mkn 421.

The flux is found to be constant over the period of observation except on one day when it went up by 6σ, after which it recovered back to normal within two days. Thus the possibility that Mkn 501 is as variable as Mkn 421 in the TeV energy region cannot be ruled out.

Of the 35 AGNs examined by the Whipple group, Mkn 421 and Mkn 501 are the only two objects that have shown emission of TeV gamma rays. It is not clear whether TeV gamma ray emission is an intrinsic feature of a sub-class of AGNs (of which these two objects are the first identified members) or whether the non-detection of TeV emission from other, more distant, AGNs is due to intergalactic absorption. A deep search for TeV gamma rays from AGNs at intermediate distances can shed some light on this question.

References

1. I. S. Schklovsky, *Dokl. Akad. Nauk* **90** (1953) 983.
2. G. Cocconi, *Proc. Moscow Cosmic Ray Conference II* (1960) 309.
3. J. Wdowczyk et al., *J. Phys.* **A5** (1972) 1419.
4. P. V. Ramana Murthy, in *Non-solar Gamma Rays* ed. R. Cowsik and R. D. Wills (Pergamon Press, Oxford and New York, 1990).
5. T. C. Weekes, *Phys. Rep.* **160** (1988) 1.
6. T. C. Weekes, *Space Sci. Rev.* **59** (1992) 315.
7. Polla V. Ramana Murthy and Arnold W. Wolfendale, *Gamma Ray Astronomy* (Cambridge University Press, Cambridge, 1986).
8. P. N. Bhat et al., *Astron. Astrophys. Lett.* **81** (1980) L3.
9. A. M. Hillas, *Proc. 19th International Cosmic Ray Conference, La Jolla* **3** (1985) 445.
10. M. V. S. Rao and S. Sinha, *J. Phys. G: Nucl. Phys.* **14** (1988) 811.
11. A. M. Hillas and J. R. Patterson, *J. Phys. G: Nucl. Phys.* **16** (1990) 1271.
12. P. R. Vishwanath et al., *Towards a Major Atmospheric Cerenkov Detector III* ed. T. Kiffune, (Universal Academy Press, Inc, Tokyo, Japan, 1994) 65.
13. E. J. Fenyves, *Proc. of Workshop on Techniques in Ultra High Energy GAmma Ray Astronomy, La Jolla* (Eds. R. J. Protheroe and S. A. Stephens, 1985) 124.
    A. M. Hillas, *Proc. 19th International Cosmic Ray Conference, Moscow*, **2** (1987) 362.
    T. Chung and P. K. Mackeown, *Il Nuovo Cimento* **11C** (1988) 193.
14. B. S. Acharya et al., *J. Phys. G: Nucl. Phys.* **19** (1993) 1053.
15. B. S. Acharya et al., *Nucl. Instr. Meth. in Phys. Res.* **A365** (1995) 224.
16. D. E. Alexandreas et al., *Nucl. Instr. Meth.* **A311** (1991) 350.
17. R. A. Ong et al., *Nucl. Phys. B (Proc. Suppl.)* **14A** (1990) 273.
18. A. E. Chudakov et al., *Proc. 8 th International Cosmic Ray Conference, Jaipur* **4** (1963) 199.
19. M. Samorski W. Stamm, *Astrophys. J.* **268** (1983) L17.
20. M. V. S. Rao and B. V. Sreekantan, *Current Science* **62** (1992) 617.
21. G. G. Fazio et al., *Ap. J. Lett.* **175** (1972) L117.
22. J. B. Mukanov et al., *Izv. Krym. Astrofiz. Obs.* **67** (1983) 55.
23. M. F. Cawley et al., *Proc. 19th International Cosmic Ray Conference, La Jolla* **1** (1985) 131.
24. T. C. Weekes et al., *Astrophys. J.* **342** (1989) 379.
25. M. J. Lang et al., *Proc. 21st International Cosmic Ray Conference, Adelaide* **2** (1990) 139.
26. G. Vacanti et al., *Astrophys. J.* **377** (1991) 467.

27. C. Akerlof et al., *Proc. 21st International Cosmic Ray Conference, Adelaide* **2** (1990) 135.
28. O. T. Tumer et al., *Nuclear Physics B (Proc. Suppl.)* **14A** (1990) 176.
29. O. T. Tumer et al., *Proc. 21st International Cosmic Ray Conference, Adelaide* **2** (1990) 155.
30. T. Dzikowski et al., *Proc. 18th International Cosmic Ray Conference, Bangalore* **2** (1983) 132.
31. T. Dzikowski et al., *Phil. Trans. R. Soc. A.* **301** (1983) 641.
32. I. N. Kirov et al., *Proc. 19th International Cosmic Ray Conference, La Jolla* **1** (1985) 135.
33. J. Boone et al., *Astrophys. J.* **285** (1984) 264.
34. N. Hayashida et al., *Proc. 17th International Cosmic Ray Conference, Paris* **9** (1981) 9.
35. A. A. Watson, *Proc. 19th International Cosmic Ray Conference, La Jolla* **9** (1985) 111.
36. M. A. B. Craig et al., *Proc. 17th International Cosmic Ray Conference, Paris* **1** (1981) 3.
37. S. C. Corbato et al., *Proc. 21st International Cosmic Ray Conference, Adelaide* **2** (1990) 159.
38. R. J. Gould, *Phys. Rev. Lett.* **15** (1965) 511.
39. J. E. Grindlay and J. A. Hoffman, *Ap. J. Lett.* **8** (1971) L209.
40. O. C. De Jager and A. K. Harding, *Astrophys. J.* **396** (1992) 161.
41. K. S. Cheng et al., *J. Phys. G: Nucl. Part. Phys.* **16** (1990) 1115.
42. K. S. Cheng, C. Ho and M. Ruderman, *Astrophys. J.* **300** (1986) 522.
43. P. W. Kwok and K. S. Cheng, *Bull. of A. P. S.* **34 (4)** (1989) 1225.
44. J. Wdowczyk and A. W. Wolfendale, *J. Phys. G: Nucl. Phys.* **13** (1987) 411.
45. E. Pare et al., *Phys. Lett.* **B242** (1990) 531.
46. A. Borione et al., *Proc. 24th International Cosmic Ray Conference, Roma* **2** (1995) 350.
47. J. E. Grindlay, *Ap. J. Lett.* **174** (1972) L9.
48. S. K. Gupta et al., *Astrophys. J.* **221** (1978) 268.
49. P. R. Vishwanath, *Proc. International Workshop om Very High Energy Gamma Ray Astronomy, Ootacamund* (eds. P. V. Ramana Murthy and T. C. Weekes), (1982) 21.
50. I. A. Gibson et al., *Nature* **296** (1982) 833.
51. P. N. Bhat et al., *Nature* **319** (1986) 127.
52. P. R. Vishwanath, *J. Astrophys. Astr.* **8** (1987) 69.
53. J. C. Dowthwaite et al., *Ap. J. Lett.* **286** (1984) L35.
54. O. T. Tumer et al., *Proc. 19th International Cosmic Ray Conference, La Jolla* **1** (1985) 139.

55. M. J. Lang et al., *Proc. 22nd International Cosmic Ray Conference, Dublin* **1** (1991)204.
56. S. K. Gupta et al., *Astron. and Astrophys.* **245** (1991) 141.
57. V. V. Alexeenko et al., *J. Phys. G: Nucl. Part. Phys.* **18** (1992) L83.
58. T. P. Li and Y. Q. Ma, *Astrophys. J.* **272** (1983) 317.
59. B. S. Acharya et al., *Nature* **347** (1990) 364.
60. B. S. Acharya et al., *AIP Conference Proceedings 220, High Energy Gamma Ray Astronomy, Ann Arbor* (1991) 137.
61. M. Aglietta et al., *Europhys. Lett.* **15** (1991) 81.
62. A. E. Chudakov, *AIP Conference Proceedings, High Energy Gamma Ray Astronnomy, Ann Arbor* (ed. James Mathews), (1990) 132.
63. W. Ochs and L. Stodolsky, *Phys. Rev.* **D33** (1986) 1247.
M. Drees and F. Halzen , *Phys. Rev. Lett.* **61** (1988) 275.
A. Subramanian, *Phys. Lett.* **244** (1990) 127.
64. G. Domokos and S. Kovesi-Domokos, *Phys. Rev.*, **D38** (1988) 2833.
65. R. N. Manchester and J. H. Taylor, *Pulsars* (Freman, San Fransisco, CA) (1977).
66. R. D. Wills et al., *Nature* **296** (1982) 723.
67. M. E. Ozel, *Europhys. Lett.* **14** (1991) 275.
68. S. V. Bogovalov, *Sov. Astron. Lett.* **16** (1990) 362.
69. S. V. Bogovalov and Yu. D. Kotov, *JETP Lett.* **52** (1990) 1.
70. F. R. Harnden and F. D. Seward, *Astrophys. J.* **283** (1984) 279.
71. J. E. Grindlay et al., *Astrophys. J.* **201** (1975) 82.
72. P. N. Bhat et al., *Astron. Astrophys.* **178** (1987) 242.
73. K. T. S. Brazier et al., *Proc. 21st International Cosmic Ray Conference, Adelaide* **2** (1990) 304.
74. P. N. Bhat et al., *Astron. Astrophys.* **236** (1990) L1.
75. T. Tanimori (CANGAROO collaboration), *Towards a Major Atmospheric Cerenkov Detector III* ed. T. Kiffune, (Universal Academy Press, Inc, Tokyo, Japan, 1994) 83.
76. R. C. Lamb et al., *Ap. J. Lett.* **212** (1977) L63.
77. A. M. Galper et al, *Pisma v Astron. Zh.* **2** (1976) 524.
78. Y. I. Neshpor et al., *Astrophys. & Space Sci.* **61** (1979) 349.
79. A. A. Stepanian et al., *Astrophys. & Space Sci.* **84** (1982) 347.
80. D. B. Mukanov, *Izv. Krym. Astrofyz. Obs.* **63** (1981) 151.
81. S. Danaher et al., *Nature* **289** (1981) 568.
82. R. C. Lamb et al., *Nature* **296** (1982) 543.
83. J. C. Dowthwiate et al., *Astron. Astrophys.* **126** (1983) 1.
84. K. T. S. Brazier et al., *Astrophys. J.* **350** (1990) 745.
85. M. F. Cawley and T. C. Weekes, *Preprint 1943, Centre for Astrophysics, Cambridge,*

USA.

86. J. Lloyd-Evans et al., *Nature* **305** (1983) 784.
87. M. van der Klis and J. M. Bonnet-Bidaud, *Astron. Astrophys.* **95** (1981) L5.
88. D. R. Parsignault et al., *Ap. J. Lett.* **209** (1976) L73.
89. C. L. Bhat et al., *Astrophys. J.* **306** (1986) 587.
90. N. C. Rana et al., *Astron. Astrophys.* **141** (1984) 394.
91. S. C. Tonwar et al., *Ap. J. Lett.* **330** (1988) L107.
92. V. S. Berezensky, *Nature* **334** (1988) 506.
93. G. L. Cassiday et al., *Phys. Rev. Lett.* **62** (1989) 383.
94. D. J. Fegan, *Proc. 21st International Cosmic Ray Conference, Adelaide* **11** (1990) 23.
95. N. Teshima et al., *Phys. Rev. Lett.* **64** (1990) 1628.
96. M. L. Marshak et al., *Phys. Rev. Lett.* **54** (1985) 2079.
97. G. Battisoni et al., *Phys. Lett. B* **155** (1985) 465.
98. R. J. Protheroe, *Proc. ASA* **6** (1986) 280.
99. A. M. Hillas, *Nature* **312** (1984) 50.
100. S. A. Stephens and R. P. Verma, *Nature* **308** (1984) 828.
101. A. M. Hillas, *Proc. 19th International Cosmic Ray Conference, La Jolla* **2** (1985) 296.
102. J. E. Gunn and J. P. Ostriker, *Phys. Rev. Lett.* **22** (1969) 728.
103. J. Wdowczyk and A. W. Wolfendale, *Nature* **306** (1983) 347.
104. J. C. Dowthwaite et al., *Nature* **309** (1984) 691.
105. P. R. Vishwanath et al., *Astrophys. J.* **342** (1989) 489.
106. R. C. Lamd et al., *Ap. J. Lett.* **328** (1988) L13.
107. L. K. Resvanis et al., *Ap. J. Lett.* **328** (1988) L9.
108. B. L. Dingus et al., *Phys. Rev. Lett.* **61** (1988) 1906.
109. S. K. Gupta et al., *Ap. J. Lett.* **354** (1990) L13.
110. D. Eichler and W. T. Vestrand, *Nature* **318** (1985) 345.
111. K. S. Cheng and M. Ruderman, *Ap. J. Lett.* **337** (1989) L77.
112. P. Slane and W. F. Fry, *Astrophys. J.* **342** (1989) 1129.
113. M. Punch et al., *Nature* **358** (1992) 477.
114. G. Mohanty et al., *Proc. 23rd International Cosmic Ray Conference, Calgary* **1** (1993) 440.
115. A. D. Kerric et al., *Ap. J. Lett.* **438** (1995) L59.
116. T. Takahashi et al., *IAU Circ. No.* 5993 (1994).
117. T. Haines, *Proc. XXIII International Cosmic Ray Conference, Invited, Rapporteur & Highlight Papers, Calgary* (1993) 341.

118. M. Catanese et al., *Towards a Major Atmospheric Cerenkov Detector IV* ed. M. Cresti (1995) 348.

Chapter 9
NEUTRINO ASTRONOMY

While the exciting field of neutrino astronomy has already opened up with the detection of solar neutrinos and the neutrinos from the Supernova 1987 A in the MeV to GeV range of energies, the prospects of Ultra High Energy Neutrino astronomy is currently engaging the attention of both experimentalists and theorists [1].

The Active Galactic Nuclei (AGNs) which have been recognized as possible sources of cosmic rays where particles can be accelerated to $\sim 10^{19}$ eV are now recognized as possible sources of very high energy neutrinos. The active galactic nuclei have luminosities in the range 10^{42}–10^{48} ergs/s and are believed to have Giant Black Holes of mass 10^4–10^{10} Solar masses. The Whipple group [2] have established beyond doubt that Markarian 421 is a source of TeV gamma rays, while the Compton Gamma Ray Observatory has discovered GeV emission from several AGNs, Mkn 421 being one of them [3]

The TeV gamma rays could in principle arise from the inverse photon scattering of high energy electrons in the AGN environments. Wherever a mechanism exists for acceleration of electrons, there should also be acceleration of protons if protons are available for acceleration. The protons being more massive can get accelerated to higher energies since the energy losses are much smaller.

The neutrino spectra generated in the central region of 3C 273 calculated by Stecker et al. [4], Szabo and Protheroe [5] and by Manheim and Biermann [6] are shown in Fig. 9.1. While the neutrinos are the products of the decay of $\pi^+ \to \mu^+ + \nu_\mu$ and $\mu^+ \to e^+ + \bar{\nu}_\mu + \nu_e$, the pions themselves are produced in the collisions of the accelerated protons with the photon background (the UV bump and X-ray photons extending to MeV range). The differences in the flux calculations are attributed to different astrophysical assumptions in the calculations [1].

The Diffuse Neutrino background extending up to 10^{19}eV from all the AGNs (unidentified) calculated by Szabo and Protheroe [5] and Stecker et al. [4] is shown in Figs. 9.2 and 9.3.

These neutrinos have to be detected either through the high energy muons they produce or through the cascades that result from the high energy electrons produced in the interactions of the neutrinos.

Fig. 9.4 gives the fluxes of muons. The dotted line in the figure gives the contribution from the neutrinos produced in the Earth's atmosphere from cosmic ray

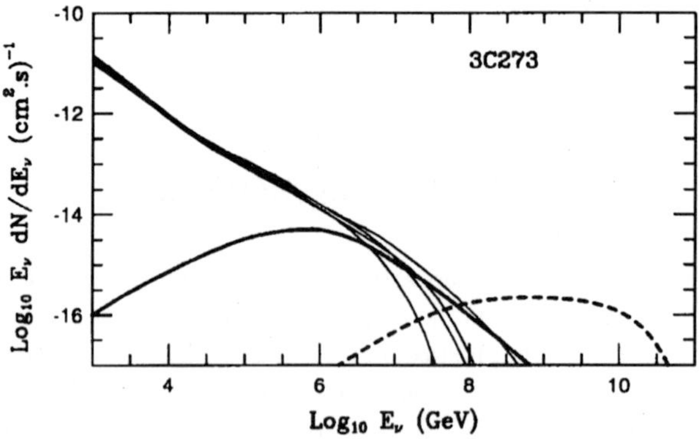

Figure 9.1 Neutrino spectra generated in the central region of 3C273 in the models of Stecker et al. (1991) (—) and of Szabo and Protheroe (1992) (—) and in its jet (Manheim and Biermann (1989) (— —)).

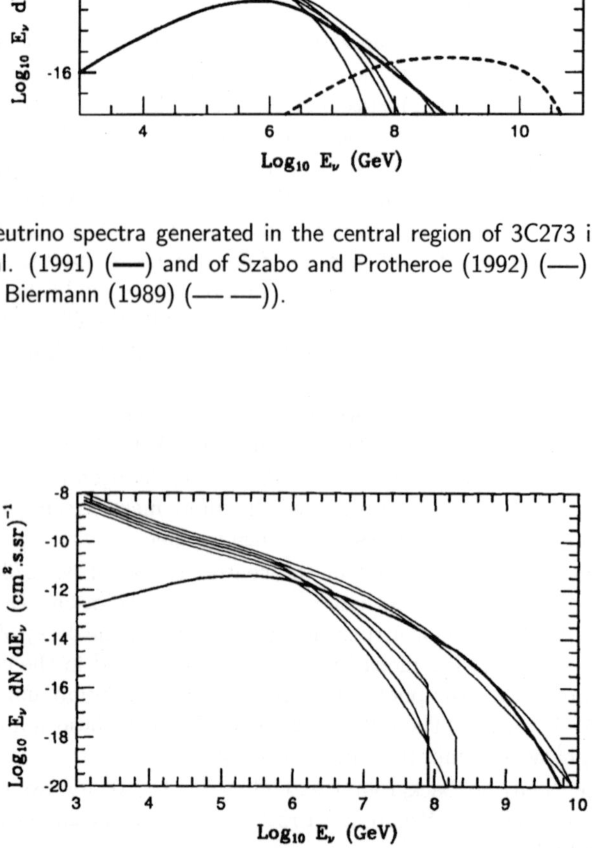

Figure 9.2 Diffuse neutrino background from undefined AGN: — selected models of Szabo and Protheroe (1992): — the model of Stecker et al. (1991).

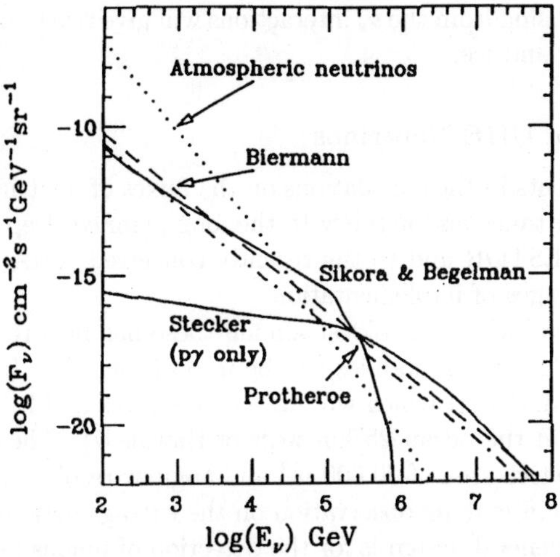

Figure 9.3 Predicted neutrino fluxes from Active Galactic Nuclei and from the earth's atmosphere.

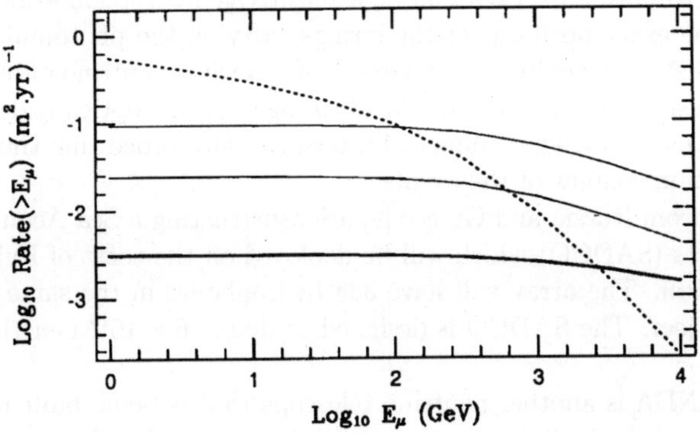

Figure 9.4 Fluxes of upward going muons as a function of the muon threshold energy predicted by the bracketing high and low models of Szabo and Protheroe (1992) (—) and Stecker et al. (1991) (**—**). The atmospheric neutrino background is shown with a dotted line.

interactions. It is seen that it is only beyond 10^{12} eV that the AGN contribution starts dominating over the atmospheric muons.

The electrons arising from the ν_e interactions will give rise to high density cascades in media like water and ice.

9.1 Detection of UHE Neutrinos

The new developments in the calculations on the fluxes of neutrinos from AGNs have added further importance and urgency to the long planned deep ocean detectors like DUMAND[a] and NESTOR and to the recently conceived AMANDA projects which are all in various stages of implementation.

The DUMAND which is the oldest among these has now reached a stage where three strings with 72 optical modules will be in position soon. It is now proposed to expand this to DUMAND II which will have nine strings and 216 optical modules at a depth of 4.8 km in the ocean 25 km west of Hawaii [7]. The nine string octagon will provide an effective area of 20,000 m^2 and angular resolution of $\sim 1°$ and will be just sufficient to begin making observation on the extragalactic neutrinos at energies above 3 TeV. This general design is for the detection of muons produced by neutrino interactions in water. The interaction of an electron neutrino in water will give rise to an ultra high energy electron that will initiate an electromagnetic cascade shower with a thick ionization column along the core. The high density high temperature core will heat the water locally and the resulting adiabatic expansion will generate cylindrical acoustic waves. The detection of these acoustic waves offers a novel method of detecting the ultra high energy neutrinos. It so happens that the DUMAND II has been planned to have an array of 53 high sensitivity hydrophones for the purpose of monitoring the exact positions of the strings carrying the photomultiplier modules. The same acoustic system has the capability of recording neutrino induced showers [8]. The combined set up has the additional advantage that the optically detected neutrino events can be used to trigger enhanced acoustical data processing thus adding to the total information content of the events.

Scientists from Russia and Greece [9] are constructing a Sea Acoustic Detector of Cosmic Objects (SADCO) which will be deployed off the coast of Pylos, Greece at a depth of 4–6 km. The array will have 384 hydrophones in the same location as the NESTOR project. The SADCO is designed to detect 6×10^{15} neutrinos in 1 km^3 of water.

The AMANDA is another neutrino telescope that is being built in the Antarctic Ice making use of the high transparency of the ice at depths of more than a kilometer and serving as an ideal Cerenkov detector. The first phase of AMANDA [10] will consist of nine strings surrounding a central string; 20 down-looking optical modules separated by 10 m are positioned on each string. The outer strings are separated by 30 meters from the central string. The effective area of the telescope is $\sim 10^4$ m^2 for

[a]The DUMAND has since been discontinued.

detecting TeV neutrinos. The effective area is 25,000 m^2 at 100 TeV.

9.1.1 Radio Detection of Neutrino induced Showers

Askar'yan [11] had proposed, as early as 1965, the possibility of detecting neutrino induced showers through the radio waves emitted. Cold ice in the Antarctic is highly transparent to the propagation of microwaves. The proposal of Provorov and Zheleznykh [12] from Russia, is to use the antarctic ice as the target material for the generation of cascades by the electrons produced in neutrino interactions and detect the microwaves by a system of 100 antennae (broad horn TEM) looking down into the antarctic ice. The spacing between the antennae will be 100 m and the total effective area is 1 km^2. This will detect neutrinos of energy greater than 100 TeV. The calculated rate of detection of neutrinos from AGNs is about 400–500 per year.

9.1.2 Horizontal Air Showers

If h_0 is the amount of matter in the vertical direction, the amount of matter at a zenith angle θ is $h = h_0/\cos\theta$. In the case of sea level, $h_0 = 10$ m.w.e. and the limiting value in the horizontal direction is 300 m.w.e. Consequently, the frequency of showers of a given size observed as a function of zenith angle falls off very steeply. Practically all the components of showers are absorbed except the very high energy muons, the number of which however is very small. These high energy muons do give rise occasionally to showers of smaller size through the process of bremsstrahlung deep in the atmosphere. Clearly the 300 m.w.e. of matter in the horizontal direction provides adequate matter for the interaction of ultra high energy neutrinos.

The frequency of horizontal air showers recorded by the Akeno air shower array as a function of shower size is compared [13] with the rate expected from the AGN neutrinos and the upper limits for contributions from NGC4151 and Mrk421 in Fig. 9.5. Clearly the background due to bremsstrahlung from high energy muons agrees with the data well and neutrino induced showers can be detected above this background only above a shower size of 10^6. What is interesting at the moment is that Akeno [14] has not recorded so far any horizontal shower of size greater than 10^5.

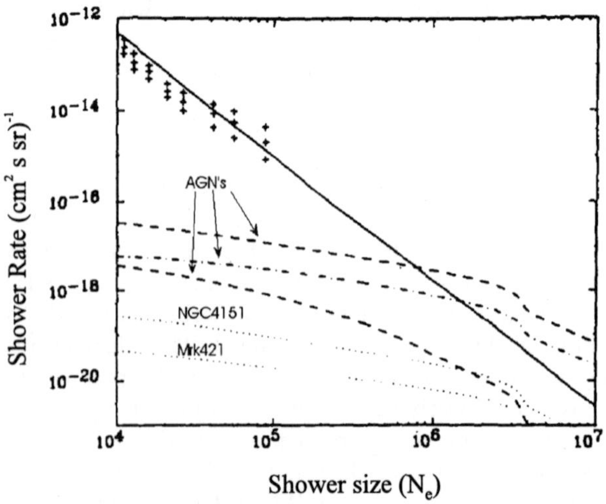

Figure 9.5 Integral horizontal shower flux initiated by neutrinos produced in AGN. Dashed lines are the upper and lower predictions from Szabo and Protheroe (1992). The dash-dotted line is the prediction from Stecker et al. (1991). The two dotted lines correspond to the upper predictions from Szabo and Protheroe (1992) for NGC4151 and Mrk421. The data points are the horizontal air shower fluxes measured by Nagano et al. (1971) and the solid line shows the expected flux due to bremsstrahlung from high energy muons.

References

1. T. Stanev, *Proc. 23rd International Cosmic Ray Conference, Calgary,* **Invited Rapporteur & Highlight Papers** (1993) 503.
2. M. Punch et al., *Nature* **358** (1992) 477.
3. C. E. Fichtel et al., *AIP Conference Proceedings 280, Compton Gamma -ray Observatory, St. Louis* (1992) 461.
4. F. W. Stecker et al., *Phy. Rev. Lett.* **66** (1991) 2697.
5. A. P. Szabo and R. J. Protheroe, in *Proc. of the Workshop on High Energy Neutrino Astronomy, Honolulu* ed. V. J. Stenger, J. G. Learned, S. Pakvasa and X. Tata (World Scientific, Singapore 1992) p. 24.
6. K. Manheim and P. L. Biermann, *Astron. Astrophysics* **22** (1989) 211.
7. C. M. Alexander et al., *Proc. 23rd International Cosmic Ray Conference, Calgary* **4** (1993) 515.
8. J. G. Learned et al., *Proc. 23rd International Cosmic Ray Conference, Calgary* **4** (1993) 538.
9. S. Karavesky et al. *Proc. 23rd International Cosmic Ray Conference, Calgary* **4** (1993) 550.
10. S. Tilay et al., *Proc. 23rd International Cosmic Ray Conference, Calgary* **4** (1993) 561.
11. G. A. Askar'yan, *Zh. Expt. Theor. Fiz.* **41** (1965) 616.
12. A. L. Provorov and I. M. Zheleznykh, *Preprint INR–825/93* July (1993).
13. E. Zas et al., *Proc. 23rd International Cosmic Ray Conference, Calgary* **4** (1993) 434.
14. M. Nagano et al., *J. Phys. Soc. Japan* **30** (1971) 33.

Chapter 10
COSMIC RAY SOURCES

Even after 80 years of cosmic ray research it is rather strange that not even one source of the major components of primary cosmic rays, namely protons and heavy nuclei, has been unambiguously identified. What is gratifying, however, is that in the TeV domain one galactic source, the Crab Nebula Pulsar and two extragalactic sources, the AGNs— Markarian 421 and Markarian 501—have been established as sources of gamma rays and therefore of electrons of comparable energy in the sources. If mechanisms exist in the source regions for acceleration of electrons to these energies, then the same mechanism should also operate for the protons as well.

10.1 Summary of Energy Spectrum and Composition

Let us first summarize the basic features that have been experimentally established on the primary energy spectrum and composition of cosmic rays at high energies. These have been arrived at from

1. nuclear emulsion flights of the JACEE type carried out at stratospheric altitudes
2. deep underground experiments on the intensity and angular distributions of muons and
3. extensive air shower experiments.

Examining the overall behavior of the spectrum over the wide energy range $10^{11} - 10^{21}$ eV shown in Fig. 10.1 [1] and guided by the indications, though crude, of the composition at various energies it is expedient, from the point of view of analyzing the possible sources and acceleration mechanisms, to make the following broad classifications:

1. cosmic rays below 10^{14} eV
2. cosmic rays in the region $10^{14} - 10^{16}$ eV
3. cosmic rays in the region $10^{16} - 10^{19}$ eV
4. cosmic rays $\sim 10^{19}$ eV.

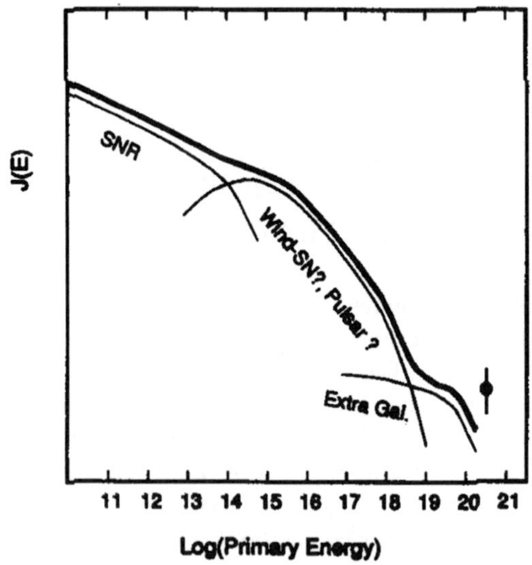

Figure 10.1 A schematic figure of the cosmic ray total energy spectrum. First component can be explained by the acceleration in the supernova shock expanding in the interstellar medium. Second one may be the galactic heavy component which originates from different sources from the first one. Third component is possibly the extragalactic protons.

5. highest energy cosmic rays ($> 10^{20}$ eV)

In Chapter 5 we have discussed the summary of the current status of the primary spectrum from balloon-borne emulsion chamber and other experiments in the energy range below 10^{14} eV by Simon Swordy [2] in his Rapporteur talk at the Calgary Cosmic Ray Conference. The spectra of proton and the He in the energy range 10 GeV/n to few times 10^{14} eV/n and of CNO, Ne-S, and Fe in the same energy range based on a variety of satellite and balloon experiments are given in Figs. 5.2 and 5.3.

At energies of interest to those concerned with air shower phenomena ($> 10^{12}$ eV), the two interesting trends seen from Fig. 5.2 are

1. the relative proportion of He compared to protons increases with energy and
2. a possible spectral cut-off of protons around 10^{14} eV (a sudden steepening in the spectrum). An important question that arises is whether this steepening[a]

[a]The JACEE group, which claimed the steepening, did not claim it any more at the International Cosmic Ray Conference at Durban in 1997.

is connected with the knee of the cosmic ray spectrum. We shall discuss this later on.

Fig. 5.3 shows that as far as the heavy components are concerned, the spectra tend to flatten at higher energies—the differential spectrum changes from -2.75 to more like $-2.5 \sim -2.6$ at energies of 10^{13} eV/n. The mean escape length of cosmic rays has been observed to decline with increasing particle energy up to at least ~ 100 GeV/n. If this decline were to become less severe around ~ 1 TeV/n the spectra would be expected to become flatter in this region. Thus, this flattening may be connected with the expected escape of particles from the galaxy at energies of $\sim 10^{12}$ eV/n.

In Chapter 6 we discussed Swordy's estimate of the composition of primary cosmic rays at the crucial energy of 10^{14} eV/ particle, which is very important for linking up the estimate with air shower observations. His estimates are given in Table 6.1. He remarks that the total estimated flux of 6.36×10^4 (particles per (m^2 s sr) $GeV^{1.75}$) is in good agreement with the proton satellite measurements of Grigorov et al. [3].

Let us now examine how these emulsion results join up with the observations in the knee region and beyond. This can be seen from Fig. 10.2 which is a compilation of the spectra from direct measurements and from air shower experiments [1]. Beyond a

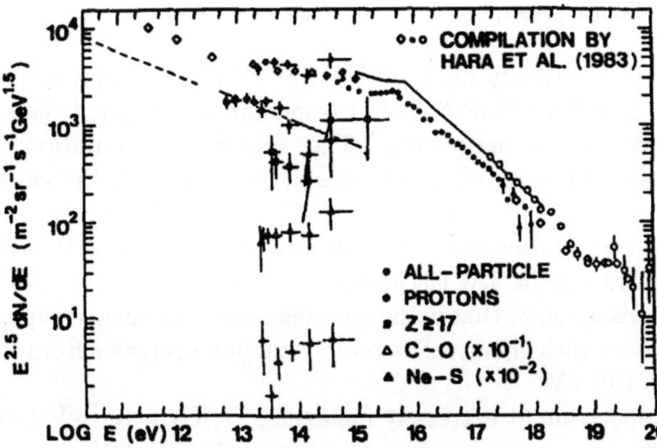

Figure 15: JACEE Energy spectrum

Figure 10.2 Compilation of the primary energy spectrum from direct as well as air shower measurements (Teshima, 1993).

few times 10^{14} eV, there are no direct measurements of the composition. The BASJE (Bolivian Air Shower Joint Experiment) group using the method of equal intensity cuts arrived at the longitudinal profiles and derived the chemical composition and spectrum on the assumption that the primaries are just protons and iron. The proton fraction as a function of energy arrived at on this basis is shown in Fig. 10.3.

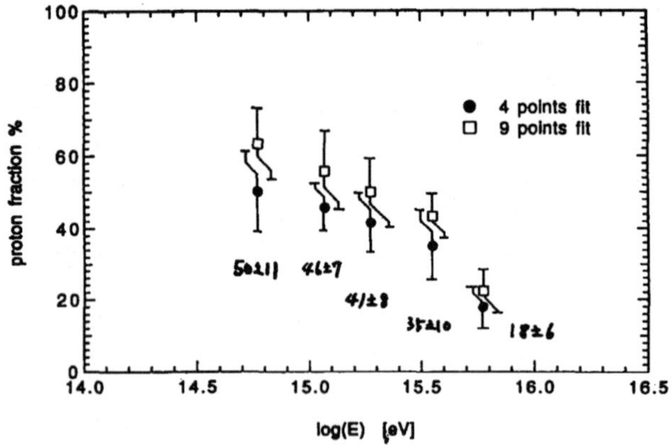

Figure 10.3 The fraction of protons as a function of primary energy determined using the longitudinal developement curves obtained from equi-intensity cut method by the BASJE group (Teshima,1993).

The trend that in the energy range $10^{15} - 10^{16}$ eV the proton fraction is decreasing is quite clear. The EAS-TOP-MACRO combination has good data on the muon content of showers in this energy range. They find that the variation of the observed muon number as a function of shower size agrees well with the calculations made assuming that over the shower size range of $10^{4.5}$ to 10^6 at the EAS-TOP level in the atmosphere, the composition and spectrum remains the same as that observed in the direct measurements of the JACEE group.

These comparisons show that in the knee region the chemical composition becomes increasingly heavier with energy. The proton fraction decreases from $\sim 60\%$ at $10^{14.5}$ eV to $\sim 18\%$ at 10^{16} eV.

The energy spectrum at extremely high energies, $E_0 > 10^{17}$ eV (see Chapter 7), shows the following:

1. a pronounced dip in the spectrum at 2×10^{18} eV,

2. flattening of the spectrum at 10^{19} eV.

From a measurement of the $X_{\max}$, the position in the atmosphere at which the shower reaches its maximum and the elongation rate which is the variation of this $X_{\max}$ with energy, the Fly's Eye group [4] have come to the following important conclusions on the composition:

1. the composition in the energy range 10^{17} to 10^{19} eV is not consistent with either a pure or mixed composition over the entire range.

2. the composition changes from a heavy to a predominantly light one in the range 3×10^{17} eV to 10^{19} eV. In the energy range 10^{17} to 3×10^{17} eV. The composition is constant and heavy.

The AGASA results do not quite support this conclusion yet [5].

The spectral change and the composition change around 10^{18} eV are perhaps interconnected and may be indicating the onset of a new cosmic ray component mostly light beyond this energy which could be consistent with extragalactic component.

If the spectrum had continued with the same slope up to energies beyond 10^{20} eV, without being influenced by the Greisen-Zatsepin-Kuzmin (GZK) cut-off due to the microwave background, then in the 30 years of total operation of Haverah Park, Yakutsk, Fly's Eye and Akeno arrays there should have been at least 22 showers of energy $> 10^{20}$ eV. However, the observed number is only eight which seems to suggest that GZK cut-off is operative. The problem with these three showers, as discussed separately, is that they should have originated in sources within a distance of 30 Mpc. There are no candidate sources within this range.

10.2 Origin and Acceleration Mechanisms

The current status of theories of acceleration mechanisms are available in the series of articles by Luke O'C Drury [6], J. R. Jokipii and G. Morfill [7], W. H. Ip and W. I. Axford [8], D.C. Ellison [9], Fumio Takahara and Toshio Terasawa [10], P. L. Biermann [11] and P. Mészáros and B. Rudak [12] in the proceedings of the International Symposium on "The Astrophysical Aspects of the Most Energetic Cosmic Rays" held in Kofu, Japan in November 1990, published by the University of Tokyo.

As far as acceleration of particles up to $\sim 10^{14} - 10^{15}$ eV is concerned, the most popular and one on which there is a general consensus is the diffusive shock acceleration in the supernova shock waves which propagate up to distances of 20–50 pc in the interstellar medium. This mechanism explains both the intensity, the spectral features, and composition of the primary radiation. The maximum energy that can be reached in such processes is about 10^{14} eV when the magnetic field is parallel to the shock normal. This is because the acceleration time is much longer than the diffusion time scale. It has been pointed out by Jokipii [13] that in highly oblique or perpendicular shocks significantly higher energies are possible ($\sim 10^{16}$ eV). Biermann and Volk [14] have examined the effect of the propagation of the supernova blast wave into a wind and find that particles can be accelerated to 10^{19} eV in case the wind is associated with a compact binary. This however creates the problem of matching the spectrum and intensities observed at the lower energies.

Jokipii and Morfill [7] take the view that since bulk of the cosmic rays are produced in the supernova blast waves propagating into interstellar medium, the origin of the ultra high energy particles must also be closely related to processes in the supernova and interstellar medium themselves. They propose the acceleration in the terminal shock of a galactic wind and perhaps with similar winds at neighboring galaxies. (Small Magellanic Clouds, Sculptor, Draco, Ursa Minor, Fornax, Leo I and II). They show that protons can be accelerated up to $\sim 10^{19}$ eV and Iron nuclei up to $\sim 10^{20}$ eV.

Reacceleration of particles with an initial energy of $\backsim 10^{14}$ eV, through their interactions with Pulsar wind terminal shocks has been proposed recently by Bell [15] and Berezhko [16]. In the Pulsar wind like that of the Crab Nebula, acceleration of particles up to 10^{17} eV is feasible. In the surroundings of pulsars of shorter rotational period acceleration of protons to 10^{18} eV and of iron to 10^{20} eV are quite likely.

Yet another new mechanism that is proposed is related to the role of the reverse shock wave that is invariably produced in all supernova explosions (Ming Zhang [17]. Fig. 10.4 shows the scenario following the SN explosion. Zhang maintains that if

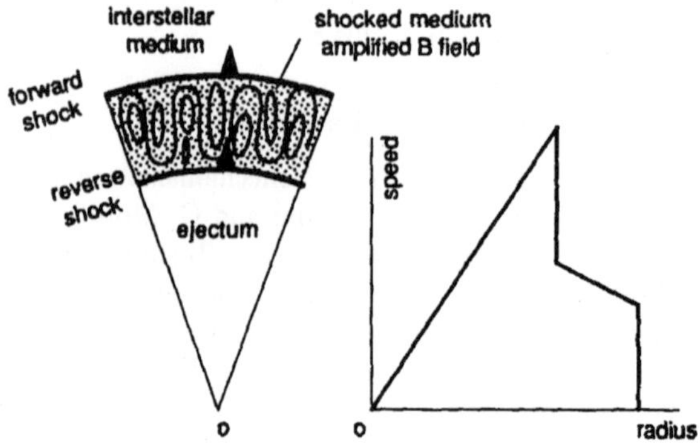

Figure 10.4 A view of young SNR structure and the gas speed profile (Ming Zhang, 1993).

a few percent of the plasma thermal energy is converted into magnetic field, then there will be a large amplification of the ambient magnetic field in the SN shell and particles can get accelerated to $10^{17} - 10^{18}$ eV/charge (see Fig. 10.5.). Among the extragalactic objects the most promising locations for acceleration of particles to the super high energies seem to be the Radio jets. Ip and Axford [8] have examined as an example the case of Cyg A, which has a double radio jet system (see Fig. 10.6). The diffuse shock acceleration takes place in the inner shock region and the accelerated electrons and nuclei get channeled into extended envelope which appears as the radio

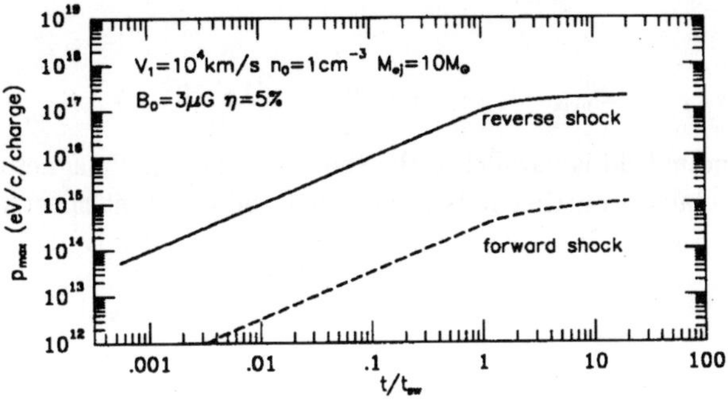

Figure 10.5 The maximum momentum of particles accelerated by shocks (Ming Zhing, 1993).

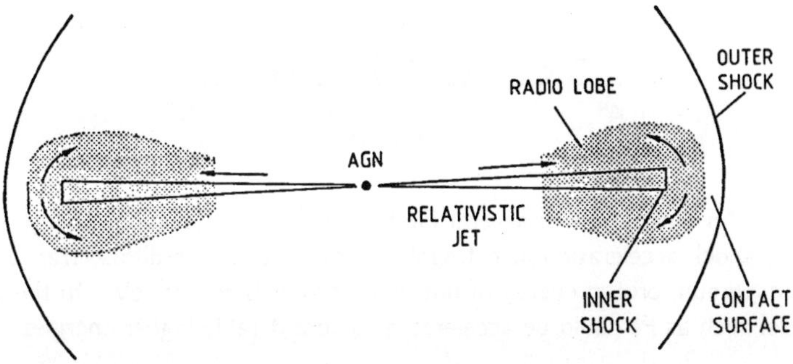

Figure 10.6 A schematic view of an extended radio system. The piston of the supersonic jet creates an inner shock and an outer shock in the intergalactic medium. These two shocks are separated by a contact surface. The inner shock is identified as the diffusive shock region where particles might be accelerated to 10^{20} eV. The accelerated electrons and nuclei will be channelled into the extended envelope appearing as the radio lobes. In this scenario the hot spot observed in synchrotron radiation emission is most likely located between the inner shock and the contact surface.

lobe. It is shown by the authors that

$$E''_{\max}\,(\text{proton}) = (3 \sim 9) \times 10^{18}\ \text{eV}$$

$$E''_{\max}\,(\text{Fe}) = 7.5 \times 10^{19} \sim 2.3 \times 10^{20}\ \text{eV}$$

when the magnetic field is parallel to the shock normal. The magnetic field assumed is $\backsim 3 \times 10^{-5}$ Gauss. Further it is shown that the resultant spectrum would be a power law (see Fig. 10.7). A very important prediction of this model is that at the

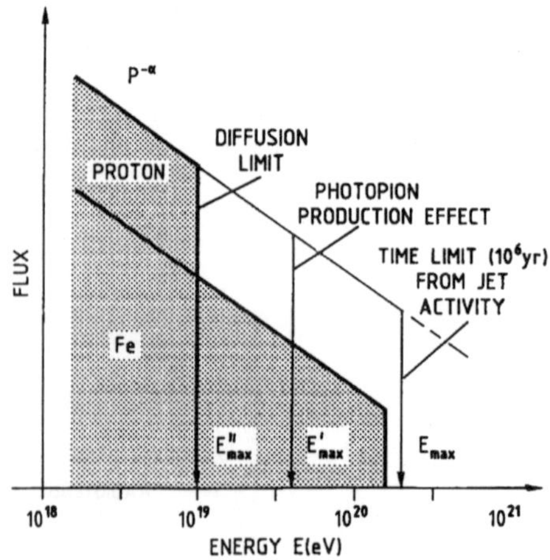

Figure 10.7 An illustration of several limitations on the maximum energy derivable from diffuse shock acceleration at extragalactic radio jets. The diffuse transport process permits a maximum proton energy of not much higher than 10^{19} eV. On the other hand heavy nuclei such as Fe could be accelerated to considerably higher energies.

highest energies (10^{19} eV) the dominant component would be heavies. In the case of radio jets in AGNs within the local cluster, the transit time of cosmic rays would be less than 10^8 years and therefore GZK cut-off would not be effective.

W. H. Sorrel [18] has considered the various scenarios in which particles are accelerated to energies beyond 10^{19} eV by Electric Dynamo action. Neutron stars with millisecond spin rates and magnetic field strength of $B = 10^{13}$ Gauss, can give rise to an electric voltage drop of $10^{20} - 10^{21}$ volts across the polar caps. Even though protons can be accelerated to 10^{21}eV, the energy losses due to curvature radiation would bring down the energy to 10^{17} eV. This mechanism therefore is not effective for such high energies.

Sorrel considers the pregalactic black holes hypothesis of Lacey and Ostriker [19] to explain the dark matter in the halos of spiral galaxies. Each black hole accretes gas and the turbulence and the magnetic fields in the accreting disk would set up a dynamo resulting in an accretion disc voltage of 9×10^{20} volts and would drive twin beams of protons of energy 9×10^{20} eV. It is shown that while the energy losses due to photomeson production and e^+ e^- pair creation would be small, curvature radiation would bring down this energy to $E_{max} \sim 3 \times 10^{20}$ eV. While this appears to be an attractive hypothesis, the main objection is that such accreting black holes in the halo would produce intense infrared sources (> 100 Jy) for which there is no evidence from the IRAS data.

The quasars and Seyfert galaxies have enormous radiation luminosities ($\mathcal{L} = 10^{46}-10^{48}$ ergs/s). Much of it is in the UV and IR regions of the electromagnetic spectrum. These high radiation densities prevent the escape of high energy particles even if they are accelerated in the central compact regions. Also the distances of these objects are so large that even the universal 3° radiation degrades their energy through repeated interactions. Therefore, it is unlikely that these can be sources of cosmic rays observed on the earth.

A non-acceleration scenario for the production of ultra high energy particles through the decay of massive particles (of mass characteristic of grand unification $\sim 10^{16}$ GeV) released in the collapse or annihilation of topological defects (TDs) such as monopoles, cosmic strings, domain walls formed in the early symmetry breaking phase transitions, has been proposed by P. Bhattacharjee [20]. Such processes can lead to the production of particles up to $\sim 10^{24}$ eV without the need of any acceleration mechanism.

The calculated spectrum for the particular process of the collapse of a cosmic string is shown in Fig. 10.8. It is seen that this process is unlikely to account for the flux of particles below an energy 10^{19} eV observed by the various experiments. However, the author observes that at energies above this range the contribution from this non-accelerator process may become significant. This is because the spectrum extends up to 10^{24} eV beyond the GZK cut-off for protons at 6×10^{19} eV and there is now evidence for finite flux beyond 10^{20} eV. A specific prediction of this model is that the primaries can only be fundamental particles (protons, neutrons and other anti-particles) and not iron nuclei. Discovery of particles beyond 10^{22} eV may provide us with an unique signature of exotic processes and the existence of exotic objects signalled by Grand Unification Theories (GUT).

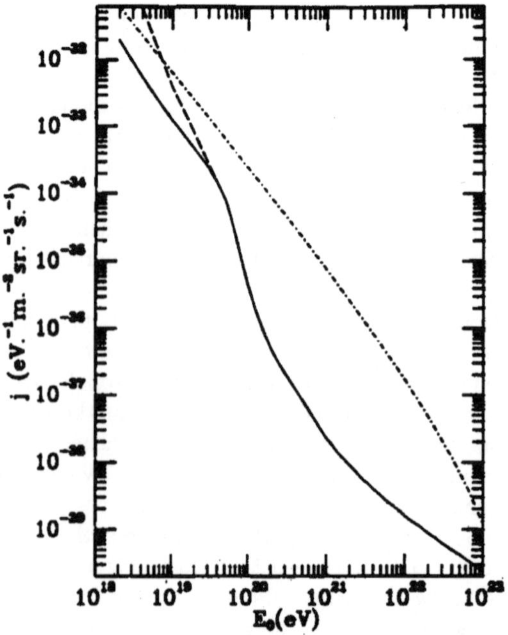

Figure 10.8 The estimated energy spectrum of protons from the decay of massive particles ($\sim 10^{16}$ GeV) from the collapse of comic strings (Bhattacharjee, 1991): — including energy losses due to expansion of the universe as well as pair production and photopion production from interactions with the microwave background radiation and requiring that the fluxes remain below the observed spectrum; – · – including energy losses due to expansion of the universe only; – – – best fit power law spectrum from the Fly's Eye experiment.

References

1. M. Teshima, *Proc. XXIII International Cosmic Ray Conference, Calgary, Invited, Rapporteur & Highlight Papers* (1993) 257.
2. Simon Swordy, *Proc. XXIII International Cosmic Ray Conference, Calgary, Invited, Rapporteur & Highlight Papers* (1993) 243.
3. N. L. Grigorov et al., *Proc. 12th International Cosmic Ray Conference, Hobart* **5** (1971) 1746.
4. D. J. Bird et al., *Proc. 23rd International Cosmic Ray Conference, Calgary* **2** (1993) 38.
5. N. Hayashida et al., *Proc. Tokyo Workshop on Techniques for the Study of Extremely High Energy Cosmic Rays, Tokyo* (1993) 63.
6. Luke O'C. Drury, *Proc. ICRR International Symposium, Astrophysical Aspects Of The Most Energetic Cosmic Rays,* Ed. M. Nagano and F. Takahara, (World Scientific, 1991) 252.
7. J. R. Jokipii and G. Morfill, *Proc. ICRR International Symposium, Astrophysical Aspects Of The Most Energetic Cosmic Rays,* Ed. M. Nagano and F. Takahara, (World Scientific, 1991) 261.
8. W. H. Ip and W. I. Axford, *Proc. ICRR International Symposium, Astrophysical Aspects Of The Most Energetic Cosmic Rays,* Ed. M. Nagano and F. Takahara, (World Scientific, 1991) 273.
9. D. C. Ellison, *Proc. ICRR International Symposium, Astrophysical Aspects Of The Most Energetic Cosmic Rays,* Ed. M. Nagano and F. Takahara, (World Scientific, 1991) 281.
10. Fumio Takahara and Toshio Terasawa, *Proc. ICRR International Symposium, Astrophysical Aspects Of The Most Energetic Cosmic Rays,* Ed. M. Nagano and F. Takahara, (World Scientific, 1991) 291.
11. Peter L. Biermann, *Proc. ICRR International Symposium, Astrophysical Aspects Of The Most Energetic Cosmic Rays,* Ed. M. Nagano and F. Takahara, (World Scientific, 1991) 301.
12. P. Mészáros and B. Rudak, *Proc. ICRR International Symposium, Astrophysical Aspects Of The Most Energetic Cosmic Rays,* Ed. M. Nagano and F. Takahara, (World Scientific, 1991) 311.
13. J. R. Jokipii, *Astrophys. J.* **313** (1987) 842.
14. P. Biermann and H. Volk, *Astrophys. J.* **333** (1987) L65.
15. A. R. Bell, *Proc. 22nd International Cosmic Ray Conference, Dublin* **2** (1991) 420.
16. E. G. Berezhko, *Proc. 22nd International Cosmic Ray Conference, Dublin* **2** (1991) 436.
17. Ming Zhang, *Proc. 23rd International Cosmic Ray Conference, Calgary* **4** (1993) 374.
18. W. H. Sorrel, *Proc. ICRR International Symposium, Astrophysical Aspects Of The

Most Energetic Cosmic Rays, Ed. M. Nagano and F. Takahara, (World Scientific, 1991) 329.

19. C. G. Lacey and J. P. Ostriker, *Ap. J.* **299** (1985) 633.
20. P. Bhattacharjee, *Proc. ICRR International Symposium, Astrophysical Aspects Of The Most Energetic Cosmic Rays,* Ed. M. Nagano and F. Takahara, (World Scientific, 1991) 382.

Chapter 11
CONCLUSIONS

The primary concern of cosmic ray research for the past several decades has been the determination of the spectrum, composition and anisotropy in the arrival directions of the primary radiation with a view to figuring out the sources of the radiation and the mechanisms of acceleration of the particles to such high energies and over such a wide band. A secondary aspect has continued to be the discernment of the characteristics of high energy interactions beyond the range of accelerator energies. We have seen that in both these studies the method of extensive air showers has played a very important role and beyond a certain energy is the only method that is available. However the method being indirect has limitations. These are gradually being overcome through a variety of innovations, by deployment of more and more composite arrays of detectors with higher spatial and temporal resolutions, that enable recording of simultaneous information on a variety of parameters of the showers and through the liberal usage of computer simulations based on the parameters of collisions determined in accelerator experiments up to the highest energies available.

Regarding the primary spectrum, the most interesting and important feature that has emerged is that while the spectrum in an overall way is a power law spectrum with a negative exponent, there are significant and pronounced structures. The integral spectrum steepens around 10^{15} eV (the exponent changes from ~ -1.6 to ~ -2.0) then flattens around 10^{19} eV (exponent ~ -1.6) with a dip around 10^{18} eV, with a clear indication that there is a tendency for a cut off beyond a few times 10^{19} eV. However it has to be stressed that there is clear evidence of particles accelerated to energies beyond 10^{20} eV even though the numbers are an order of magnitude lower than expected on the basis of extrapolation of the spectrum in the energy range 10^{17}–10^{19} eV.

The situation regarding primary composition at the air shower energies is still not clear and unambiguous. While some experimenters have claimed a change from the normal composition observed at lower energies in the knee region of the spectrum around 10^{15} eV, others report no change. The uncertain results on the change of composition are compounded further by the fact that the direct measurements of the composition by balloon born emulsion payloads seem to indicate a steepening of the proton spectrum around 10^{14} eV itself, well below the knee region, with a tendency for the heavy component to dominate beyond. Again in the energy region 10^{17}–10^{19}

eV, there are conflicting reports on the composition. Some groups have claimed a change from a heavy dominant to a lighter composition while others do not quite agree with this.

Regarding the anisotropy of the radiation, while there is indication that the amplitude of the first harmonic in Right Ascension increases with energy near the knee there is no evidence at all for any pronounced anisotropy at higher energies. Whether the structure that is seen in the primary spectrum is connected with the onset of drastic changes in the characteristics of interactions remains a moot point. There is some evidence from emulsion chamber and hybrid experiments which do point to changes, but these have not been taken seriously enough in the interpretation of air shower data by most of the groups.

In the domain of TeV astronomy, what is encouraging is that at least one source in the galaxy, i.e. the Crab Nebula remnant, and two extragalactic sources, Mkn421 and Mkn501, have been unambiguously established as sources of gamma rays of energy $> 10^{12}$ eV. The pulsed nature of its radiation is still in dispute.

In the PeV domain, while no steady source has been identified yet, transient emission has been claimed from the Crab, Her X-1 and Cyg X-3.

Thus many of the problems concerning the primary radiation continue to be open, requiring further experiments of a more extended and sophisticated nature for solution. It is encouraging that several new arrays are coming up for this purpose. A new field that is opening up is the field of very high energy neutrino astronomy. The impetus has come from the calculated fluxes of very high energy neutrinos from extragalactic compact objects like Active Galactic Nuclei which are powered by giant black holes in their centre. The deep underwater installations and the deep Antarctic Ice installations hold great promise in this regard.

The way technology is advancing, entirely newer methods of detection especially suited for the recording of very high energy showers may come up.

Appendix A
MONTE CARLO TECHNIQUE

Monte Carlo technique is used to generate individual events or values of a parameter according to a given distribution with the help of random numbers. Random numbers generated by computers are uniformly distributed between 0 and 1. The technique involves mapping the random numbers on to the parameter space. If the probability density distribution function of a parameter x is $f(x)\,dx$ in the range $a \leq x \leq b$ such that

$$\int_a^b f(x)\,dx = 1 \tag{A.1}$$

then the problem consists of assigning a corresponding value x for each random number R in the range $0 \leq R \leq 1$ so that the distribution of the resultant values of the parameter approach the function $f(x)$. This is achieved by equating the differential probability of occurrence of the parameter to that of the random number

$$f(x)\,dx = dR. \tag{A.2}$$

which on integration gives

$$\int_a^x f(x)\,dx = \int_0^R dR = R. \tag{A.3}$$

If the function is integrable Eq. A.3 can be solved for x. If the result of integration is a transcendental equation, iterative procedures have to be used. For example, the probability of interaction of a particle at x in dx with an interaction mean free path λ is $e^{-x/\lambda}\frac{dx}{\lambda}$ and the equation becomes

$$\int_0^x e^{-x/\lambda}\frac{dx}{\lambda} = R \tag{A.4}$$

which gives

$$x = -\lambda \ln(R-1) = -\lambda \ln(R)\,. \tag{A.5}$$

Since $1-R$ is also a random number uniformly distributed between 0 and 1 it can be replaced by R. On the other hand, in the case of transverse momentum distribution the solution has to be obtained by iterative procedure. The transverse momentum distribution is given by

$$f(x)\,dx = xe^{-x}dx \tag{A.6}$$

where $x = p_t/p_0$. The equation in this case is

$$\int_0^x xe^{-x}dx = R \tag{A.7}$$

which gives

$$x = \ln\left(\frac{x+1}{R}\right). \tag{A.8}$$

This is solved by iterative procedure, e.g. using Newton-Raphson method.

If the function is not integrable Neumann's method is used. The function is normalized to its maximum value in the range $[a,b]$ of interest so that its peak value in this range is 1. A random number R_1 is chosen and the value of x is taken to be $x = a + (b-a)R_1$. This value is accepted if $f(x) \leq R_2$ where R_2 is another random number. If $f(x) > R_2$ the value is rejected and the whole procedure is repeated. When a sufficiently larger number of values are thus generated, their distribution approaches $f(x)$. This procedure is schematically illustrated in Fig. A.1.The efficiency of this

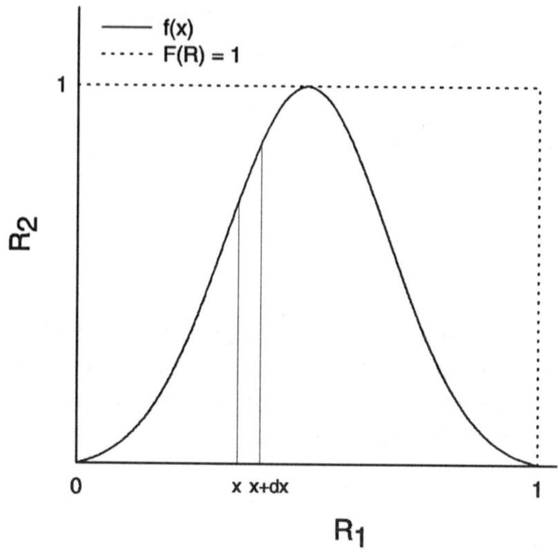

Figure A.1 Neumann's method of Monte Carlo generation.

method depends on the ratio of the area under the curve to unity. If the curve is too narrow and steep the method is very inefficient and takes too many trials.

Another method that is generally used is to create a look-up table of values of the parameter in the required range such that a cell picked with a random number gives the value. This method necessarily yields discrete values and the steps have to be chosen carefully so that the tail of the function is also sampled adequately. Though this method is fast, it requires large storage space.

A function which is very often used is the Gaussian function. This function can be generated by choosing two random numbers R_1 and R_2 from which two independent values of the parameter can be obtained as

$$\begin{aligned} x_1 &= \sqrt{-2\ln R_1}\,\sin(2\pi R_2) \\ x_2 &= \sqrt{-2\ln R_1}\,\cos(2\pi R_2)\,. \end{aligned} \tag{A.9}$$

Both the values x_1 and x_2 are normally distributed.

Appendix B
MONTE CARLO SIMULATION OF EXTENSIVE AIR SHOWERS

The backbone of the Monte Carlo simulation of air shower cascades is the hadronic interaction model. A variety of interaction models have been used by various authors to check their validity at air shower energies. Here, we shall briefly describe the simulation procedure adopted by the KGF group [1] using the scaling model of high energy interactions with inelastic cross sections increasing with energy.

B.1 The Interaction Model

The inelastic cross section, σ_{p-air}, of protons on air nuclei is given by

$$\begin{aligned} \sigma_{p-air}(E) &= 280 \quad \text{mb} \quad \text{for } E < 100\,\text{GeV} \\ &= 280 + 2.5\left[\ln(E/100)\right]^{1.8} \quad \text{mb for } E > 100 \text{ GeV} \end{aligned} \tag{B.1}$$

and for pions and kaons

$$\sigma_{\pi-air}(E) = \sigma_{K-air}(E) = \frac{2}{3}\sigma_{p-air}(E)\,. \tag{B.2}$$

The corresponding interaction mean free path for $p - air$ collisions is

$$\lambda_{p-air} = \frac{2.45 \times 10^4}{\sigma_{p-air}}, \tag{B.3}$$

where σ_{p-air} is in units of mb, with similar expressions for pion and kaons. Apart from pions, production of kaons and baryons is also included in the model. The energy dependence of the average multiplicity, $< n >$, is given by

$$\begin{aligned} \langle n \rangle &= A + B\ln S + C/\sqrt{S} && \text{for } \pi^+,\ \pi^-,\ K^+ \text{ and } K^-, \\ &= A + \left(B + C/\sqrt{S}\right)\ln S && \text{for } \overline{p} \end{aligned} \tag{B.4}$$

where S is the square of the center of mass energy and the constants A, B and C are given in Table B.1. The multiplicity of neutral pions is half the charged pion multiplicity and for neutral kaons it is equal to the sum of charged kaons. The multiplicities of p, n and $\overline{n}$ are same as that of $\overline{p}$. The multiplicity distribution is

Table B.1 Constants in the expressions for the energy dependence of multiplicities of different kinds of particles.

Particle	A	B	C
π^+	−1.7	0.84	1.0
π^-	−2.6	0.87	2.7
K^+	−0.5	0.13	0.65
K^-	−0.52	0.11	0.80
$\overline{p}$	−1.2	0.15	1.1

assumed to obey KNO scaling [2]. The scaling variable [3] $\eta = n\,(n-1) / \langle n\,(n-1)\rangle$ with the scaling function

$$\chi\,(\eta) = e^{-\eta} \tag{B.5}$$

is used. Using the dispersion, D, of the multiplicity distribution given by

$$D = \sqrt{\langle n^2\rangle - \langle n\rangle^2} = 0.548\,\langle n\rangle - 0.44 \tag{B.6}$$

the denominator in the definition of η is obtained as

$$\langle n\,(n-1)\rangle = 1.3\,\langle n\rangle^2 - 1.482\,\langle n\rangle + 0.194. \tag{B.7}$$

Thus knowing η and $\langle n\,(n-1)\rangle$ the multiplicity in individual interactions is obtained as the positive solution of the equation

$$n = 0.5 \pm 0.5\sqrt{1 + 4\eta\,\langle n\,(n-1)\rangle}. \tag{B.8}$$

The invariant cross section in terms of the variables rapidity Y and transverse momentum p_t can be factorized as

$$\frac{d^2\sigma}{dY\,dp_t} = f\,(Y)\,g\,(p_t) \tag{B.9}$$

where

$$Y = \frac{1}{2}\ln\left(\frac{E + p_l}{E - p_l}\right) \tag{B.10}$$

and

$$\begin{aligned} f\,(Y) &= ke^{-m(z+Y)^{-l}} \\ g\,(p_t) &= p_t e^{-bp_t}. \end{aligned} \tag{B.11}$$

The constants in the functions are given in Table B.2. Having generated Y and p_t, the longitudinal momentum p_l is obtained from

$$p_l = 0.5\sqrt{p_t^2 + m^2}\frac{e^{2Y} - 1}{e^Y} \tag{B.12}$$

Table B.2 Constants in the expressions for rapidity and transverse momentum distributions

Particle	k	m	l	z	b
π^+	188.2	66.81	4.33	2.0	6.5
π^-	245.1	45.29	3.48	2.0	7.0
K^+	10.56	3.96	3.26	0.65	5.0
K^-	10.45	6.53	2.49	0.65	5.0
$\overline{p}$	4.51	4.65	2.24	0	4.0

where m is the mass of the particle. The azimuthal angles are obtained form a uniform distribution.

The incident and target particles are treated separately as far as their rapidity is concerned. No explicit function is used. Instead, their momenta and azimuthal angles are determined in individual interactions by balancing momentum and energy after all the produced particles are generated. A charge exchange probability of 0.25 is assumed for leading nucleons.

B.2 Simulation of Individual Interactions

All interactions are treated in the center of mass (CM) system of the colliding particles and the energies and production angles of the particles are converted to the laboratory system after all the conservation laws are satisfied.

The number of particles of each kind is generated first from the average multiplicities of various kinds of particles and the multiplicity distribution. Then each particle is assigned a p_t value from the corresponding $g(p_t)$ such that the vector sum of all p_ts is zero. Finally the rapidity, Y, and from this p_l is generated for each particle from the corresponding $f(Y)$ such that momentum and energy are simultaneously conserved.

B.2.1 Simulation of Individual Particle Multiplicities

First the average total multiplicity $\langle n \rangle$ at the energy of the incident particle is calculated as the sum of the individual average multiplicities of the different kinds of particles as given by Eq. B.4. Then the scaling variable η of the multiplicity distribution is generated from random number R using the relation $\eta = -\ln R$. The total multiplicity n in that individual interaction is then calculated form Eqs. B.7 and B.8. Using random numbers, baryons are selected out of the n particles with a probability equal to the ratio of the average multiplicity of baryons to the average total multiplicity. Protons and neutrons are selected from these baryons with equal probability. Then an equal number of anti-baryons are selected to conserve the baryon number and out of these anti-protons and anti-neutrons are selected with equal probability.

The target nucleon is assigned to be a proton or neutron with equal probability since the target air nucleus contains equal number of protons and neutrons. The leading nucleon is assigned a charge with a charge exchange probability of 0.25.

Out of the remaining particles, charged mesons are then selected and charges assigned in a similar way. To conserve charge, the number of charged mesons is increased or decreased by 1 with equal probability. Out of these charged mesons, pions and kaons are then selected. The remaining particles, which are neutral mesons, are assigned to be pions or kaons in proportion to their multiplicities. This procedure assures charge conservation, which is necessary to study the charged to neutral ratio of hadrons.

B.2.2 Generation and Conservation of Transverse Momentum

The transverse momentum p_t of each particle is assigned according to the corresponding distribution as described in Appendix A. The equations to be satisfied for the transverse momentum to be conserved are

$$\begin{aligned} \sum p_{t_i} \sin\varphi_i &= 0 \\ \sum p_{t_i} \cos\varphi_i &= 0. \end{aligned} \tag{B.13}$$

The following procedure is adopted to ensure conservation:

1. p_t is generated for each particle

2. All particles except two are assigned a φ with a uniform distribution.

3. Equations B.13 are solved for these two values of φ.

4. If the two values of φ thus obtained are not real, then different combinations of the four particles having the highest values of p_t are selected and step 3 is repeated.

5. If none of the combinations satisfy the above equations, then step 2 to step 4 are repeated.

6. If conservation is not possible even after repeating step 5 5 times, then step 1 to step 5 are repeated until p_t is conserved.

B.2.3 Generation and Conservation of Longitudinal Momentum and Energy

Rapidity Y for each produced particle is generated from the corresponding rapidity distribution $f(Y)$ using Neuman's method as described in Appendix A. The longitudinal momentum p_l is calculated using Eq. B.12 and a positive (forward hemisphere) or negative (backward hemisphere) sign is assigned with equal probability.

The equations to be satisfied to conserve p_l and E are

$$\begin{aligned} \sum p_{l_i} &= 0 \\ \sum \sqrt{p_{l_i}^2 + p_{t_i}^2 + m_i^2} &= \sqrt{S}. \end{aligned} \tag{B.14}$$

Conservation is assured by solving these two equations for the longitudinal momenta of the two leading particles. The following procedure is adopted to ensure conservation.

1. A p_l for each produced particle is generated as mentioned above.

2. If the sum of the energies of all the produced particles in the CM system is $> \sqrt{S}$, go to step 1.

3. Solve Eqs. B.14 for the values of p_l of the two leading particles. If the both values are real, go to step 5.

4. If any of the solutions is not real, then the four particles having the highest values of p_l are alternately placed in the forward and backward hemispheres and step 3 is repeated. If any of the solutions is still not real, then steps 1 to 3 are repeated.

5. The two real solutions are accepted as the p_ls of the two leading particles.

It has been verified that enforcing the procedures used for conserving charge, energy and momenta do not alter the original distributions.

B.3 Simulation of the Air Shower Cascade

The air shower cascade simulation is based on the method of successive generations in which each particle in a generation is followed through until it interacts or decays or reaches the observation level. If it interacts or decays, the produced particles are stored as particles of next generation. Once all particles in a particular generation are followed, the particles in the next generation are considered. This procedure is repeated until all the particles reach the observation level or their energy falls below a predetermined value (5 GeV in the present case). All muons above 1 GeV are considered and their energy loss in the atmosphere and their decay in flight are taken into account.

Primary energy E_0 is sampled from a spectrum of the form $E^{-\gamma}dE$. In a particular case, E_0 is given by

$$E_o = \left[E_1^{-\gamma} - R\left(E_1^{-\gamma} - E_2^{-\gamma}\right)\right]^{-\frac{1}{\gamma}} \tag{B.15}$$

where the sampling is done in the energy interval between E_1 and E_2. The point of interaction t is given by

$$t = -\lambda \ln R \quad \text{g cm}^{-2} \tag{B.16}$$

where λ is the interaction mean free path in units of g cm^{-2}. The interaction products and their energy and momenta are obtained as described in the previous section. For unstable particles such as pions and kaons, the point of interaction as well as decay are obtained from random numbers and whichever is closer to the point of its production, that process is taken to occur. All particles reaching the observation level are accumulated and the total numbers, energy spectra and lateral distributions are formed for different kinds of particles.

Neutral pions are assumed to decay into two gamma rays of equal energy and their contribution to the shower size and lateral distribution of electrons at the observation level are calculated from Greisen's formula for the longitudinal development of electromagnetic cascades (Eq. 2.14) and NKG function (Eq. 2.22) respectively. Contribution due to electrons, which may come form the decay of kaons, is taken to be half that due to a gamma ray of twice the energy starting one cascade unit above the production point of the electron.

Showers initiated by heavy nuclei are treated assuming the superposition model. A simplified flow chart of the simulation program is shown in Fig. B.1.

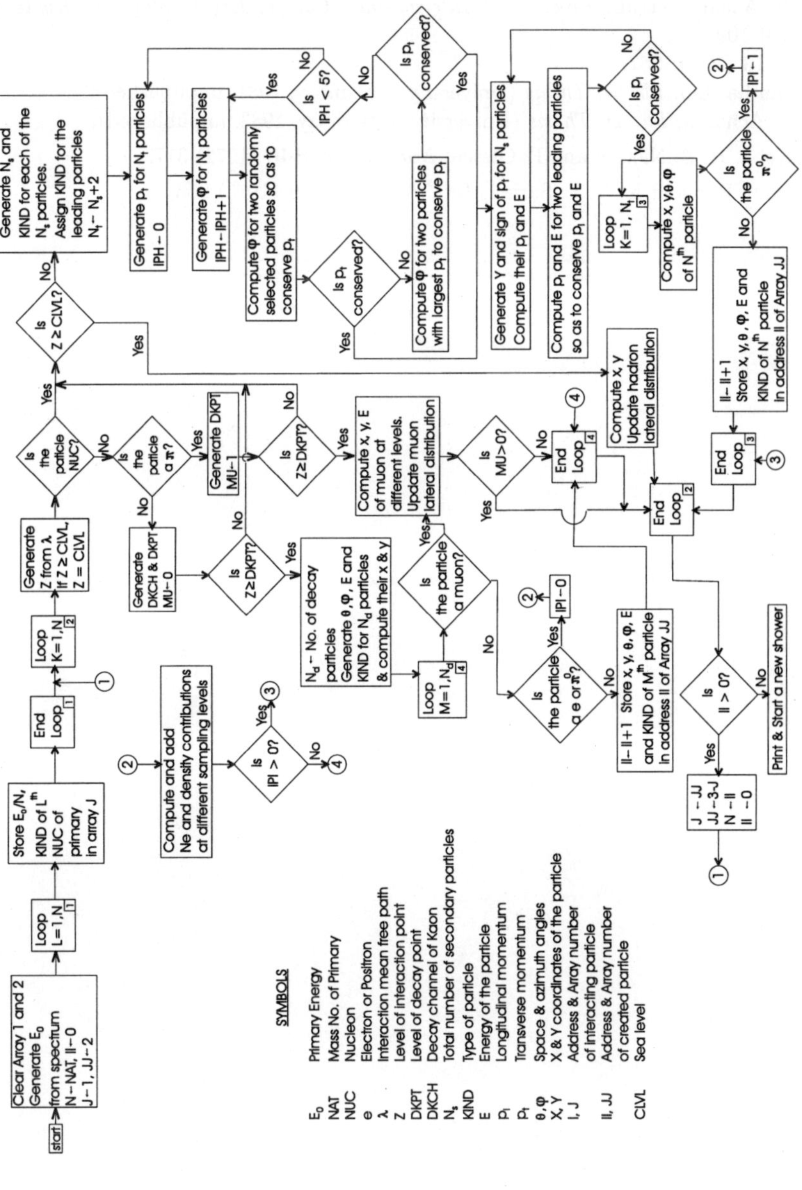

Figure B.1 Simplified flow chart of the air shower simulation program.

References

1. B. S. Acharya et al., *Proc. 16th International Cosmic Ray Conference, Kyoto* **9** (1979) 109.
   ibid. **13** (1979) 287.
   Srikantha Rao, *Ph. D. Thesis* (University of Bombay, 1981, unpublished).
   B. S. Acharya, *Ph. D. Thesis* (University of Bombay, 1983, unpublished).
2. Z. Koba, H. B. Nielsen and P. Olesen, *Nucl. Phys.* **B40** (1972) 317.
3. I. Rama Rao and K. V. L. Sarma, *Lett. Nuo. Cem.* **8** (1973) 963.

Author Index

F

G

H

I

J

K

L

M

W

Y

Z

Subject Index